Tactile Sensors for Robotic Applications

Tactile Sensors for Robotic Applications

Editor

Salvatore Pirozzi

MDPI • Basel • Beijing • Wuhan • Barcelona • Belgrade • Manchester • Tokyo • Cluj • Tianjin

Editor
Salvatore Pirozzi
Università degli Studi della
Campania "Luigi Vanvitelli"
Italy

Editorial Office
MDPI
St. Alban-Anlage 66
4052 Basel, Switzerland

This is a reprint of articles from the Special Issue published online in the open access journal *Sensors* (ISSN 1424-8220) (available at: https://www.mdpi.com/journal/sensors/special_issues/tactile_sensors_robotics).

For citation purposes, cite each article independently as indicated on the article page online and as indicated below:

LastName, A.A.; LastName, B.B.; LastName, C.C. Article Title. *Journal Name* **Year**, *Volume Number*, Page Range.

ISBN 978-3-0365-0424-7 (Hbk)
ISBN 978-3-0365-0425-4 (PDF)

Contents

About the Editor . . . . **vii**

Salvatore Pirozzi
Tactile Sensors for Robotic Applications
Reprinted from: *Sensors* **2020**, *20*, 7009, doi:10.3390/s20247009 . . . . **1**

Marco Costanzo, Giuseppe De Maria, Ciro Natale andSalvatore Pirozzi
Design and Calibration of a Force/Tactile Sensor forDexterous Manipulation
Reprinted from: *Sensors* **2019**, *19*, 966, doi:10.3390/s19040966 . . . . **5**

Shan-Qian Ji, Ming-Bao Huang and Han-Pang Huang
Robot Intelligent Grasp of Unknown Objects Based on Multi-Sensor Information
Reprinted from: *Sensors* **2019**, *19*, 1595, doi:10.3390/s19071595 . . . . **29**

Hiep Xuan Trinh, Yuki Iwamoto, Van Anh Ho and Koji Shibuya
Localization of Sliding Movements Using Soft Tactile Sensing Systems with
Three-axis Accelerometers
Reprinted from: *Sensors* **2019**, *19*, 2036, doi:10.3390/s19092036 . . . . **59**

Vinicius Prado da Fonseca, Thiago Eustaquio Alves de Oliveira and Emil M. Petriu
Estimating the Orientation of Objects from Tactile Sensing Data Using Machine Learning
Methods and Visual Frames of Reference
Reprinted from: *Sensors* **2019**, *19*, 2285, doi:10.3390/s19102285 . . . . **79**

Shunsuke Nagahama, Kayo Migita and Shigeki Sugano
Soft Magnetic Powdery Sensor for Tactile Sensing
Reprinted from: *Sensors* **2019**, *19*, 2677, doi:10.3390/s19122677 . . . . **99**

Ryuichiro Koike, Sho Sakaino and Toshiaki Tsuji
Hysteresis Compensation in Force/Torque Sensors Using Time Series Information
Reprinted from: *Sensors* **2019**, *19*, 4259, doi:10.3390/s19194259 . . . . **111**

Alireza Mohammadi, Yangmengfei Xu, Ying Tan, Peter Choong and Denny Oetomo
Magnetic-Based Soft Tactile Sensors with Deformable Continuous Force Transfer Medium
for Resolving Contact Locations in Robotic Grasping and Manipulation
Reprinted from: *Sensors* **2019**, *19*, 4925, doi:10.3390/s19224925 . . . . **125**

Seondae Kim, Yeongil Ryu, Jinsoo Cho, and Eun-Seok Ryu
Towards Tangible Vision for the Visually Impaired through 2D Multiarray Braille Display
Reprinted from: *Sensors* **2019**, *19*, 5319, doi:0.3390/s19235319 . . . . **139**

Francisco Pastor, Juan M. Gandarias, Alfonso J. García-Cerezo and
Jesús M. Gómez-de-Gabriel
Using 3D Convolutional Neural Networks for Tactile Object Recognition with
Robotic Palpation
Reprinted from: *Sensors* **2019**, *19*, 5356, doi:10.3390/s19245356 . . . . **167**

Pascal Weiner, Caterina Neef, Yoshihisa Shibata, Yoshihiko Nakamura and Tamim Asfour
An Embedded, Multi-Modal Sensor System for Scalable Robotic and Prosthetic Hand Fingers
Reprinted from: *Sensors* **2020**, *20*, 101, doi:10.3390/s20010101 . . . . **183**

Lukas Merker, Sebastian J. Fischer Calderon, Moritz Scharff, Jorge H. Alencastre Miranda, Carsten Behn
Effects of Multi-Point Contacts during Object Contour Scanning Using a Biologically-Inspired Tactile Sensor
Reprinted from: *Sensors* **2020**, *20*, 2077, doi:10.3390/s20072077 . . . . **205**

Oliver Ozioko, William Navaraj, Marion Hersh and Ravinder Dahiya
Tacsac: A Wearable Haptic Device with Capacitive Touch-Sensing Capability for Tactile Display
Reprinted from: *Sensors* **2020**, *20*, 4780, doi:10.3390/s20174780 . . . . **223**

About the Editor

Salvatore Pirozzi (Prof.) is currently an Associate Professor at the University of Campania "Luigi Vanvitelli", Italy. He has authored or co-authored more than 80 international journal and conference papers. His research interests include the modeling and control of smart actuators for active noise and vibration control and the design and modeling of innovative sensors, in particular tactile solutions, as well as the interpretation and fusion of data acquired from developed sensors. He is currently an Associate Editor for the IEEE Transactions on Control Systems Technology and a member of the Editorial Board for MDPI's journal *Sensors*.

Editorial

Tactile Sensors for Robotic Applications

Salvatore Pirozzi

Dipartimento di Ingegneria, Università degli Studi della Campania "Luigi Vanvitelli", Via Roma, 29, 81031 Aversa (CE), Italy; salvatore.pirozzi@unicampania.it; Tel.: +39-081-501-0433

Received: 19 November 2020; Accepted: 30 November 2020; Published: 8 December 2020

In recent years, tactile sensing has become a key enabling technology to implement complex tasks by using robotic systems. For example, the successful execution of robotic grasping and manipulation tasks is strongly dependent on the knowledge of objects' geometrical and physical characteristics, especially when objects are deformable and can change their shapes depending on their interaction with the environment. To this aim, robotic systems are more and more frequently equipped with sensorized grippers, which estimate the object's features by using tactile sensors. Moreover, a safe and efficient physical Human Robot Interaction (pHRI) requires the knowledge of interaction forces and contact locations in order to perform cooperation and co-manipulation tasks and to limit damage from accidental impacts. This crucial information can be obtained through direct measurements by using an artificial sense of touch.

Very often, in grasping tasks, the object features can also be estimated by combining the tactile sensors with additional sensors, e.g., the vision systems and the six-axis force/torque sensors. Vision data are used more frequently due to the efficiency in data collection, but the vision alone may not be an efficient solution due to the difficulties in extracting the image features from a complex background. Many researchers have been working on integrating vision, force/torque and tactile data for object recognition for more than 30 years. Recent papers on this field concern object pose and shape estimation, combination of visual and tactile exploration procedures, estimation of surface features, match tactile features to visual maps, reconstruct contact force/torque from tactile data and object recognition by using cross-modal approaches.

It is evident that the number of different contexts in which the sense of touch, alone or in combination with other sensors, can be fundamental for the robotic systems of the future is high and growing. The aim of this Special Issue is to present robotic applications for which tactile sensing together with alternative sensing systems represent solutions that allow clear improvements for task automation.

Summary of Special Issue

In [1], authors present the design and calibration of a new force/tactile sensor for dexterous manipulation. After the calibration, the device is able to estimate contact forces, torsional moment, object slip and object geometry: fundamental features when objects are fragile and deformable. Grasping experiments are reported to validate the proposed solution.

The combination of multi-sensor systems and control algorithms is another approach frequently used for robotic applications. The study in [2] proposes an intelligent real-time grasping system for the handling of objects with unknown properties in cluttered scenes, by combining different sensing system and control algorithms. Experiments with a five-finger robot hand are reported.

The estimation of slippage events, the sliding avoidance and stability of a grasp are fundamental aspects for advanced manipulation tasks. Many researchers, such as the authors of [3], continue to work on the development of technology solutions which allow improvements for these aspects. In particular, they proposed a solution based on a three-axis accelerometer embedded in a silicone chamber combined with a silicone rubber base. The design, the working principle and the fabrication procedure are detailed, and experiments are presented to demonstrate the effectiveness of the proposed approach in presence of sliding movements.

The authors of [4] propose the fusion of tactile data and vision data for estimating the pose of in-hand objects grasped with an underactuated robotic hand. These kind of data are very often combined with the aim to reproduce the human approach in manipulation. The data are used together with fuzzy logic controllers suitably designed to obtain stable grasps. Experiments are reported to demonstrate system capabilities.

New technologies are continuously investigated for the development of innovative tactile sensors. In [5], the authors present a novel type of soft resistive tactile sensor called "soft magnetic powdery sensor" (soft-MPS), which comprises ferromagnetic powder, immobilized in a liquid resin after orienting in a magnetic field. The proposed technology allows the realization of sensor units in any shape and to detect collisions in robot hands/arms or in ultra-sensitive touchscreen devices.

The authors of [6] exploit machine a learning approach to compensate hysteresis in a six-axis force sensor, in order to achieve high-precision measurements. The proposed approach allows to extension to multiple axis the hysteresis compensation typically applied only to one-axis sensors. Experimental results show the effectiveness of the proposed solution.

Many robotic tasks request the estimation of contact location with high precision. The paper [7] presents a design and characterization procedure for magnetic-based soft tactile sensors with the aim to locate a contact force application point. This procedure provides conditions under which it is possible to achieve the desired performance for the tactile sensor in identifying the contact location. An illustrative example demonstrates the efficacy of the proposed design procedure.

Development of human–machine interfaces, also for impaired people, is another research field that has been expanding recently. The paper [8] presents two methodologies for delivering multimedia content to visually impaired people with the use of a haptic device and braille display. The proposed approach uses 2D multiarray braille display to represent media content contour, illustrations and figures through quantization and binarization.

The exploitation of tactile data represents an open research field, very interesting in recent years. In paper [9], authors propose a novel method for active tactile perception based on 3D neural networks on the basis of tactile data available from a high-resolution tactile sensor installed on a robot gripper. An exploration procedure is performed in order to estimate information about both the external shape of the grasped object and internal features. A new 3D representation of tactile data is presented, by introducing an appropriate tensor used to feed a 3D Convolutional Neural Network, called 3D TactNet.

Manipulation by using anthropomorphic robotic and prosthetic hands still represents a research challenge. A fundamental aspect concerns the integration of a sensory system for control purposes in the limited space available. The authors of [10] present a scalable design model of artificial fingers, which combines mechanical design and a multi-modal sensor system for the measurement of normal and shear force, distance, acceleration, temperature, and joint angles. The design is fully parametric, allowing automated scaling of the fingers to arbitrary dimensions in the human hand spectrum. Different physical demonstrators are presented to demonstrate the effectiveness of the proposed approach.

Taking a cue from the animal world alternative solutions to the touch are represented by vibrissae. Animals like mice and rats use vibrissae in order to detect different features, e.g., object-distances and object-shapes. In [11], the authors deal with the effect of multi-point contacts in a specific scanning scenario, where an artificial vibrissa is swept along an object contours. They propose a model to simulate vibrissae during an object scanning and also experiments to validate the simulation results.

The author of [12] presents a dual-function wearable device (Tacsac) with capacitive tactile sensing and integrated tactile feedback capability to enable communication among deafblind people. Tacsac comprises two main modules: the touch-sensing module and the vibrotactile module. The Tacsac device has been tested for independent sensing and actuation as well as a dual sensing-actuation mode.

A mobile application was also developed to demonstrate the application of Tacsac for communication between deafblind person wearing the device and a mobile phone user who is not deafblind.

Funding: This research received no external funding.

Acknowledgments: Thanks to all authors for their valuable contributions to this Special Issue. All published papers have been rigorously reviewed and hence high gratitude is also owed to the international reviewers for their diligence and hard work, fundamental to obtain high quality papers. Thanks also to employees of MDPI Sensors Editorial Office, who supported me during the Special Issue management.

Conflicts of Interest: The author declares no conflict of interest.

References

1. Costanzo, M.; De Maria, G.; Natale, C.; Pirozzi, S. Design and Calibration of a Force/Tactile Sensor for Dexterous Manipulation. *Sensors* **2019**, *19*, 966. [CrossRef] [PubMed]
2. Ji, S.; Huang, M.; Huang, H. Robot Intelligent Grasp of Unknown Objects Based on Multi-Sensor Information. *Sensors* **2019**, *19*, 1595. [CrossRef] [PubMed]
3. Trinh, H.; Iwamoto, Y.; Ho, V.; Shibuya, K. Localization of Sliding Movements Using Soft Tactile Sensing Systems with Three-axis Accelerometers. *Sensors* **2019**, *19*, 2036. [CrossRef] [PubMed]
4. Prado da Fonseca, V.; Alves de Oliveira, T.; Petriu, E. Estimating the Orientation of Objects from Tactile Sensing Data Using Machine Learning Methods and Visual Frames of Reference. *Sensors* **2019**, *19*, 2285. [CrossRef] [PubMed]
5. Nagahama, S.; Migita, K.; Sugano, S. Soft Magnetic Powdery Sensor for Tactile Sensing. *Sensors* **2019**, *19*, 2677. [CrossRef] [PubMed]
6. Koike, R.; Sakaino, S.; Tsuji, T. Hysteresis Compensation in Force/Torque Sensors Using Time Series Information. *Sensors* **2019**, *19*, 4259. [CrossRef] [PubMed]
7. Mohammadi, A.; Xu, Y.; Tan, Y.; Choong, P.; Oetomo, D. Magnetic-based Soft Tactile Sensors with Deformable Continuous Force Transfer Medium for Resolving Contact Locations in Robotic Grasping and Manipulation. *Sensors* **2019**, *19*, 4925. [CrossRef] [PubMed]
8. Kim, S.; Ryu, Y.; Cho, J.; Ryu, E. Towards Tangible Vision for the Visually Impaired through 2D Multiarray Braille Display. *Sensors* **2019**, *19*, 5319. [CrossRef] [PubMed]
9. Pastor, F.; Gandarias, J.; García-Cerezo, A.; Gómez-de-Gabriel, J. Using 3D Convolutional Neural Networks for Tactile Object Recognition with Robotic Palpation. *Sensors* **2019**, *19*, 5356. [CrossRef] [PubMed]
10. Weiner, P.; Neef, C.; Shibata, Y.; Nakamura, Y.; Asfour, T. An Embedded, Multi-Modal Sensor System for Scalable Robotic and Prosthetic Hand Fingers. *Sensors* **2020**, *20*, 101. [CrossRef] [PubMed]
11. Merker, L.; Fischer Calderon, S.; Scharff, M.; Alencastre Miranda, J.; Behn, C. Effects of Multi-Point Contacts during Object Contour Scanning Using a Biologically-Inspired Tactile Sensor. *Sensors* **2020**, *20*, 2077. [CrossRef] [PubMed]
12. Ozioko, O.; Navaraj, W.; Hersh, M.; Dahiya, R. Tacsac: A Wearable Haptic Device with Capacitive Touch-Sensing Capability for Tactile Display. *Sensors* **2020**, *20*, 4780. [CrossRef] [PubMed]

Article

Design and Calibration of a Force/Tactile Sensor for Dexterous Manipulation

Marco Costanzo *, Giuseppe De Maria, Ciro Natale and Salvatore Pirozzi

Dipartimento di Ingegneria, Università degli Studi della Campania Luigi Vanvitelli, Via Roma, 29, 81031 Aversa, Italy; giuseppe.demaria@unicampania.it (G.D.M.); ciro.natale@unicampania.it (C.N.); salvatore.pirozzi@unicampania.it (S.P.)

* Correspondence: marco.costanzo@unicampania.it; Tel.: +39-081-5010-271

Received: 24 January 2019; Accepted: 20 February 2019; Published: 25 February 2019

Abstract: This paper presents the design and calibration of a new force/tactile sensor for robotic applications. The sensor is suitably designed to provide the robotic grasping device with a sensory system mimicking the human sense of touch, namely, a device sensitive to contact forces, object slip and object geometry. This type of perception information is of paramount importance not only in dexterous manipulation but even in simple grasping tasks, especially when objects are fragile, such that only a minimum amount of grasping force can be applied to hold the object without damaging it. Moreover, sensing only forces and not moments can be very limiting to securely grasp an object when it is grasped far from its center of gravity. Therefore, the perception of torsional moments is a key requirement of the designed sensor. Furthermore, the sensor is also the mechanical interface between the gripper and the manipulated object, therefore its design should consider also the requirements for a correct holding of the object. The most relevant of such requirements is the necessity to hold a torsional moment, therefore a soft distributed contact is necessary. The presence of a soft contact poses a number of challenges in the calibration of the sensor, and that is another contribution of this work. Experimental validation is provided in real grasping tasks with two sensors mounted on an industrial gripper.

Keywords: distributed force/tactile sensing; dexterous manipulation; sensor calibration

1. Introduction

The ability of modern service robots to grasp and manipulate objects in a dexterous way is still very far from human manipulation skills. Although complex artificial hands, even anthropomorphic [1–5], have been constructed, the manipulation technology is still at its infancy, not only due to current limitations of both actuation and sensing systems integrated into such complex devices, but also due to the lack of proper control algorithms that should allow the robot to perform manipulation tasks. A detailed review of tactile sensing systems for dextrous robot hands appears in [6]. In particular, only few tactile sensors are able to measure the full 6-D wrench, moreover, sensors that are able to measure also torque (e.g., the 6-axis version of OptoForce) uses an array of 3-D sensors coupled with a rigid surface that limits the frictional torque that the sensor can apply to the grasped object.

One of the main features that any grasping device should possess is the ability to grasp any kind of object as firmly as possible avoiding its slippage even in the presence of external disturbances applied to the object. Such features requires the possibility of modulating the grasping force to allow the robot to manipulate both rigid objects and fragile ones that have to be grasped with the minimum force required to hold them without causing breakage. In multifingered hands, controlling the grasping force requires the proper measurement of contact forces and moments as well as the contact locations at each finger so that external and internal forces can be estimated and properly controlled. These kinds of measurements can be performed with the combined use of tactile and force sensors [7] or by resorting to integrated force/tactile sensors [8,9]. Recently, this kind of sensor has been integrated into commercial parallel grippers for controlling both the linear and rotational slippage of rigid objects of parallelepiped shape [10], demonstrating that it allows the safe grasping of objects under uncertain conditions, namely unknown weight, center of mass and friction, as well as subject to unknown external forces. Underactuated soft robotic hands have been recently used to safely grasp both rigid and fragile objects [11,12]; they exploit the intrinsic compliance of the apparatus to grasp objects with different shapes and stiffnesses.

Grasping is an issue, but how about in-hand manipulation? At a first sight, one might think it can be performed only by multifingered hands. However, currently available anthropomorphic robotic hands still have limited reliability and high complexity, whereas, underactuated soft robotic hands have a limited number of controlled degrees of freedom hence they are mostly used for compliant grasping. That is why manipulation applications mostly adopt simple grippers. Parallel jaw grippers are by far the most widespread, owing to their reliability, low cost and ease of control and integration into standard industrial robots. Such kind of grippers have a limited dexterity, nevertheless researchers have recently demonstrated that they can be used to perform dexterous manipulation actions exploiting the so-called *extrinsic dexterity* concept [13]. In practice, dexterity is provided not only by the degrees of freedom of the grasping device but also by external aids such as gravity or environmental constraints. Examples of in-hand manipulation with these simple devices can be found in [14,15], where a visual feedback was used in an adaptive control algorithm to allow a grasped object rotate in-hand to achieve a given orientation. The same task has been executed by resorting only to force/tactile feedback in [16], where the measurement of both normal and tangential force components together with the torsional moment demonstrated to be effective without the need of any additional external sensors. These experiments, however were carried out only with rigid objects of parallelepiped shape.

In the present paper the complete design and calibration of a new and upgraded version of that force/tactile sensor is presented. The new sensor design starts from the main requirement to manipulate objects of generic shape avoiding both linear and rotational slippage. This goal leads to the need of a soft contact surface so that significant torsional moments can be held by the sensor. To allow application of slipping avoidance algorithms when a soft sensor pad interacts with an object with generic shape, estimation of the contact geometry is essential for the determination of friction model parameters [17], and thus a tactile map with suitable spatial resolution should be designed based on the accuracy requirement on the contact geometry estimation. In this work, it is assumed that objects interacting with the sensor have a curvature radius larger than the one of the sensor pad, therefore estimating contact geometry means an estimation of the normal direction to the contact surface.

Since the tactile map is the sensor output from which all measurements have to be derived, gray-box approaches like in [18] could be adopted. However, physics-based methods were revealed to be effective only in special cases, i.e., interaction with objects with parallelepiped shape, hence a more general calibration procedure has to be sought for. Calibration is of paramount importance in robotics not only for the perception system but also for the robot system itself [19,20]. Furthermore, estimation of the torsional moment poses some constraints on the geometry of the sensor pad. First, the effect of the torsional warping of the deformable layer can be exploited to extract the torsional moment only if its geometry is not axisymmetric. Then, a suitable trade-off between surface curvature and extension of the contact surface must be achieved to ensure proper estimation of force tangential components and torsional moment.

The novel calibration procedure is aimed at enhancing the estimation accuracy especially of the torsional moment component of the contact wrench. It is based on a machine learning approach and, in detail, on the training of a multi-layer feed-forward neural network (FF-NN), and special attention had to be given to the construction of the training set, as the dimension of the input space is 25 while the dimension of the target space is only 6. Therefore, a twofold approach has been followed to ensure proper coverage of both input and target spaces. First, a dedicated graphical user interface (GUI) has been designed to aid the user during the manual calibration of the sensor. The GUI displays in real-time the data acquired by a reference force/torque sensor and those of the force/tactile sensor to be calibrated. Such data are then recorded only if they are considered properly acquired. Specific metrics based on the so-called *limit surface* concept [21] have been purposefully devised to distinguish between admissible and spurious samples. Second, a novel data decimation algorithm in combination with a data compression technique based on the principal component analysis (PCA) has been conceived to ensure the most uniform sampling of both input and target spaces, to avoid unnecessary overfitting of the FF-NN. An important issue in the training of the FF-NN is the large difference in the dimensionality of input and target spaces. In fact, on one hand, accurate estimation of the contact geometry requires high spatial resolution and at least an array of 4×4 taxels. On the other hand, the contact wrench has only six components. Therefore, the data compression pre-processing step is crucial to ensure the minimal complexity of the network architecture. The validation of the calibration procedure is demonstrated by specific validation experiments performed with two sensors mounted on an industrial gripper equipping a robotic arm to grasp different objects. To carry out experiments on the robotic arm, the sensor was interfaced to the control PC via a USB connection in order to avoid the limitations of the gripper communication interface. To reduce the cabling problems, the sensor PCB design is compatible with a wi-fi interface too.

2. Design of the Force/Tactile Sensor

This section firstly recalls the requirements and the working principle of a force/tactile sensor for dexterous manipulation applications. Then a discussion about the generalization of the design procedure is presented, which allows the acquisition of an optimal and detailed design of the force/tactile sensor for the integration into commercial grippers.

2.1. Requirements for Dexterous Manipulation

The dexterous manipulation is a very challenging robotic task, especially when the objects to manipulate have mechanical properties (e.g., weight, shape, stiffness, friction coefficient) very different from each another and/or not a priori known. In these cases, the use of exteroceptive sensors, able to provide data concerning object properties, is necessary to implement suitable control strategies. The more information that is available, the greater the likeliness of a successful completion of a dexterous manipulation task. The major objective that can be pursued is to design a sensor as similar as possible to human touch, i.e., able to supply different types of information about the manipulated objects at the same time: A tactile map, an estimate of contact forces and moments. Moreover, it should also allow the robot to handle fragile objects without breaking them. To this aim, a first requirement concerns the mechanical interface between the sensing components and the manipulated objects. The use of a rigid interface does not allow manipulation of fragile objects and adaptation to different object shapes. Moreover, a rigid interface allows the measurement of pure torsional moments only by using a flat extended surface and by ensuring that the contact occurs with this flat surface perfectly parallel with the object. As a consequence, a solution with a soft pad as contact interface becomes a mandatory requirement. The design of this soft pad has to provide a domed shape to improve the sensor adaptation to different object shapes. By using a Finite Element Model, in [8] the authors demonstrated how the deformations of the soft cap are related to external force components and that a trade-off between a domed shape with a high and a low curvature radius provides good sensitivity for all force components. The second requirement concerns the transduction method. Whereas the objective is not only the estimation of contact force and moment, the classic solution of a force/torque sensor with mechanical frames with bonded sensing elements (usually strain gages or capacitive sensors) has to be overcome. A distributed measurement of the soft pad deformation due to a contact appears to be the best solution to reconstruct the maximum amount of information about the manipulated object properties. Since this solution is not implementable, the alternative is to measure the deformation of the soft pad in a discrete number of points by using sensing elements spatially distributed on a plane positioned on the bottom side of the pad. These distributed measurements are correlated to the contact state and allow, after a suitable calibration, the reconstruction of several mechanical properties of the object. In particular, in [17] the authors demonstrated how the distributed measurements at the bottom of the soft pad can be used to reconstruct the contact plane pose and consequently the force components in the contact plane, from which physical properties for the contact can then be estimated, such as the friction. Concerning the spatial resolution of the sensing elements, the paper [18] discusses how the estimation of the contact properties depend on this parameter. Moreover, for specific applications (e.g., wires manipulation) the number of sensing elements can also be optimized as shown in [22]. Another requirement is related to both the soft pad shape and the spatial distribution of the sensing elements. In particular, in order to guarantee a sensitivity to external torsional moments applied to the sensor, the soft pad has to be designed with a shape able to transduce the torsional moment into a deformation measurable by the sensing elements. To this aim, the torsional warping phenomenon [23] can be used, by recalling that only the torsion of a non axisymmetric structure allows its generation. As a consequence, the soft pad with a domed top as contact interface should have a non axisymmetric bottom (e.g., a square base). The sensing elements have to be spatially distributed on the area covered by the soft pad base. Additional characteristics to define, common to all sensors, are the working range and the sampling frequency. As deeply discussed in previous papers of the same authors, the force and moment measurement range depends on the hardness of the material used for the realization of the soft pad. Instead, the sampling frequency depends on the number of sensing elements, the interrogation strategy and the communication interface with the main controller [9,24]. For dexterous manipulation tasks the slipping detection and avoidance are fundamental and all techniques presented in literature work much better with higher sampling frequency.

2.2. The Working Principle and the Technology

The proposed sensor is based on the working principle firstly presented in [9]. On the basis of requirements described above, the basic idea foresees a suitably designed deformable layer positioned above a discrete number of sensible points (called "taxels"), in order to transduce the external force and moment, applied to the sensor, into deformations, which are measured by the taxels. The taxels, spatially distributed below the deformable layer, provide a set of signals corresponding to a distributed information (called "tactile map") about the sensor deformations. The whole tactile map allows, after a calibration procedure, to estimate contact force and moment together with information about the orientation of the contact surface and object properties. The taxels have been developed by using optoelectronic technology, and in particular each sensing point is constituted by an emitter and a receiver, mounted side by side, working in reflection mode. The soft pad has been realized by using the silicone molding technology with the molds made with a high resolution 3D printing manufacturing process.

2.3. Detailed Design of the Rigid-Flex PCB

Similarly to the initial prototype in [9], the developed sensor is mainly constituted by three components: A Printed Circuit Board (PCB), a rigid grid and a deformable cap. However, for this work, the design of all these components has been optimized on the basis of the requirements discussed in Section 2.1. The first improvement with respect to [9] concerns the type of integrated taxels. In detail, for each taxel, the emitter/receiver couple is here constituted by a unique optoelectronic component: A Surface Mount Technology (SMT) photo-reflector, manufactured by New Japan Radio Co. (San Jose, CA, USA), with part number NJL5908AR. This device integrates in the same package the emitter, an infrared Light Emitting Diode (LED), with a peak wavelength at 920 nm and the receiver, a PhotoTransistor (PT), with a peak wavelength at 880 nm. The surface encumbrance of a single device is $1.06 \times 1.46\ \text{mm}^2$. These devices allow the realization of the PCB with a standard robotized pick-and-place procedure, which guarantees the minimization of uncertainties on the taxel positioning and on the relative orientation among the emitter and the receiver of a taxel. These uncertainties negatively affected the sensor performance when separated components were used. Differently from [9], the number of taxels have been increased in order to obtain a sensitive area sufficiently wide to manipulate a larger number of objects. In particular, the optoelectronic section of the PCB integrates 25 taxels, organized in a 5×5 matrix. The device positioning on the PCB has been made in order to obtain a grid of photo-reflectors, with spacings both vertically and horizontally of 3.55 mm among their optical axes. The same distance has been considered on the edges of the optical component matrix, by obtaining a total area to cover with the deformable layer equal to $21.3 \times 21.3\,\text{mm}^2$, as shown in Figure 1. The PCB design also foresaw specific holes for the mechanical assembly between the electronic layer and the other sensor components. Figure 1 shows the holes designed to mechanically connect the board to the finger case (via M2 screws) and the holes used to connect the grid and the deformable layer as described in the following. As in previous prototypes, for each taxel, the conditioning electronics is constituted by two resistors: One to drive the LED and a second to transduce the photocurrent measured by the PT into a voltage directly compatible with an Analog-to-Digital (A/D) converter. The same 12-bit A/D converters (manufactured by Analog Devices, with part number AD7490) with 16 channels and a Serial Peripheral Interface (SPI), used in previous works, has been integrated in the PCB design. In this case two converters are needed for the conversion of the 25 taxel signals. A preliminary version of this solution has been presented in [18], where a standard rigid PCB has been realized only with the optoelectronic and the A/D conversion sections. Preliminary tests have been carried out by interrogating the prototype with an external board via the SPI interface. In order to integrate the tactile sensor in a standard parallel gripper, for this work, a microcontroller-based section has been integrated into the PCB design. In particular, an interfacing section constituted by the microcontroller PIC16F1824, manufactured by Microchip Technology, has been added on a separate rigid board, connected to the previously described part via a flexible section. The integration of the microcontroller allows to obtain a fully integrated sensor with a programmable device used to interrogate the sensor via a standard serial interface already available in most commercial grippers. The board is completed by a standard low-noise voltage regulator with an input voltage range up to 12 V (typical range of supply voltage available on commercial grippers) and an output voltage equal to 3.3 V to supply the whole PCB. The use of the rigid-flex technology allows the integration of the PCB into a finger case compatible with the mechanical connection of standard grippers. Figure 2 reports some pictures of the whole rigid-flex PCB, with the dimensions and the description of all components. In the realized prototype, the connector compatible with the sensor port available on the commercial grippers WSG-series, manufactured by Weiss Robotics, has been integrated, in order to provide the 5 V voltage supply to the sensor and for the physical implementation of the serial interface.

Figure 1. Details of the sensitive area on the Printed Circuit Board (PCB).

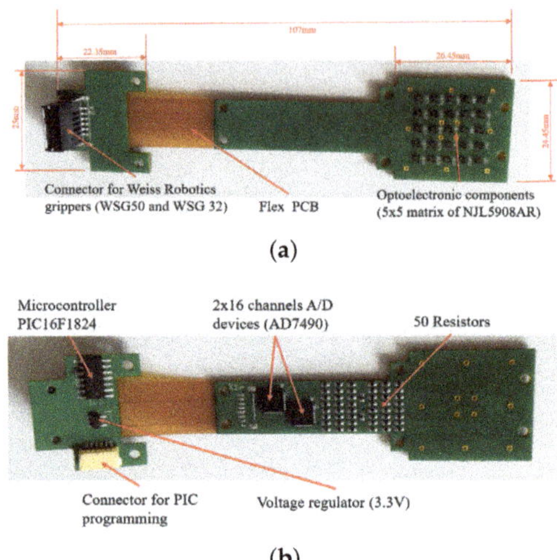

Figure 2. The tactile sensor PCB: a top view with dimensions (**a**) and a bottom view (**b**) with the components highlighted.

2.4. Detailed Design of the Deformable Pad

A mechanical structure constituted by the deformable layer and the rigid grid is connected above the PCB. The deformable layer is mainly made of white silicone with a domed top side and a square base, as shown by the picture in Figure 3a. As discussed above, the use of a non axisymmetric shape implies, in presence of torsional moments applied to the deformable layer, the generation of the torsional warping effect, which is measurable by the tactile map and allows the reconstruction of the applied moment. The mechanical properties of the silicone determine the full-scale and the sensitivity of the sensor. The realized prototype uses a shore hardness of 26 A, which corresponds to the working range reported in Section 3. Figure 3b shows the bottom side of the deformable layer, where there are the twenty-five empty cells, which present the ceilings (which in the final assembly are positioned in front of photo-reflectors) made of white silicone, while the walls among the taxels are black, to avoid cross-talk effects. According to the working priciple explained in Section 2.2, when external forces and/or moments are applied to the deformable layer, they produce vertical displacements of the white ceilings for all cells. The distances between the top of photo-reflectors and the white surfaces change, by producing variations of the reflected light and, accordingly, of the voltage signals measured by the PTs.

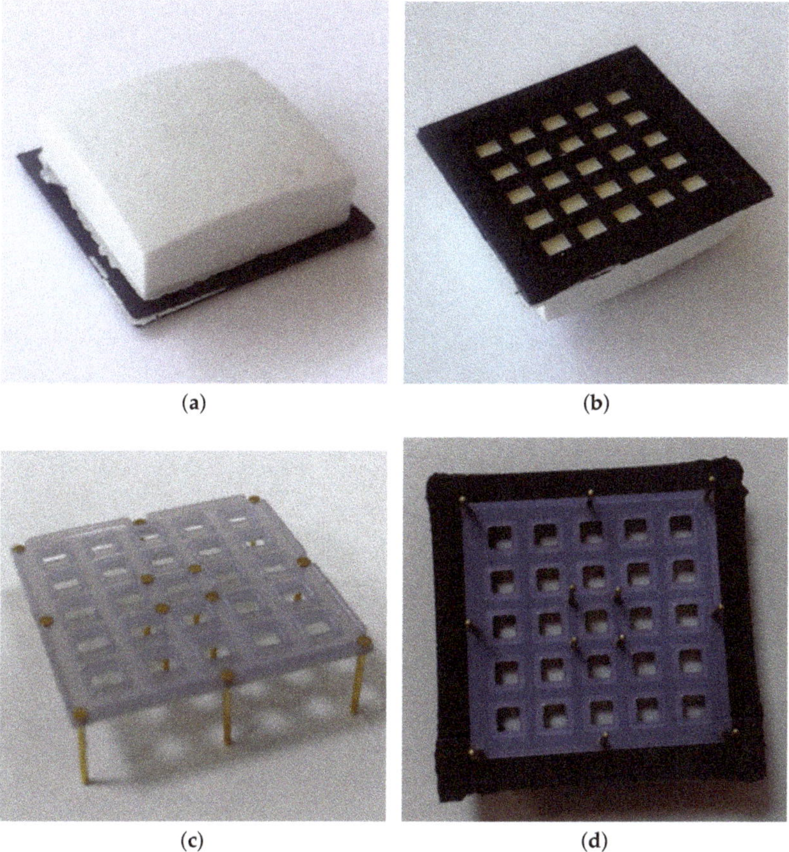

(a) (b) (c) (d)

Figure 3. Pictures of deformable layer and rigid grid: Top view (**a**) and bottom view (**b**) of the deformable layer, grid with bonded pins (**c**) and deformable layer assembled with the grid (**d**).

The addition of the third component (i.e., the rigid grid) became necessary due to the electromechanical characteristic of the optical components. In particular, the NJL5908AR photo-reflector has a non-monotonic characteristic (see Figure 4), which relates the measured voltage to the distance of a reflecting surface positioned in front of the component. As a consequence, the rigid grid has to ensure that the reflecting surface never reaches distances from the component that fall into the non-monotonic area, highlighted by the red bars in Figure 4. Note that this part of the characteristic curve is related to the optical behaviour of the photodetector and it has been obtained without the deformable sensor pad. Taking into account that the height of a component is 0.5 mm, the rigid grid has been designed with a thickness of 0.8 mm. With this choice the minimum reachable distance between a reflecting surface and a photo-reflector is $d_m = 0.3$ mm. On the other side, the silicone layer have been designed so that, in rest condition, the sum of the grid thickness and of the cell walls fixes the white ceilings at an initial distance $d_0 = 1$ mm from the emitting surface of the optical components. The integrated design of these two components allows to force the photo-reflectors to work in the monotonic working area, highlighted by the green bars in Figure 4. Considering the high definition needed for the grid realization, a 3D printing manufacturing process based on the PolyJet technology has been selected with a resolution of 16 μm. The grid design foresees holes suitable for housing rigid pins. Figure 3c shows a grid assembled with the pins, which are bonded via a cyanoacrylate-based glue. This assembled grid is then bonded to the deformable layer, by using the same glue, by obtaining the final mechanical cap reported in Figure 3d. The rigid pins that come out of the assembled cap are used to align the cells with the optical components on the PCB, thanks to the mechanical holes available on the board (see Figure 1). After the alignment, the same pins are used to mechanically connect the cap and the PCB, by soldering the rigid pins to the bottom side of the board (see Figure 2b).

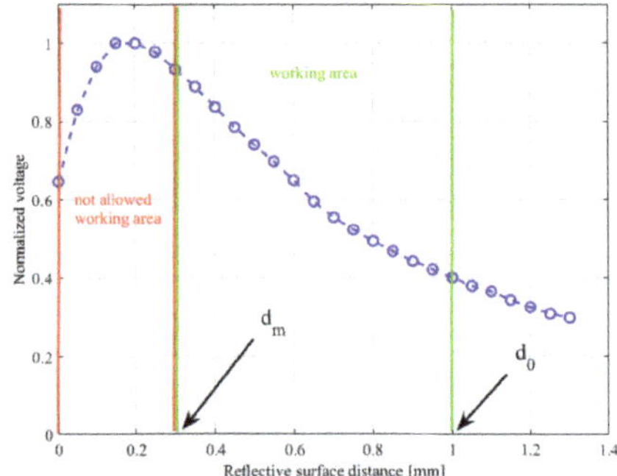

Figure 4. Characteristic for a single taxel: Normalized voltage vs. reflective surface distance.

The distance-to-voltage characteristic in Figure 4 is not the only source of nonlinearity in the sensor. The hemispherical geometry of the pad yields a nonlinear relationship between the normal force applied to the sensor and the deformation of the sensor pad, due to the variation of the contact area with the force. Figure 5 reports the normal force versus the displacement of the sensor pad tip, when applied with a rigid plane parallel to the sensitive board. The figure shows how a quadratic regression accurately fits the sampled data. Another source of nonlinearity is the material hysteresis that, however, is quite limited (about 5%) due to the specific silicone material selected, as shown in Figure 6, which reports the de-biased voltage of the central taxel versus the applied force.

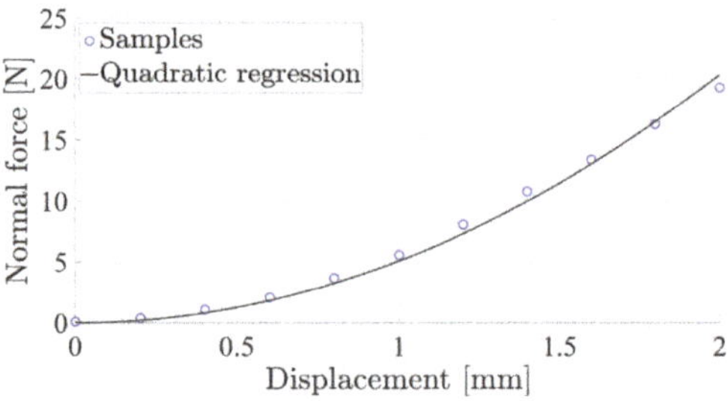

Figure 5. Force–displacement characteristic curve of the sensor pad.

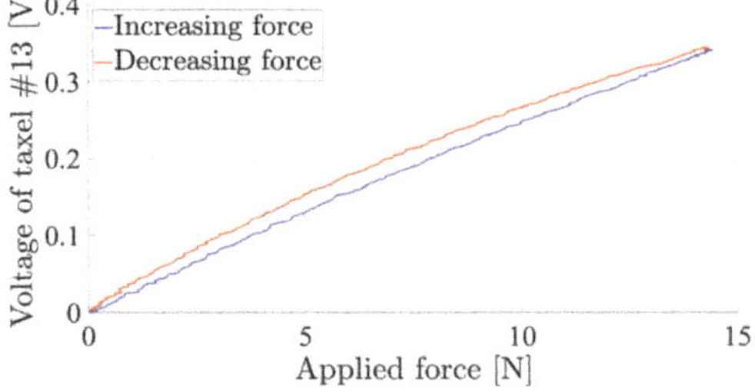

Figure 6. Force–voltage hysteresis curve of a sensor taxel.

2.5. Integration of the Sensor into a Commercial Gripper

The assembled force/tactile sensor is finally fixed inside an aluminum case suitably designed to house the sensor and for the mechanical connection to the WSG-series flange. Figure 7 reports a picture of the sensorized finger fully integrated with the gripper. The microcontroller section available on the PCB allows two possible connections to exchange data with the main PC. In the fully integrated version, the PCB takes the voltage supply directly from the sensor port available on the WSG-series flange. The same port is used to implement a standard serial communication between the gripper and the sensor. The microcontroller interrogates the A/D converters via an SPI interface and transmits the raw data (2 bytes for each taxel, for a total of 50 bytes) via its serial port. The gripper is programmable by using the LUA programming language, that is an interpreted language suitably designed for embedded applications. This language allows to interrogate the serial port with a maximum baudrate equal to 115,200 bps. The conditioning electronics integrated into the gripper, together with the interpreted language, allows to reach a maximum sampling frequency for all 25 taxels equal to 50 Hz, even though the connection between the gripper and the main PC is implemented by using an Ethernet interface. The second possible connection from the microcontroller to the robot control PC foresees the use of a standard USB-to-serial converter with an external cable, that directly connects the microcontroller to the main PC. In this case, the power supply and the serial transmission are implemented directly from the PC. With this solution the baudrate of the serial port can reach a maximum baudrate equal to 500,000 bps (in this case limited by the microcontroller). This baudrate together with the latency time of the serial port used on the control PC allows to reach a sampling frequency for all 25 taxels equal to 333 Hz. Figure 8 reports a scheme of possible connections currently available. The microcontroller is ready to be interfaced with a serial-to-WIFI adapter, in order to use a wireless connection directly with the PC. This solution will allow the avoidance of limitations related to the serial port latency time with an expected sampling frequency up to 1 kHz. On the control PC, two different ROS nodes have been developed: One to interact with the gripper, if the first solution is selected, and another one to directly interact with the microcontroller in the second case. In both cases the ROS nodes receive raw data (i.e., the 50 bytes acquired by the A/D converters) and the first elaboration consists in the reconstruction of actual voltage values, which are published to be available for the whole ROS network.

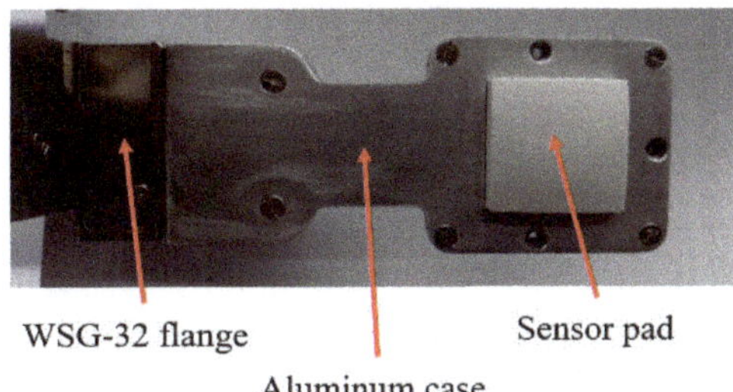

Figure 7. Picture of the sensorized finger fully integrated with the WSG-32 gripper.

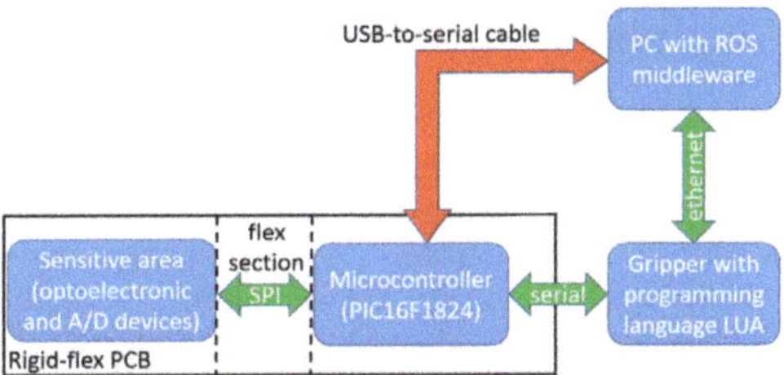

Figure 8. Data flow scheme of possible connections from the sensor to the control PC.

3. Sensor Calibration

This section illustrates the calibration procedure of the sensor. In [18] the sensor was calibrated with a gray-box model deduced by a FEM analysis. The main limitations of the previous approach are the necessity of an equalizing scaling factor for each voltage and the assumption of a contact plane orientation of $\theta = 0$. The equalization is needed to counteract unavoidable different taxel sensitivities due to different gains and operating points of the phototransistor. Moreover, the previous algorithm was able to estimate only the forces and not all the wrench components.

In this paper a new calibration procedure, based on a FF-NN, is proposed. This approach is able to overcame the limitations of the previous procedure, i.e., the parameters are implicitly learned by the FF-NN; the contact plane can be oriented in different positions; the moment vector can be estimated by including it in the target data.

The critical point of the machine learning-based approach is the training data collection. The objective is to estimate the wrench in all possible combinations in a large interval of the contact plane orientation. The dimensionality of the problem is large, so there is a significant risk of missed wrench/orientation combinations in the training set.

Moreover, the dimensionality and the correlation among the inputs, whose number (25 for the sensor in this paper) is significantly larger than the dimension of the target set (6), and the large number of samples acquired during the calibration phase can slow down the training phase and can easily cause unnecessary overfitting.

3.1. Construction of the Training Set

In order to collect the training set data, the sensor is mounted on a reference force/torque sensor, the Robotous RFT40, as in Figure 9. Σ_{sens} is the reference frame of both sensors; Σ_{sph} is a frame placed in the center of the undeformed silicone sphere; Σ_{CoP} is a frame placed in the center of pressure (CoP) of the contact area with the z axis normal to the contact plane.

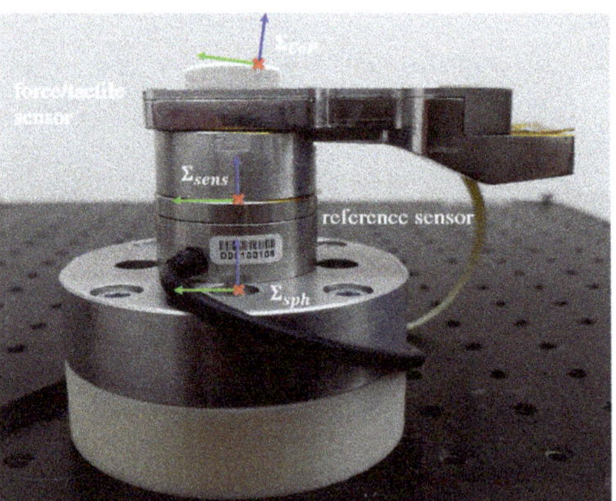

Figure 9. Testbench for sensor calibration.

Data generation is made by an operator who applies forces and moments by touching the sensor with an object. The target wrench and the input tactile voltages are recorded synchronized through the ROS network (see Section 2.5).

In order to ensure a good training set, the input space (and consequently the target space) has to be properly covered. Furthermore, bad data should be avoided, e.g., samples during slipping of the object on the sensor pad surface or during the relaxing phase of the deformable layer.

These issues are tackled by resorting to a dedicated MATLAB GUI (Figure 10). The user interface displays in real-time the calibration data acquired. The visualization of the samples is carried out using the limit surface (LS) theory [21], which is an extension of the Coulomb friction model to the case of roto-translational slippage. The LS gives information about the maximum force and torsional moment that can be applied before a slippage occurs, it is a surface defined in the 3D space of the two tangential force components and the torsional moment (the component of the contact moment along the direction normal to the contact surface). When the wrench is inside this surface no slippage occurs, otherwise, there is relative motion between the two contacting surfaces. The maximum pure tangential force (that is the component of the force tangential to the contact surface) and torsional moment are given by:

$$f_{t\max} = \mu f_n; \tag{1}$$

$$\tau_{n\max} = \alpha f_n^{\gamma+1}; \tag{2}$$

where f_n is the component of force normal to the contact frame, μ is the classical Coulomb friction coefficient, α and γ are parameters of the maximum torque model [25]. All parameters are experimentally estimated through the procedure described in [16].

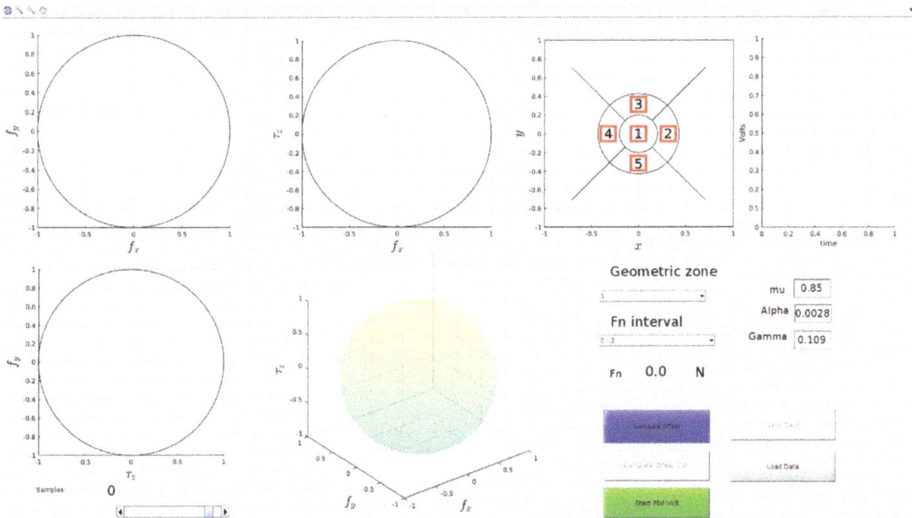

Figure 10. Graphical user interface (GUI) used in the calibration procedure.

The GUI visualizes the 3D space of tangential force and torsional moment normalized with respect to $f_{t\max}$ and $\tau_{n\max}$, respectively, using four different plots: A 3D plot and three separate plots one for each view from each coordinate axis. In this normalized space, the LS is approximated as a unit sphere centered in the origin drawn in the 3D plot. This method is useful to discard samples acquired in any slipping phase, namely, samples outside the LS are not included into the training set.

This decision about data inclusion cannot be directly taken based on the measured wrench referred to the sensor frame Σ_{sens}. In fact, the LS is defined based on the wrench referred to the Σ_{CoP} frame. The homogeneous transformation matrix expressing the pose of the CoP frame with respect to the sensor frame is estimated considering the tactile map. First of all, the centroid of the tactile map is calculated as

$$x_C = \frac{\sum_{i=1}^{25} x_i \Delta v_i}{\sum_{i=1}^{25} \Delta v_i}; \tag{3}$$

$$y_C = \frac{\sum_{i=1}^{25} y_i \Delta v_i}{\sum_{i=1}^{25} \Delta v_i}; \tag{4}$$

where (x_i,y_i) are the coordinates of the ith taxel and Δv_i is the difference between the actual voltage value and the voltage value in rest conditions. The centroid is also plotted in the GUI in a separate plot to help the user understand where he/she currently is touching the sensor. The CoP is considered located in the point on the contact surface corresponding to neglecting the deformation of the sphere (consider that this computation is simply aimed at helping the operator in the calibration procedure). Hence, the coordinates of the CoP with respect to the Σ_{sph} frame are

$$\boldsymbol{p}_{CoP}^{sph} = \begin{bmatrix} x_C \\ y_C \\ \sqrt{R^2 - x_C^2 - y_C} \end{bmatrix} \tag{5}$$

where $R = 50\,\text{mm}$ is the sphere radius. Given the distance between Σ_{sph} and Σ_{sens} (20 mm) it is trivial to find the coordinates of the CoP with respect to the sensor frame ($\boldsymbol{p}_{CoP}^{sens}$).

The orientation of the contact plane is basically given by the normal vector to the contact plane. Since the GUI is just an aid for the operator, the contact plane can be well approximated as tangent to the sphere. So the normal unit vector can be calculated with respect to the sphere frame as

$$\hat{\boldsymbol{n}}_{CoP}^{sph} = \frac{1}{R}\boldsymbol{p}_{CoP}^{sph}. \tag{6}$$

Choosing the sphere frame aligned to the sensor frame, this normal vector has the same components in the sensor frame and it is selected as the z-axis of the contact frame. The x and y axes of the contact frame can be trivially choosen as the projection of the same axes of sensor frame on the contact plane (conveniently normalized). The computed axes can be organized into a rotation matrix $\boldsymbol{R}_{CoP}^{sens}$ and, finally, the homogenouse transformation matrix of the contact frame is

$$\boldsymbol{T}_{CoP}^{sens} = \begin{bmatrix} \boldsymbol{R}_{CoP}^{sens} & \boldsymbol{p}_{CoP}^{sens} \\ \boldsymbol{0}^T & 1 \end{bmatrix}. \tag{7}$$

Finally, by inverting the last matrix, it is possible to find the force and moment vectors in the contact frame

$$\boldsymbol{f}^{CoP} = \boldsymbol{R}_{sens}^{CoP}\,\boldsymbol{f}^{sens}, \tag{8}$$

$$\boldsymbol{\tau}^{CoP} = \boldsymbol{R}_{sens}^{CoP}\,\boldsymbol{\tau}^{sens} + \boldsymbol{p}_{sens}^{CoP} \times \boldsymbol{f}^{CoP}. \tag{9}$$

With the LS aid, the operator can visualize only the tangential forces and the torsional moment. It is not possible to see variations in the normal force and in the contact plane orientation. Moreover, it is impossible to include such information in the plot because it is already a 3D plot and plots with higher dimensions are impossible to easily visualize. To overcome this problem, in the GUI the operator can select a target interval for the normal force among a set of predefined intervals. In addition, the GUI shows the base surface of sensor divided in polar areas (see Figure 10). Given the centroid position, a polar area is uniquely defined. In the same manner, given the normal force value, an interval of forces is defined. In this way it is possible to define various 3D spaces, one for each possible combination of the normal force interval and polar area. The task of the operator is to cover all these 3D spaces with samples, and the program will automatically discard bad samples.

3.2. Training Set Pre-Processing

The pre-processing step has a twofold aim. It first reduces the dimension of the input space, that is significantly larger (25) than the target space dimension (6), through a data compression technique, e.g., the Principal Component Analysis (PCA). Furthermore, it tries to make the sample density of the training set as uniform as possible via a new decimation algorithm.

The motivation for such an algorithm is that samples are often collected so that there are zones of the training set with a very high density compared to others. This is typical when forces are low, e.g., the operator is at the beginning of a maneuver. So there are a lot of samples that add few new information to the dataset. This can cause a useless increase in the computational load and can encourage the learning algorithm to specialize the model towards the behaviour in these high density zones. Therefore, these samples should be removed.

The number of samples is reduced through a novel bubble-based decimation algorithm described hereafter in a general case. The idea is to fix a maximum density for the samples in the input space. Let N be the total number of samples and the couple $(\boldsymbol{v}_i, \boldsymbol{w}_i)$ the ith sample with input $\boldsymbol{v}_i \in \mathbb{R}^m$ and target $\boldsymbol{w}_i \in \mathbb{R}^t$, the training set is:

$$T_s = \{(\boldsymbol{v}_i, \boldsymbol{w}_i),\, i \in \mathcal{I}_{T_s}\}. \tag{10}$$

where

$$\mathcal{I}_{T_s} = \{1, ..., N\}. \tag{11}$$

Note that in the particular case of study $\boldsymbol{v}_i \in \mathbb{R}^{25}$ and $\boldsymbol{w}_i \in \mathbb{R}^6$. The main idea is to define a bubble in the space of the inputs such that, centering the bubble in a sample, no other sample is in the bubble. In other words, the objective is to find a subset T_s^* of T_s such that

$$T_s^* = \{(\boldsymbol{v}_i, \boldsymbol{w}_i) \in T_s, i \in \mathcal{I}_{T_s^*}\}, \tag{12}$$

where

$$\mathcal{I}_{T_s^*} = \{j \in \mathcal{I}_{T_s} : \|\boldsymbol{v}_j - \boldsymbol{v}_k\| > r \;\forall k \in \mathcal{I}_{T_s},\; k \neq j\} \tag{13}$$

being r the radius of the bubble. In this way the maximum density in the input space will be of one sample per bubble.

A MATLAB function that implements the decimation algorithm is the following:

```
function [inputs,targets,mask_keep]...
        = decimation(inputs,targets,radius)
% Bubble-based decimation function,
% Ts = {(inputs_i,targets_i)}

%Initialization
radius_square = radius^2;
mask_keep = false(1,size(inputs,2));
mask_not_computed = ~mask_keep;
%Repeat until process all samples
while any(mask_not_computed)
  actual_index = find(mask_not_computed,1);
  mask_keep(actual_index) = true;
  mask_not_computed(actual_index) = false;
  mask_not_computed(mask_not_computed)=...
    sum( ( inputs(:,mask_not_computed)...
        - inputs(:,actual_index) ).^2 ...
       ) > radius_square ;
end
%select only good samples
inputs = inputs(:,mask_keep);
targets = targets(:,mask_keep);
end
```

The bubble-based decimation can be applied to an heterogeneous input space too, e.g., made of inputs of a different scale. In that case it is necessary to have a pre-normalization of the input data.

Considering that for each normal force interval and polar area the voltage map has to be rather different, this algorithm can be applied separately on the data of each 3D space defined in Section 3.1. In this way the computational load of the decimation is reduced.

The second goal of the preprocessing step is the reduction of the input space dimension and this is achieved by applying a Principal Component Analysis (PCA) technique. Figure 11 reports the plot of the singular values of the input covariance. The plot is normalized with respect to the maximum singular value and, after 15 components, the singular values are below the 0.1%. The choice made here is to take into account the first $r = 15$ components.

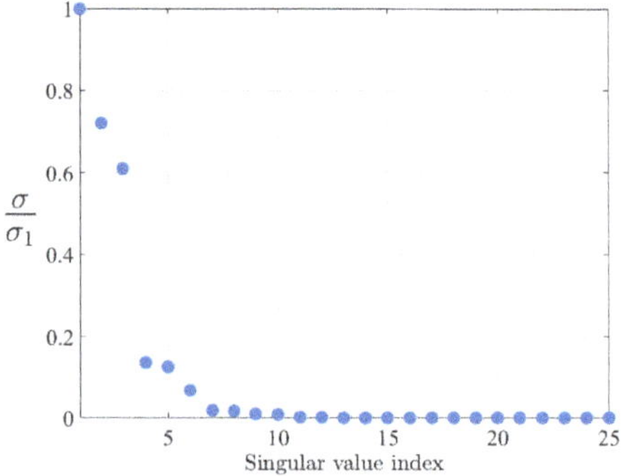

Figure 11. Normalized singular values of the Principal Component Analysis (PCA) of the training set.

Let $\boldsymbol{U} \in \mathbb{R}^{m \times r}$ be the matrix of the first r singular vectors, the ith compressed input will be

$$\boldsymbol{v}_i^* = \boldsymbol{U}^T \boldsymbol{v}_i. \tag{14}$$

Eventually, the reduced and compressed training set is used to train a FF-NN with h hidden layers, each made of n_l neurons, with $l = 1...h$.

3.3. FF-NN Training

After collecting the training set according to the procedure described in Section 3.1, the bubble-based decimation algorithm of Section 3.2 has removed the 53.45% of samples. The GUI can be used to visualize the reduced set in order to graphically view the quality of the decimation and the degree of coverage of the target space. If the decimation had removed too many samples there would be holes in one or more of the plots in the GUI. Figure 12 shows the T_s^* data in the GUI, the samples are visualized for a normal force in the interval $[8, 10]$ and for a centroid position inside zone 4. It is possible to see that, despite the data decimation, the 3D space is still well covered.

The data are finally used to train a FF-NN. The network is made of six hidden layers in each one with 30 neurons and a sigmoidal activation function, contrariwise the output layer has a linear activation function and six neurons. Figure 13 shows the FF-NN fitting on the training data in a certain range. The quality of the reconstruction of all wrench components is satisfactory.

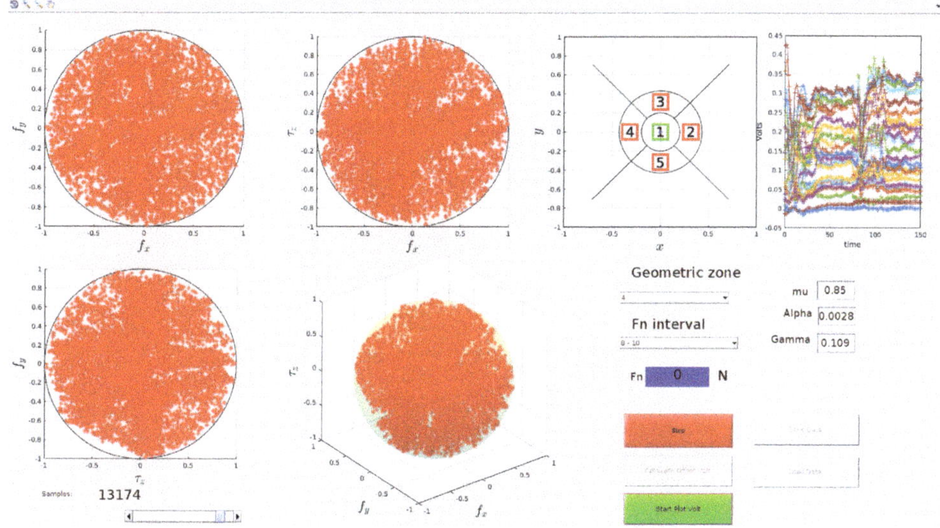

Figure 12. Training set visualized on the GUI after the bubble-based decimation.

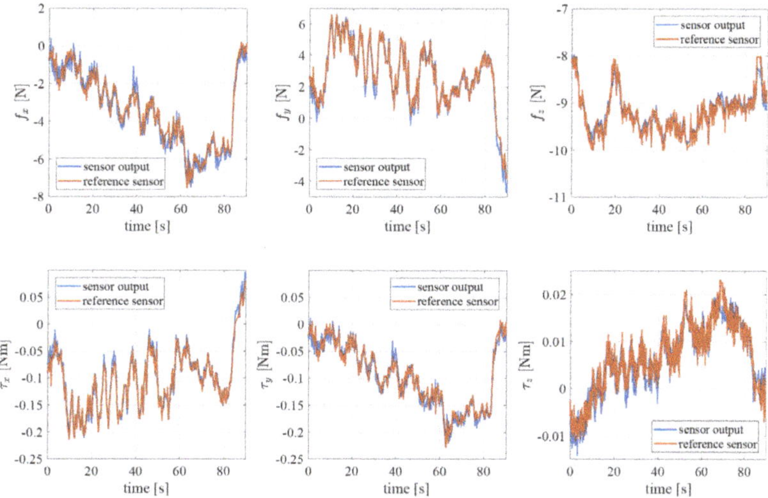

Figure 13. Result of the feed-forward neural network (FF-NN) fitting on the whole acquired training set before decimation.

4. Experimental Validation

This section presents the experimental validation of the proposed sensor and algorithms for its calibration in different experiments with two sensors mounted on the WSG-50 parallel gripper grasping an object in several configurations so as to assess the quality of calibration of all components of the wrench.

The sensors are mounted on the parallel gripper as shown in Figure 14. The external forces applied affect both sensors in a non-symmetric manner (depending on the grasped object orientation and shape). In order to measure the external forces on the object it is necessary to combine the measures of both sensors. Let Σ_{s1} and Σ_{s2} be the frame of each sensor, and Σ_{grasp} be the grasp frame that is located at the center of the fingers as shown in Figure 14. The relative position of these frames is not constant, but it depends on the position of the gripper fingers.

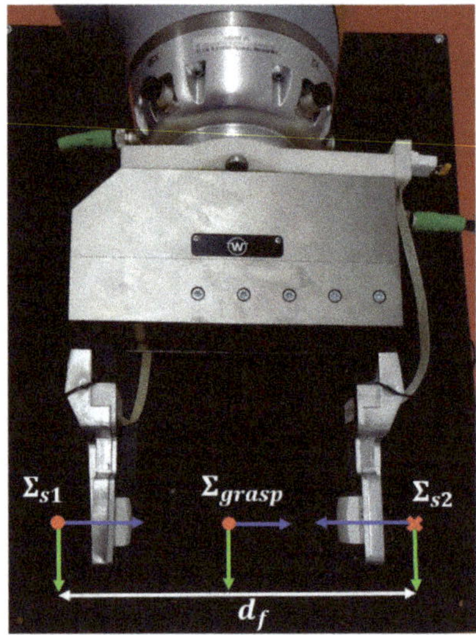

Figure 14. Definition of the grasp frame Σ_{grasp}.

Choosing the grasp frame aligned with the sensor 1 frame, the relative position of the frames is given by

$$R_{s1}^{grasp} = I_3, \tag{15}$$

$$R_{s2}^{grasp} = \begin{bmatrix} -1 & 0 & 0 \\ 0 & 1 & 0 \\ 0 & 0 & -1 \end{bmatrix}, \tag{16}$$

$$p_{s1}^{grasp} = \begin{bmatrix} 0 \\ 0 \\ -d_f/2 \end{bmatrix}, \tag{17}$$

$$p_{s2}^{grasp} = \begin{bmatrix} 0 \\ 0 \\ d_f/2 \end{bmatrix}, \tag{18}$$

where d_f is the distance between fingers which depends on gripper state. Let s_i be the ith sensor, the wrench of each finger can be expressed in the grasp frame as

$$f_{s_i}^{grasp} = R_{s_i}^{grasp} f_{s_i}^{s_i} \tag{19}$$

$$\tau_{s_i}^{grasp} = R_{s_i}^{grasp} \tau_{s_i}^{s_i} + p_{s_i}^{grasp} \times f_{s_i}^{grasp} \tag{20}$$

Finally, the external wrench in the grasp frame is simply the sum of the components of each finger in the grasp frame, in fact, the grasp matrix is simply constituted by two identity matrices owing to the so-called virtual stick concept [26], i.e.,

$$\begin{bmatrix} \boldsymbol{f}^{grasp} \\ \boldsymbol{\tau}^{grasp} \end{bmatrix} = \begin{bmatrix} \boldsymbol{I}_6 & \boldsymbol{I}_6 \end{bmatrix} \begin{bmatrix} \boldsymbol{f}_{s_1}^{grasp} \\ \boldsymbol{\tau}_{s_1}^{grasp} \\ \boldsymbol{f}_{s_2}^{grasp} \\ \boldsymbol{\tau}_{s_2}^{grasp} \end{bmatrix}. \tag{21}$$

Note that since the z-axes of the three frames are aligned to the same straight line, the variable d_f affects only the two moments τ_x and τ_y and not the others components of the wrench.

4.1. Reconstruction of the Normal Force Component

A first validation experiment is aimed at assessing the quality of the reconstruction of the normal force component. The reference force sensor is grasped by the gripper applying a chirp force signal from 0.05 Hz to 0.1 Hz in 40 s to the fingers. The estimated normal force of a finger is then compared to the normal force measured by the reference sensor. The signals and the corresponding error are reported in Figure 15, which shows a maximum error of about 0.7 N for a maximum force of 16 N.

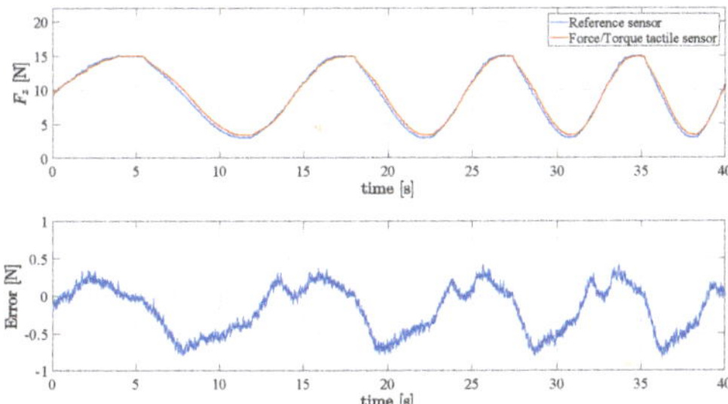

Figure 15. First experiment: Assessment of the reconstruction of the normal force component.

4.2. Reconstruction of the Tangential Force Components

Figures 16 and 17 show the result of a second experiment, where an empty aluminium box is grasped with two orientations of the gripper, one holding the object as in Figure 18 (left) and the second one holding the object as in Figure 18 (right) so as to generate tangential components along both x_s and y_s. Then, the box is filled in with weights of 0.49 N released one after the other. Figure 16 reports the ground truth weights (red bars) compared to estimated weights in terms of the total tangential force in the grasp frame (indicated with f_t in the figure legends). Taking into account that the weight of the empty box is 1.5 N, the accuracy of the sensor is quite satisfactory as the largest error is about 0.2 N and each measured step corresponds to about 0.5 N. The same experiment has been repeated by rotating of the gripper 45° and the results in Figure 17 (top) show a similar behaviour and accuracy. Figure 17 (bottom) refers to a similar trial by starting the filling process of the box previously filled in for a total initial weight of 5.9 N, so as to reach higher tangential forces. The errors keep below 0.2 N.

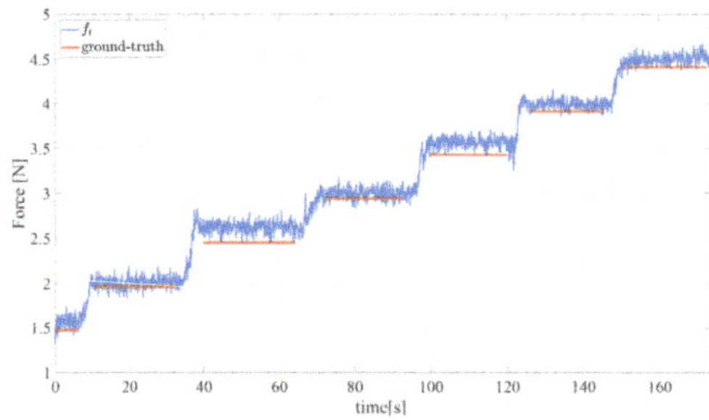

Figure 16. Second experiment: Assessment of the reconstruction of the grasped object weight—first grasp.

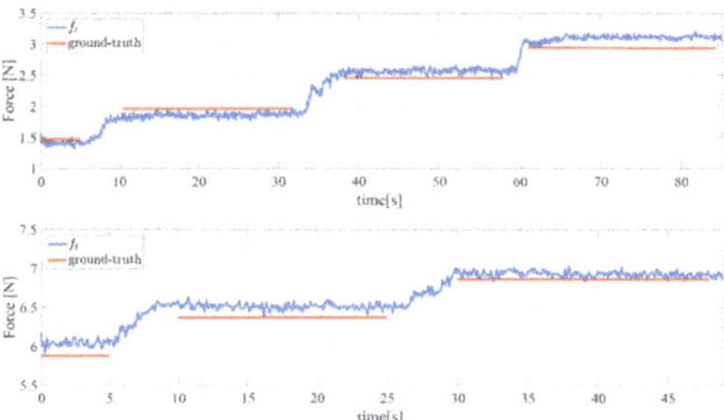

Figure 17. Second experiment: Assessment of the reconstruction of the grasped object weight—second grasp.

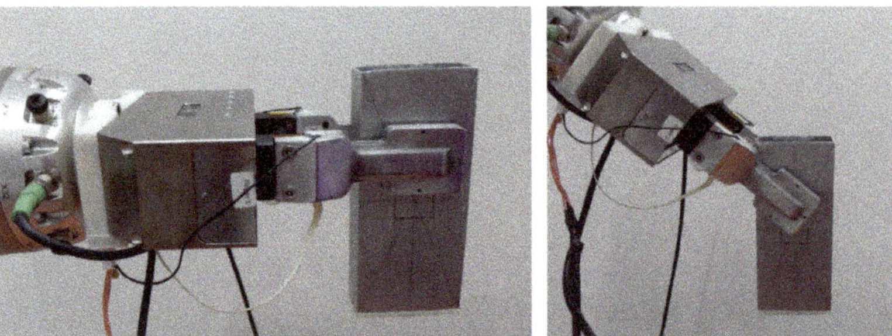

Figure 18. Second experiment: First grasp (**left**) and second grasp (**right**) for the assessment of the tangential force components accuracy.

4.3. Assessment of Sensor Sensitivity and Dynamic Range

A third experiment has been carried out to assess the sensor sensitivity and to demonstrate that it is able to measure very low forces and thus it can be exploited to grasp both light objects with low gripping forces and heavy objects that require large forces to be held. A heavy glass bottle is grasped by the gripper and the weight estimated by the sensors is compared to the actual weight in Figure 19 (top) to show how the sensor behaves close to its full scale range. To demonstrate that the sensor can actually measure low forces with an accuracy high enough to effectively grasp an object with correspondingly light gripping force, a light and empty cardboard box is grasped and its estimated weight is compared to the actual one. Figure 19 (bottom) shows the result of the test, which confirm that the light grasping force of about 0.5 N does not cause any noticeable deformation of the box. It should be highlighted that the estimated weight of about 0.4 N is computed as in (21) and this means that each sensor is able to measure half of such value.

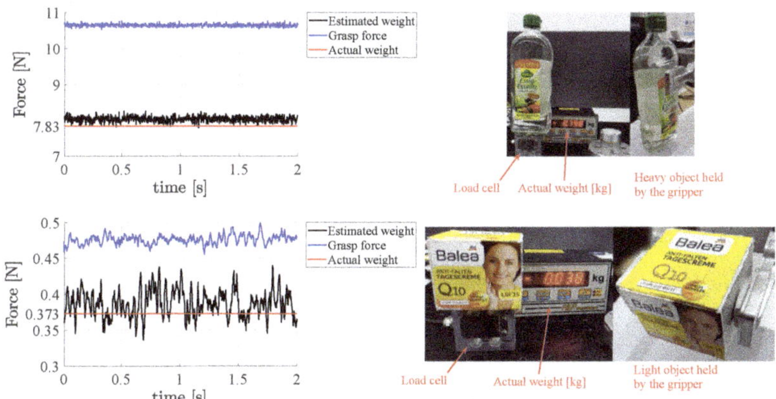

Figure 19. Third experiment: Assessment of sensor sensitivity (**bottom**) and dynamic range (**top**).

4.4. Reconstruction of the Contact Plane Orientation

A fourth experiment is devoted to evaluate the capability of the force/tactile sensor to estimate the contact force components in the CoP frame as defined in Section 3. An object with non parallel faces is grasped between the fingers as shown in Figure 20. It is clear that in the CoP frame, the tangential components should be lower than those in the sensor frame, on the contrary the normal component in the sensor frame is lower than that in the CoP frame. This expectation is confirmed by the results reported in Figure 21. To quantify the accuracy in the estimation of the contact plane orientation, the contact normal is estimated according to Equation (6) based on the centroid of the tactile map, shown in Figure 22 by spatially interpolating the taxel values. Note that the small x component of the centroid can be attributed to a slight misalignment of the gripper with respect to the horizontal table where the object is placed. The estimated angle between the normal direction and the z_{s1} direction is equal to 4.5°, compared to an actual value of 5°. It is evident that computation of the ratio between normal and tangential contact force components is significantly affected by the angle between the contact normal and the z axis of the sensor frame. Such ratio is at the basis of any strategy for slipping avoidance based on friction models.

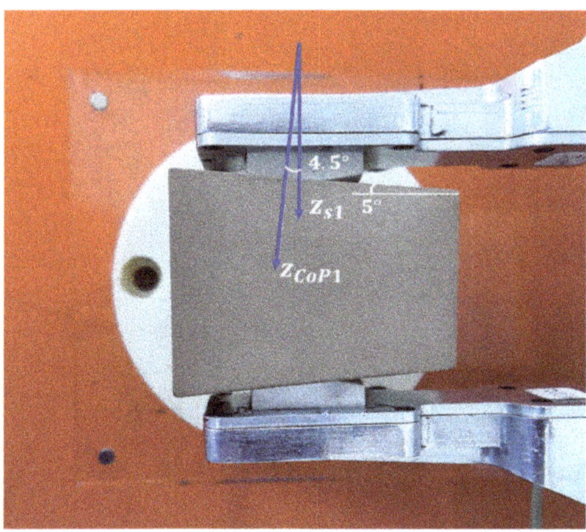

Figure 20. Fourth experiment: Grasp to validate the contact geometry estimation capability.

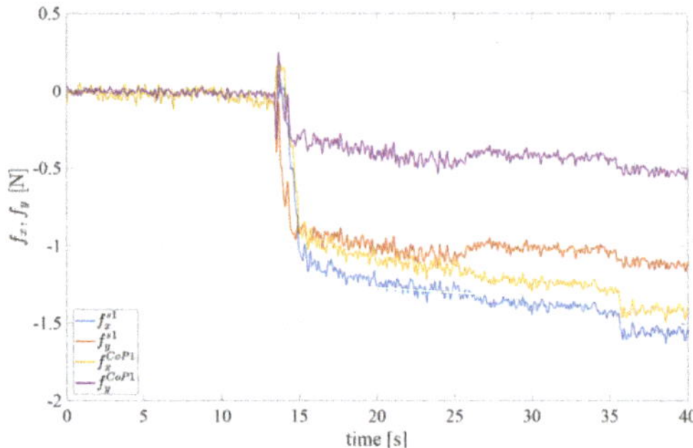

Figure 21. Fourth experiment: Force components rotated into the CoP frame.

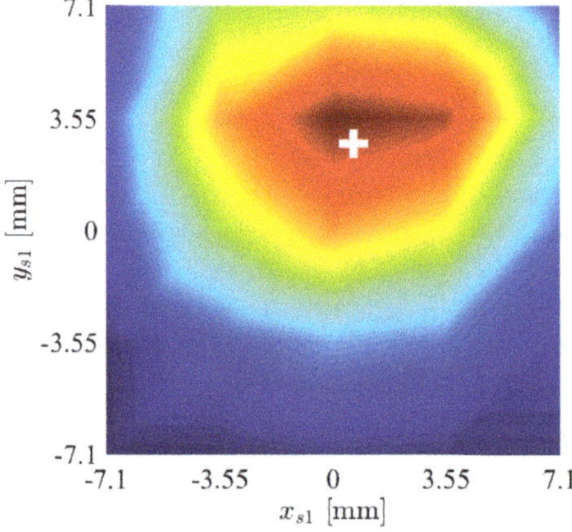

Figure 22. Fourth experiment: Tactile map and corresponding centroid (white cross).

4.5. Reconstruction of the Torsional Moment

A fifth experiment is devoted to validate the calibration algorithm for the reconstruction of the torsional moment component. An aluminium box with an initial weight of 1.22 N has been grasped as shown in Figure 23 in its center of gravity. Then, different weights have been hung at an extremity to apply a torsional moment to the sensors. The results are reported in Figure 24. The top subplot refers to a normal grasp force of 5 N and three weights each of 0.098 N have been hung one after the other corresponding to steps in the torsional moment of about 0.0068 Nm. This final condition corresponds to the image reported in Figure 23 where the 10° rotation of the aluminium box is caused by the torsional deformation of the sensor pad, and not by a slippage. In the trial shown in the middle subplot, the grasp force is increased to 15 N to allow holding larger torsional moment. Three weights of 0.29 N have then been hung to the box extremity reaching a total torsional moment of about 0.061 Nm with steps of 0.02 Nm. In the last trial reported in the bottom subplot the grasp force is reduced down to 2.5 N and after hanging a second weight of 0.098 N a slipping event occurs as demonstrated by the sudden drop of the measured torsional moment. The tangential forces and torsional moments measured in the whole fourth experiment are normalized with respect to (1) and (2), respectively, and reported together with the limit surface in Figure 25. The red and blue dots are all inside the limit surface and in fact no slipping occurs, while the green dot outside the limit surface corresponds to the slipping event indicated in the bottom subplot of Figure 24.

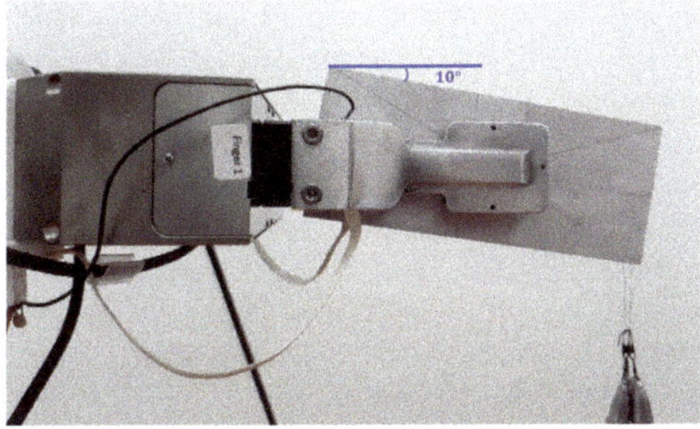

Figure 23. Fifth experiment: Grasp for the validation of the calibration of the torsional moment component.

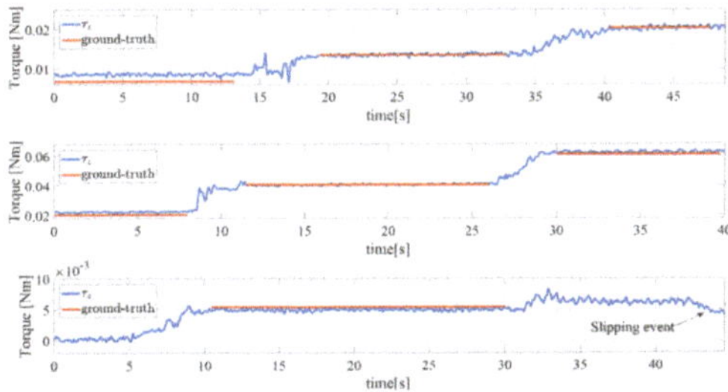

Figure 24. Fifth experiment: Validation of the calibration of the torsional moment component, with $f_n = 5\,\mathrm{N}$ (**top**), $f_n = 15\,\mathrm{N}$ (**middle**), $f_n = 2.5\,\mathrm{N}$ (**bottom**).

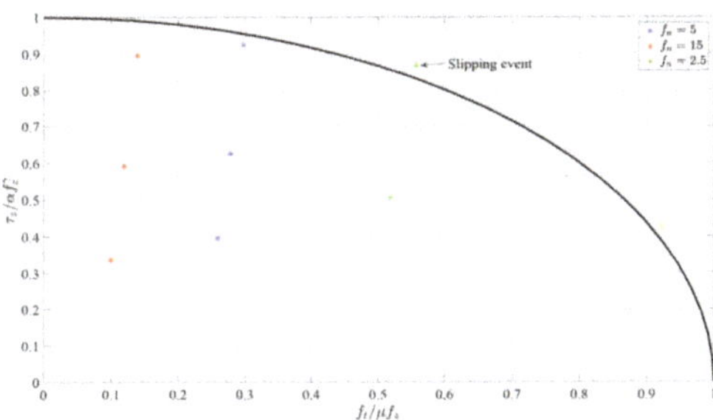

Figure 25. Fifth experiment: Tangential force and torsional moment couples with respect to the limit surface.

5. Conclusions

This paper has presented the detailed design and experimental characterization of a force/tactile sensor able to measure distributed contacts and estimate contact force and torsional moments to be used for robotic dexterous manipulation tasks. The mechanical interface of the device is a soft pad of silicone so as to adapt to different object shapes and hold high torsional moments. The sensitive part, based on optoelectronic technology, is exploited not only to estimate the total contact wrench but also to detect the orientation of the contact surface essential to correctly detect friction force, a relevant quantity in any dexterous manipulation control algorithm. The complete design of both the transducer and the interface electronics for integration into industrial grippers is discussed together with a novel calibration procedure aided by a specifically designed GUI that is aimed at ensuring a proper coverage of the training set necessary for the neural-network training selected as the calibration algorithm. The effectiveness of the calibration is experimentally validated through a number of trials carried out on a standard parallel gripper equipped with two sensorized fingers mounted on a robotic arm.

Future developments will be devoted to exploit the tactile map correlated to the force measurements for texture recognition through machine learning algorithms. This ability is of particular interest in service robotics applications where the robot has to recognize the object it is interacting with. Furthermore, an integrated force/tactile sensor is the enabling technology for executing not only grasping of unknown objects with the minimum required force to avoid slippage, but also for carrying out more complex in-hand manipulation actions, such as controlled sliding maneuvers. The possibility to control the orientation of the manipulated object by simply acting on the gripping force allows the robot to adopt only simple parallel grippers to perform such dexterous operations. This can pave the way to the adoption of robots in many application fields which require sophisticated manipulation abilities, such as logistics and household environments.

Author Contributions: M.C. developed the sensor calibration procedure with the collaboration of G.D.M.; S.P. contributed to the design of the electronic PCB of the sensor and its concept with the collaboration of C.N.; all authors equally contributed to experimental validation, analysis of results, writing and editing the manuscript.

Funding: This research was funded by Horizon 2020 Framework Programme of the European Commission under project ID 731590—REFILLS project.

Conflicts of Interest: The authors declare no conflict of interest.

References

1. Palli, G.; Melchiorri, C.; Vassura, G.; Scarcia, U.; Moriello, L.; Berselli, G.; Cavallo, A.; De Maria, G.; Natale, C.; Pirozzi, S.; et al. The DEXMART hand: Mechatronic design and experimental evaluation of synergy-based control for human-like grasping. *Int. J. Robot. Res.* **2014**, *33*, 799–824. [CrossRef]
2. Grebenstein, M.; Chalon, M.; Friedl, W.; Haddadin, S.; Wimböck, T.; Hirzinger, G.; Siegwart, R. The hand of the DLR Hand Arm System: Designed for interaction. *Int. J. Robot. Res.* **2012**, *31*, 1531–1555. [CrossRef]
3. Kawasaki, H.; Komatsu, T.; Uchiyama, K. Dexterous anthropomorphic robot hand with distributed tactile sensor: Gifu hand II. *IEEE/ASME Trans. Mech.* **2002**, *7*, 296–303. [CrossRef]
4. Shadow Robot Company. Design of a Dexterous Hand for Advanced CLAWAR Applications. In Proceedings of the International Conference on Climbing and Walking Robots, Catania, Italy, 17–19 September 2003.
5. Controzzi, M.; Cipriani, C.; Carrozza, M.C. Design of Artificial Hands: A Review. In *The Human Hand as an Inspiration for Robot Hand Development*; Balasubramanian, R., Santos, V.J., Eds.; Springer International Publishing: Cham, Switzerland, 2014; pp. 219–246.
6. Kappassov, Z.; Corrales, J.A.; Perdereau, V. Tactile Sensing in Dexterous Robot Hands—Review. *Robot. Auton. Syst.* **2015**, *74*, 195–220. [CrossRef]
7. Dahiya, R.S.; Mittendorfer, P.; Valle, M.; Cheng, G.; Lumelsky, V.J. Directions Toward Effective Utilization of Tactile Skin: A Review. *IEEE Sens. J.* **2013**, *13*, 4121–4138. [CrossRef]
8. D'Amore, A.; De Maria, G.; Grassia, L.; Natale, C.; Pirozzi, S. Silicone rubber based tactile sensor for measurement of normal and tangential components of the contact force. *J. Appl. Polym. Sci.* **2011**, *122*, 3758–3770. [CrossRef]
9. De Maria, G.; Natale, C.; Pirozzi, S. Force/Tactile Sensor for Robotic Applications. *Sens. Actuators A Phys.* **2012**, *175*, 60–72. [CrossRef]
10. Cirillo, A.; Cirillo, P.; Maria, G.D.; Natale, C.; Pirozzi, S. Control of linear and rotational slippage based on six-axis force/tactile sensor. In Proceedings of the 2017 IEEE International Conference on Robotics and Automation (ICRA), Singapore, 29 May–3 June 2017; pp. 1587–1594.
11. Deimel, R.; Brock, O. A novel type of compliant and underactuated robotic hand for dexterous grasping. *Int. J. Robot. Res.* **2016**, *35*, 161–185. [CrossRef]
12. Della Santina, C.; Piazza, C.; Grioli, G.; Catalano, G.M.; Bicchi, A. Towards Dexterous Manipulation with Augmented Adaptive Synergies: The Pisa/IIT SoftHand 2. *IEEE Trans. Robot.* **2018**, *34*, 1141–1156. [CrossRef]
13. Dafle, N.C.; Rodriguez, A.; Paolini, R.; Tang, B.; Srinivasa, S.S.; Erdmann, M.; Mason, M.T.; Lundberg, I.; Staab, H.; Fuhlbrigge, T. Extrinsic dexterity: In-hand manipulation with external forces. In Proceedings of the 2014 IEEE International Conference on Robotics and Automation (ICRA), Hong Kong, China, 31 May–7 June 2014; pp. 1578–1585.

14. Dafle, N.; Rodriguez, A. Prehensile pushing: In-hand manipulation with push-primitives. In Proceedings of the IEEE/RSJ International Conference on Intelligent Robots and System, Hamburg, Germany, 28 September–2 October 2015; pp. 6215–6222.
15. Viña, F.E.; Karayiannidis, Y.; Smith, C.; Kragic, D. Adaptive Control for Pivoting with Visual and Tactile Feedback. In Proceedings of the IEEE International Conference on Robotics and Automation, Stockholm, Sweden, 16–21 May 2016; pp. 399–406.
16. Costanzo, M.; Maria, G.D.; Natale, C. Slipping Control Algorithms for Object Manipulation with Sensorized Parallel Grippers. In Proceedings of the 2018 IEEE International Conference on Robotics and Automation (ICRA), Brisbane, QLD, Australia, 21–25 May 2018; pp. 7455–7461.
17. Maria, G.D.; Natale, C.; Pirozzi, S. Tactile data modelling and interpretation for stable grasping and manipulation. *Robot. Auton. Syst.* **2013**, *61*, 1008–1020. [CrossRef]
18. Cirillo, A.; Cirillo, P.; Maria, G.D.; Natale, C.; Pirozzi, S. Modeling and Calibration of a Tactile Sensor for Robust Grasping. In Proceedings of the 20th World Congress of the IFAC, Toulouse, France, 9–14 July 2017; pp. 7037–7044.
19. Qian, S.; Bao, K.; Zi, B.; Wang, N. Kinematic calibration of a cable-driven parallel robot for 3D printing. *Sensors* **2018**, *18*, 2898. [CrossRef] [PubMed]
20. Huang, T.; Zhao, D.; Yin, F.; Tian, W.; Chetwynd, D.G. Kinematic calibration of a 6-DOF hybrid robot by considering multicollinearity in the identification Jacobian. *Mech. Mach. Theory* **2019**, *131*, 371–384. [CrossRef]
21. Howe, R.; Cutkosky, M. TPractical force-motion models for sliding manipulation. *Int. J. Robot. Res.* **1996**, *15*, 557–572. [CrossRef]
22. Cirillo, A.; De Maria, G.; Natale, C.; Pirozzi, S. Design and Evaluation of Tactile Sensors for the Estimation of Grasped Wire Shape. In Proceedings of the IEEE International Conference on Advanced Intelligent Mechatronics (AIM2017), Munich, Germany, 3–7 July 2017; pp. 490–496.
23. Timoshenko, S.; Goodier, J. *Theory of Elasticity*, 3rd ed.; McGraw-Hill: New York, NY, USA, 1970.
24. Cirillo, A.; Cirillo, P.; De Maria, G.; Natale, C.; Pirozzi, S. An Artificial Skin Based on Optoelectronic Technology. *Sens. Actuators A Phys.* **2014**, *212*, 110–122. [CrossRef]
25. Xydas, N.; Kao, I. Modelling of contact mechanics and friction limit surfaces for soft fingers in robotics, with experimental results. *Int. J. Robot. Res.* **1999**, *18*, 941–950. [CrossRef]
26. Uchiyama, M.; Dauchez, P. Symmetric kinematic formulation and non-master/slave coordinated control of two-arm robots. *Adv. Robot.* **1992**, *7*, 361–383. [CrossRef]

Article

Robot Intelligent Grasp of Unknown Objects Based on Multi-Sensor Information

Shan-Qian Ji, Ming-Bao Huang and Han-Pang Huang *

Robotics Laboratory, Department of Mechanical Engineering, National Taiwan University, Taipei 10617, Taiwan; shanqianji@gmail.com (S.-Q.J.); d00522011@ntu.edu.tw (M.-B.H.)

* Correspondence: hanpang@ntu.edu.tw; Tel.: +886-2-3366-2700

Received: 4 March 2019; Accepted: 30 March 2019; Published: 2 April 2019

Abstract: Robots frequently need to work in human environments and handle many different types of objects. There are two problems that make this challenging for robots: human environments are typically cluttered, and the multi-finger robot hand needs to grasp and to lift objects without knowing their mass and damping properties. Therefore, this study combined vision and robot hand real-time grasp control action to achieve reliable and accurate object grasping in a cluttered scene. An efficient online algorithm for collision-free grasping pose generation according to a bounding box is proposed, and the grasp pose will be further checked for grasp quality. Finally, by fusing all available sensor data appropriately, an intelligent real-time grasp system was achieved that is reliable enough to handle various objects with unknown weights, friction, and stiffness. The robots used in this paper are the NTU 21-DOF five-finger robot hand and the NTU 6-DOF robot arm, which are both constructed by our Lab.

Keywords: contact modelling; force and tactile sensing; grasping and manipulation; grasp planning; object features recognition; robot hand-arm system; robot tactile systems; sensor fusion; slipping detection and avoidance; stiffness measurement

---

1. Introduction

Rapid technology development is enabling intelligent robots to be used in many fields, such as medicine, the military, agriculture, and industry. A robot's ability is a key function to grasp and manipulate an object that helps people with complicated tasks. In order to provide daily support by using humanoid hands and arms [1,2], robots must have the ability to grasp a variety of unseen objects in human environments [3].

In the unstructured environment, a common gripper has the limitation of not being able to grasp a great variety of objects. Therefore, studies need to be focused on using a multi-fingered robot hand to grasp objects with different shapes. Second, recognition and grasping of unknown objects in a cluttered scene have been very challenging to robots. This study attempted to develop a grasping system that is fast, robust and does not need a model of the object beforehand in order to reduce reliance on preprogrammed behaviors. The contributions of this work are summarized as follows:

- We present a method that combines three-dimensional (3D) vision and robot hand action to achieve reliable and accurate segmentation given unknown cluttered objects on a table.
- We propose an efficient algorithm for collision-free grasping pose generation; the grasp pose will be further checked for its grasp quality. Experiments show the efficiency and feasibility of our method.
- We addressed the problem associated with grasping unknown objects and how to set an appropriate grasp force. To fulfill this requirement, we adopted a multi-sensor approach to identify the stiffness of the object and detect slippage, which can promote a more lifelike functionality.

- By fusing all available sensor data appropriately, an intelligent grasp system was achieved that is reliable and able to handle various objects with unknown weights, friction, and stiffness.

2. Preliminary Knowledge

2.1. Point Cloud Processing

For a grasping task [4–7], finding the position and pose of the target object must be done first. The real environment (Figure 1a) is complex, and the raw point cloud is typically dense and noisy. This will hurt performance in object recognition (Figure 1b). Many point cloud processing methods can be used to deal with the point cloud set, and then useful information can be extracted to reconstruct the object.

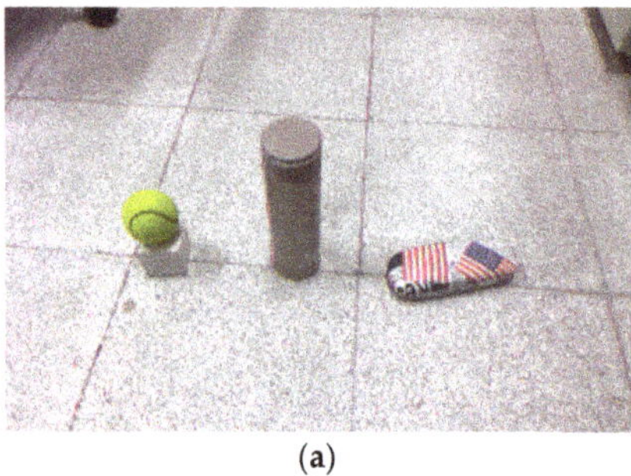

(a)

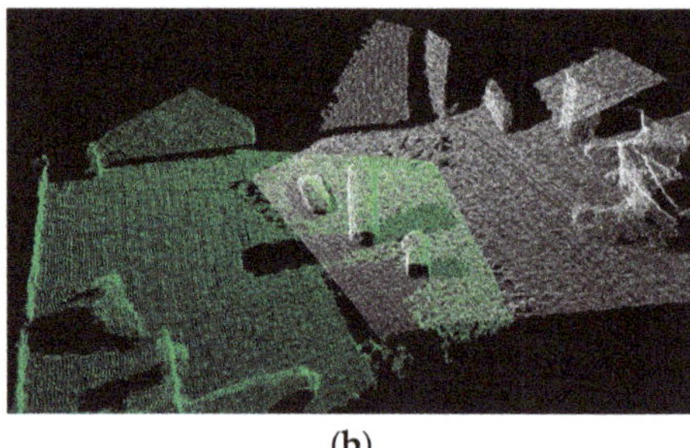

(b)

Figure 1. Real environment and point cloud: (**a**) Real environment; (**b**) Point cloud from the depth sensor.

2.1.1. Pass-Through Filter

The raw data of a point cloud contains numerous useless points. Therefore, a pass-through filter is used to set a range over the 3D space so that points within that range are kept unchanged, and points outside of the range are removed. In Figure 2, we show the point cloud after using a pass-through filter.

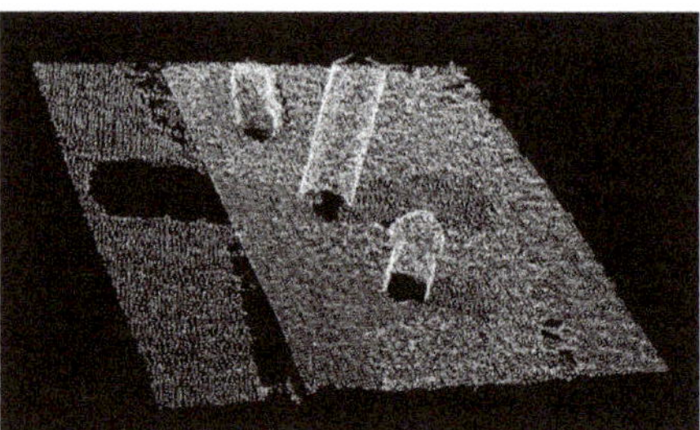

Figure 2. The output of pass-through filter.

2.1.2. Down-Sampling

A point cloud might contain points that provide no additional information due to noise or inaccuracy of the capture. On the other hand, when there are too many points in a point cloud, processing can be computationally expensive. The process of artificially decreasing the number of points in a point cloud is called down-sampling. In this study, the algorithm we used was voxel grid down-sampling. Figure 3 shows a point cloud before down-sampling and the same point cloud after down-sampling.

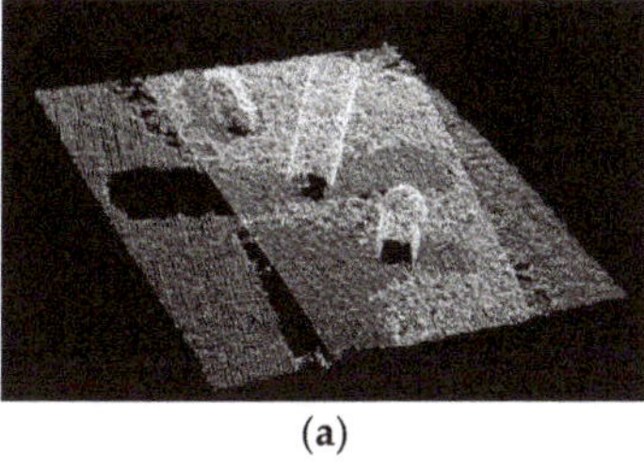

(a)

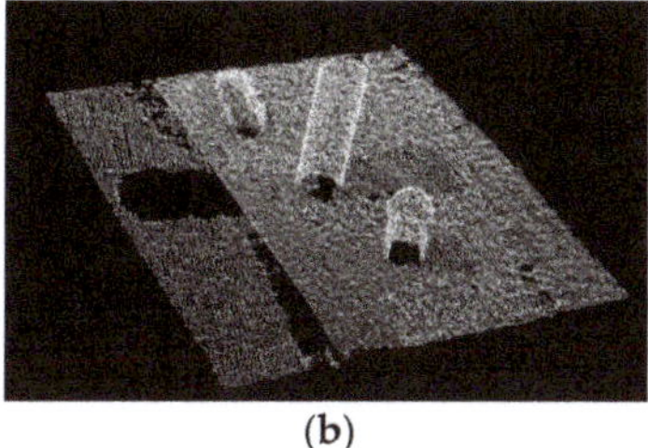

(b)

Figure 3. Down-sampling: (**a**) Original point cloud; (**b**) After down-sampling.

2.1.3. Random Sample Consensus (RANSAC)

First, we assumed objects are placed on flat surfaces. In a point cloud representation of a scene, a random sample consensus [8] can help to distinguish flat surfaces. Isolating a point cloud cluster that represents the unknown object is done by removing the entire points on the flat surface. After the planar model has been found, we can easily remove the ground and keep those objects on the ground. Figure 4 shows the objects retrieved from the original point cloud after image processing.

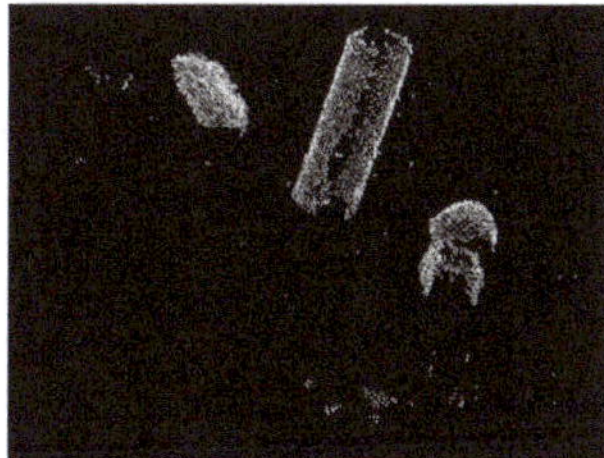

Figure 4. Objects after removing the ground.

2.1.4. Euclidean Segmentation

After separating the object from the table, the next problem was how to detect the number of objects remaining. The simple method we used was Euclidean segmentation. The result is shown in Figure 5, where we used different colors to represent different objects.

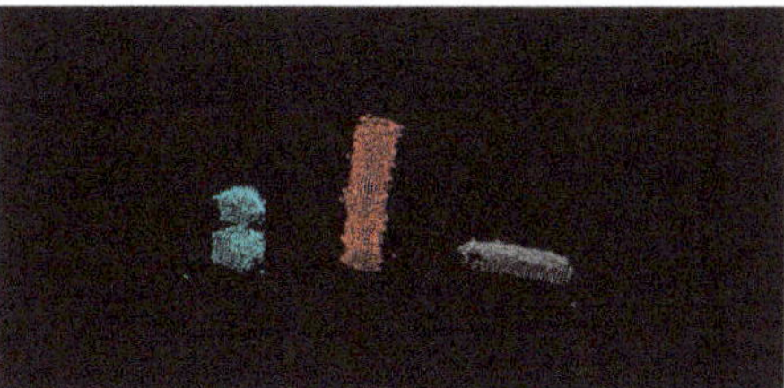

Figure 5. Point cloud after performing segmentation.

2.2. Oriented Bounding Box and Decomposition

For a common object, there are many possible grasps, and searching all possible grasps is difficult and unrealistic. In order to select a good grasp for an unknown object, we adopted an oriented bounding box (OBB) to identify the shape and pose of objects. The box is one of the most common shape primitives to represent unknown objects [6,9–11], and it can help us quickly find a suitable grasp. We assume that a hand can grasp an object if the bounding box of the object can be grasped. We created the OBB via principal component analysis (PCA). The steps involved in the OBB method are as follows:

- Compute the centroid (c0, c1, c2) and the normalized covariance.
- Find the eigenvectors for the covariance matrix of the point cloud (i.e., PCA).
- These eigenvectors are used to transform the point cloud to the origin point such that the eigenvectors correspond to the principal axes of the space.
- Compute the max, min, and center of the diagonal.

Figure 6 shows a joystick, a glass case, and a tennis ball, each with their bounding box as generated using the algorithm discussed above.

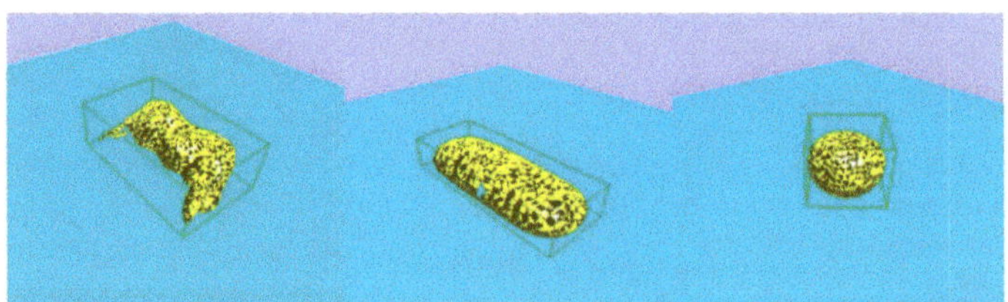

Figure 6. The bounding box of a set of objects.

In order to get better resolution of the actual models, we approximated the objects with bounding box decomposition. For this process, the number of boxes needed depends on the accuracy of approximation that is required. The principle of bounding box decomposition is that each bounding box is six-sided, and the opposing sides are parallel and symmetric. Considering the computation time, we determined that there were three valid split directions that were perpendicular to planes A, B, and C (Figure 7). The goal was to find the best split plane between each of these opposing sides.

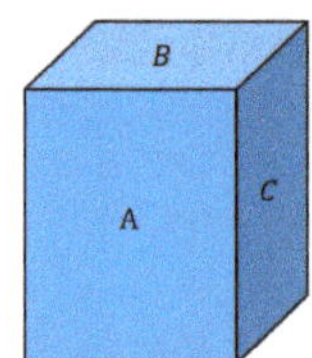

Figure 7. Parallel box cutting planes.

The split measure was defined as:

$$\alpha(d,\ x), \tag{1}$$

where d is one of the three split directions and x is the split position on this axis. For each x that cuts the target point cloud, the original point cloud can be divided into two subsets of data points. These can be used as inputs for the OBB algorithm to produce two child bounding boxes. Then we can obtain $\alpha(d,\ x)$ as the fraction between the sum volume of the two parts of the OBB and the whole volume. It is intuitive that the minimum is the best split (Figure 8). In this way, the box and the data points can be iteratively split, and the new boxes will better fit the shape.

Figure 8. Best cuts.

Additionally, for the purpose of efficiency, an iterative breaking criterion was set as:

$$\Theta^* = \frac{V(C_1) + V(C_2)}{V(P)} \geq \theta, \tag{2}$$

where P is the current (parent) box, C_1 and C_2 are the two child boxes produced by the split, V is a volume function, and θ is the threshold value.

The bounding box decomposition algorithm has two constraints. First, if the Θ is too high, the split is not valuable. Second, boxes that include a very low number of points that can be considered noise must be removed.

2.3. Object Segmentation in a Cluttered Scene

When there are objects in a cluttered scene, it is difficult for a robot to distinguish among the unknown objects and grasp them separately. However, a robot's object segmentation ability is key when it must work in human environments. In our study, the goal was to distinguish the object states by using the robot visual system and inference of object logic states through taking specific actions. This can instruct a robot about an accurate segment object point cloud in the scene without supervision.

The framework of unknown object segmentation is shown in Figure 9. First, the vision module was used to capture the 3D point cloud of the current task environment, and the unknown object was segmented through Euclidean segmentation and saved as the original object dataset. At the same time, the current visual detection results were sent to the object logic state inference module to generate the initial logic state of each object. This was done because the spatial position relationship between the objects was unclear, such as with the presence of an occlusion and superposition. The initial object dataset is a generally inaccurate representation of the current environment.

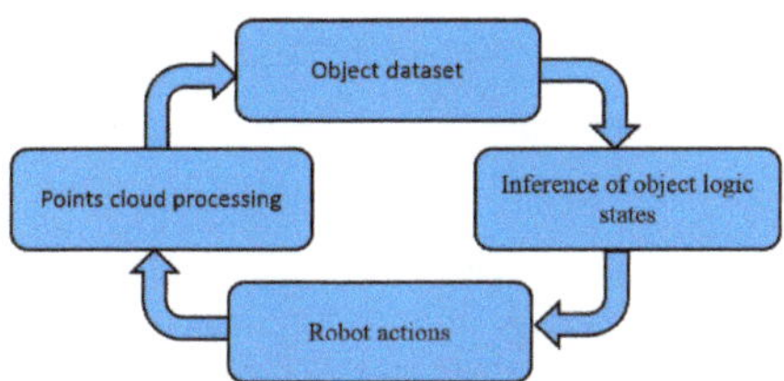

Figure 9. The framework of unknown object segmentation.

Specifically, when the robot performs a certain action, it feeds back the execution result of the action to the object logic state inference module, updates the logic state of the related object, and then updates the object point cloud. By observing the point clouds of objects that do not conform to the changed logic state, the object that is blocked or superimposed can be found, and the visual inspection result is fed to the object logic state inference module to update the object's logic state space.

First, define the object and logic state:

$$s_i = [Z_1, Z_2, Z_3], \tag{3}$$

where Z_1 is equal to 1 or 0, which indicates if object i is in or out of the robot's hand. Variable Z_2 presents the number of objects that are piled around object i. If Z_2 is equal to 0, it indicates that no objects are piled around object i. Variable Z_3 represents the number of objects under object i. If Z_3 is equal to -1, it indicates that object i is in the air. If variable Z_3 is equal to 0, it indicates that object i is on the ground.

According to the definition of object logic state, the two most important robot actions must be the focus: PickUp() and PutDown(). The action PickUp(object i) can be used to grasp object i from the table, and PutDown(object i, j) can be used to load the grasped object i into the target position j. It is assumed that only the robot's actions change the cluttered scene. The examples for inference of the object logic states based on action feedback are shown in Table 1.

Table 1. The rules for inference of object logic states.

Pre-States	Actions	Current States
$s_1 = \{0,0,0\}$	PickUp(object 1)	$s_1 = \{1,0,-1\}$
$s_1 = \{0,0,0\}$	PutDown(object 2, object 1)	$s_1 = \{0,2,0\}$
$s_1 = \{1,0,-1\}$	PutDown(object 1, object 2)	$s_1 = \{0,0,2\}$

Combining the above-mentioned logic state of objects and the basic actions of robots, the fundamental steps of object segmentation in a cluttered scene are shown in Figure 10. This method is demonstrated by the example shown in Figure 11.

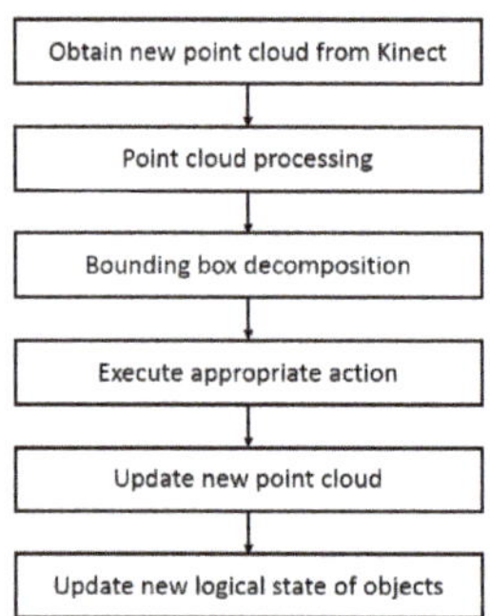

Figure 10. The flow of object segmentation.

According to the result of the bounding box decomposition (Figure 11a), box 1 and box 2 may be an integrated object or two separate objects. Thus, the reasonable action is picking up the top box.

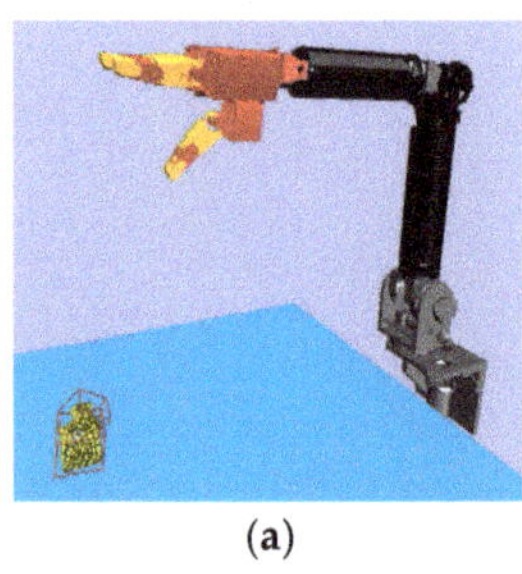

(a)

(b)

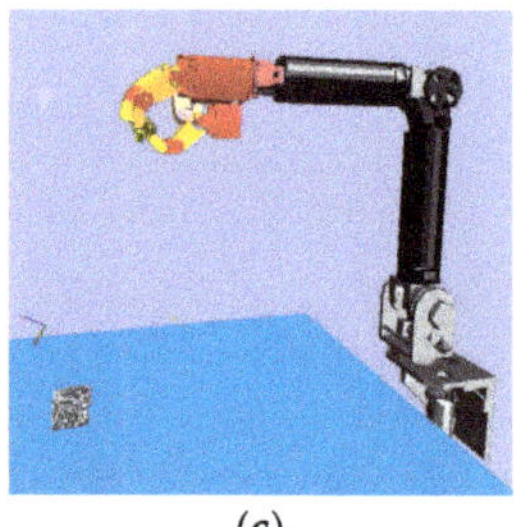

(c)

Figure 11. Object segmentation in the simulator: (**a**) Object's original position; (**b**) Previous moment; (**c**) Current moment.

After executing the PickUp() action (Figure 11b), it can be determined whether the object was blocked by observing the variation of the point cloud according to Table 2, where the first rule can be used to find an overlaid object. After object 1 was grasped, if it was an integrated object, the vision system should no longer detect the point cloud in the object's original position. If the point cloud is detected in that location, it indicates the point cloud is a new object 2, and that object 2 was overlaid with object 1 before the robot hand acted. The current pose of object 1 is determined based on the pose of the robot hand, and no visual algorithm is needed to re-detect it. The point cloud model of object 1 is updated according to the difference between the point cloud of object 1 at the previous moment and the point cloud of object 2 at the current moment, with a result shown in Figure 11c.

Table 2. The rules for segmentation point sets.

Action	Previous State	Current State	Detect Point Cloud in *Previous* Position		Detect Point Cloud in *Current* Position	
			Yes	No	Yes	No
PickUp(object 1)	$s_1 = \{0,0,0\}$	$s_1 = \{1,0,-1\}$	covered object	normal	object 1	err
PickUp(object 1)	$s_1 = \{0,0,2\}$	$s_1 = \{1,0,-1\}$	object 2	err	object 1	err
PutDown(ground)	$s_1 = \{1,0,-1\}$	$s_1 = \{0,0,0\}$	err	normal	object 1	err
PutDown(object 1, object 2)	$s_1 = \{1,0,-1\}$	$s_1 = \{0,0,2\}$	err	normal	object 1 and 2	err

3. Grasp Planning

In general, there are many pose candidates for grasping a target object. Additionally, contact points between the object and the robot need to check if it is a stable grasp, such as with the force closure property. In this section, we describe an efficient grasp planning algorithm for high-quality and collision-free grasping pose generation. The complete grasp planning approach has two main aspects: The first is checking that the grasping pose has a collision-free path for a 6-DOF arm using rapidly exploring random trees (RRTs). The next is fast grasping quality analysis in wrench space.

3.1. Grasp Strategy

Like human beings grasp things, robot grasping movements can be defined as using one of two reach-to-grasp movements: top grasp or side grasp [12], as shown in Figure 12.

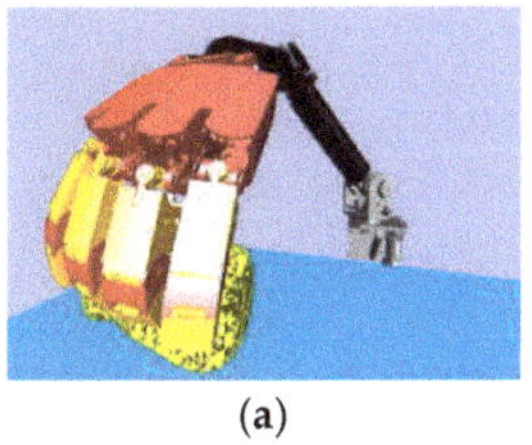

(a)

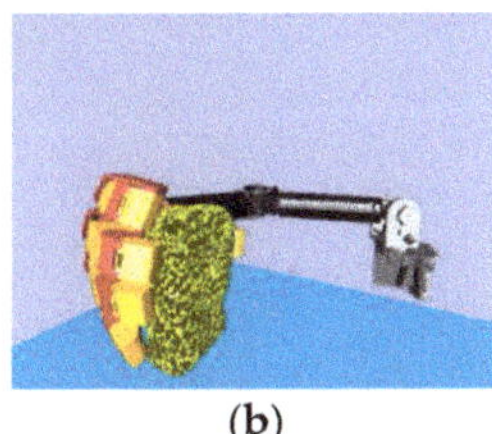

(b)

Figure 12. The robot grasping movements: (**a**) Top grasp; (**b**) Side grasp.

Humans may grasp contact points that are on the longer side of an object. Therefore, we decide the grasping direction according to the height of the bounding box of the target object. If the height is adequate, we can use a side grasp movement. Otherwise, we use a top grasp movement.

Thus, based on the OBB of the object, we can sample the position and orientation of the end-effector of the robot arm, as shown in Figure 13. Then, inverse kinematics (IK) are solved for the robot arm and a check is done for whether any collisions may occur between the robot arm system and the environment. If a collision occurs, then the grasping pose is resampled in other directions until the pose can avoid the collision. Next, we find a better grasp pose by looking at two criteria. One is the grasp pose can find a short collision-free path or the pose results in a high-quality grasp.

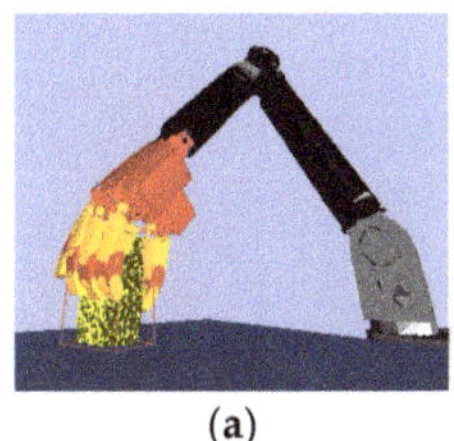

(a)

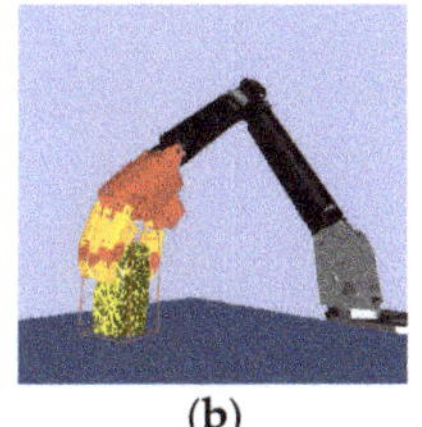

(b)

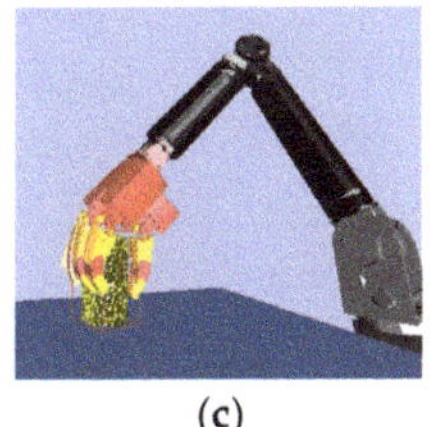

(c)

Figure 13. The sampling of the robot arm: (**a**) Top grasp movement; (**b**) A short collision-free path; (**c**) High-quality grasp.

3.2. Path Planning

Path planning involves generally searching the path in a configuration space (C-space). For the robot arm, C means the workspace of the robot arm and each $q \in C$ is a joint angle. Previous research has focused on only one goal for the purposes of motion path planning. However, there are multiple possible goals in the real world. The multi-objective RRT-Connect path planner was designed specifically for a real robot made as a high degrees of freedom (DOFs) kinematic chain [13].

The idea of multi-objective is illustrated in Figure 14. First, we set some goals in the configuration space where it is possible that some objectives are invalid, such as goal 3. Therefore, we need to check whether or not the goals are feasible according to the RRT-Connect algorithm. Let T_{init} and T_{goal} be the tree generated by q_{init} and $\{q_{goal}\}$, respectively. The multi-objective RRT-Connect algorithm is given in Table 3.

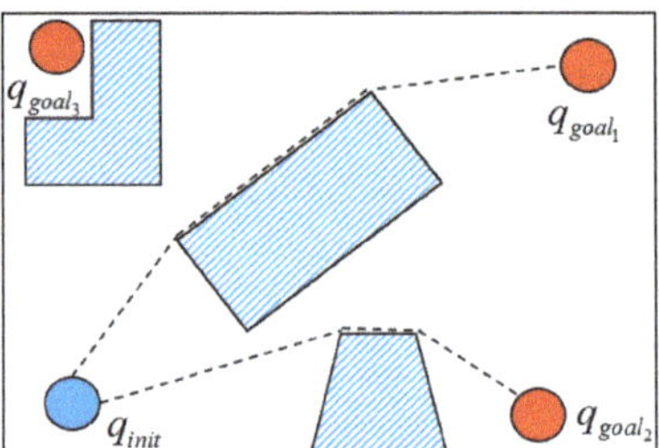

Figure 14. Multi-goal planner.

Table 3. Multi-objective RRT-connect algorithm.

1. Input $k = 1$, $r = 1$, m, $\{q_{init}\}$ and $\{q_{goal1}, \ldots, q_{goaln}\}$;
2. $q_{rand} \leftarrow randomConfig()$;
3. Find the nearest neighbor n_{init} of q_{rand} to T_{init};
4. Define k;
5. Make a segment toward q_{rand} from n_{init} with a fixed incremental distance d;
6. Define $q_{newinit}$;
7. If the segment connecting n_{init} and $q_{newinit}$ is collision free then
8.   extend T_{init} (generate by $\{q_{newinit}, T_{init}\}$);
9.   Find the nearest neighbor n_{goal} of q_{rand} to T_{goal};
10.   Make a segment toward q_{rand} from n_{goal} with fixed incremental distance d;
11.   Define $q_{newgoal}$;
12.   If the segment connecting n_{goal} and $q_{newgoal}$ is collision free then
13.   extend T_{goal} (generate by $\{q_{newgoal}, T_{goal}\}$);
14.   If T_{init} and T_{goal} are connected then
15.   Output the path, and end.
16.   Else replace k by $k + 1$, and return to Step 4
17.   Else if $r < m$ then replace r by $r + 1$, and return to Step 2
18.   Else end;
19.   Else if $r < m$ then replace r by $r + 1$, and return to Step 2
20.   Else end;

The random Config() function generates random $q_{rand} \in C$. The extend function T_{init} is illustrated in Figure 15. Each iteration attempts to extend the RRT by adding a new vertex that is biased by a randomly selected configuration. First, we find the vertex q_v that is the nearest vertex to q in tree T. Then we extend the tree toward q from q_{near} with ad-hoc variable distance d and assign a new variable $q_{newinit}$ before the collision event is checked.

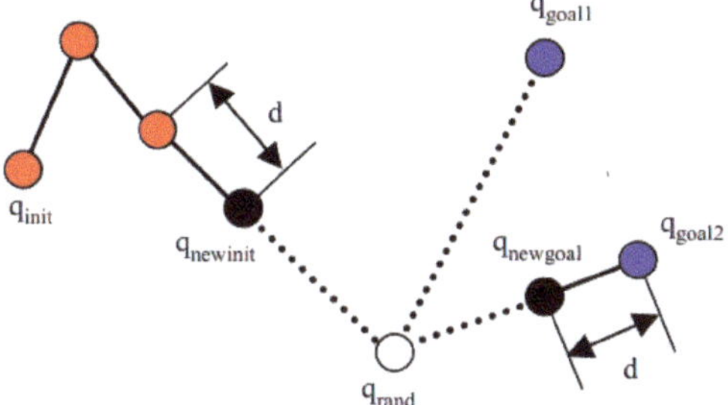

Figure 15. The extend function.

Consider the robot arm joint space path planning. Figure 16 shows that with the same start configuration but different goal configurations, there will be a different path. The yellow object is the target object, and the white object is the obstacle which the robot arm and hand must avoid. The multi-objective RRT-Connect algorithm is used to search the six-dimensional C-space of the robot arm in order to find the collision-free path. Figure 16a shows the grasping pose that had a short collision-free path.

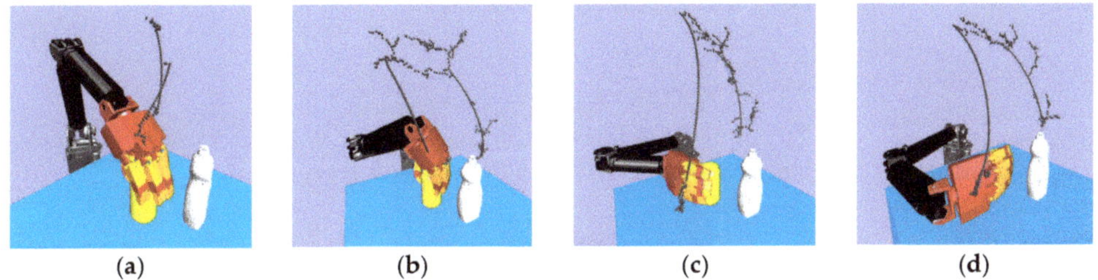

Figure 16. Multi-objective RRT-Connect in the simulator: (**a**) Path size: 59; (**b**) Path size: 238; (**c**) Path size: 191; (**d**) Path size: 231.

After searching for a path, we then needed to flatten, or prune, the path by removing any unnecessary waypoints, after which we can obtain a suitable trajectory for real robot grasping and manipulation. There are some efficient path pruning algorithms, like that shown in Ref. [14]. Even though the pruned path is not optimal, this method can remove most of the unnecessary waypoints very quickly, eliminating the chattering phenomenon in the RRT-Connect algorithm. Figure 17 shows the path before using the path pruning algorithm and the path after running the path pruning algorithm.

Figure 17. Path pruning algorithm: (**a**) Original path; (**b**) The path through path pruning algorithm.

3.3. Grasp Analysis

After path planning is complete, the robot arm moves into position with a given orientation, and the robot hand performs continuous grasping until it touches the object. Here, the robot hand was treated as a simple gripper to reduce its number of DOFs, but a multi-fingered robot hand can adapt to the object better than a simple gripper because it has five fingers that can be independently adjusted to the object's geometry. When all fingers are in contact with the object's surface, it is still not clear whether or not the robot hand can firmly grasp the target object. Thus, grasp analysis is necessary. Grasp analysis is a matter of determining, given an object and a set of contacts, whether the grasp is stable using common closure properties.

A grasp is commonly defined as a set of contacts on the surface of an object [15–17]. The quality measure is an index that quantifies the effectiveness of a grasp. Force closure is a necessary property for grasping that requires a grasp to be capable of resisting any external wrench on the object, maintaining it in mechanical equilibrium [16].

One important grasp quality measure [18] often used in optimal grasp planning [19] assesses the force efficiency of a grasp by computing the minimum of the largest wrench that a grasp can resist, over all wrench directions, with limited contact forces [20], which equals the minimum distance from the origin of the wrench space to the boundary of a grasp wrench set.

The friction cone given by use of an equation is a convex cone, which consists of all non-negative combinations of the primitive contact force set U_i, defined as follows:

$$U_i = \left\{ f_i | f_{n_i} = 1, \sqrt{f_{t_i}^2 + f_{o_i}^2} = \mu \right\}, \tag{4}$$

The set U_i is the boundary of the friction cone. The image of U_i in the wrench space $\mathbb{R}^6$ through the contact map G_i is called the primitive contact wrench set, which is expressed as:

$$W_i = G_i(U_i), \tag{5}$$

Let W_{L_1} be the union of W_i and W_{L_∞} be the union of the Minkowski sum for any choice of them:

$$W_{L_1} = \bigcup_{i=1}^{m} W_i,\ W_{L_\infty} = \bigoplus_{i=1}^{m} W_i, \tag{6}$$

The grasp wrench set is defined to be the convex hull of W_{L_1} or W_{L_∞}, which is denoted by $W_{L_1}^{co}$ or $W_{L_\infty}^{co}$:

$$W_{L_1}^{co} = \left\{ \sum_{i=1}^{m} G_i f_i | f_i \in friction\ cone,\ \sum_{i=1}^{m} f_{n_i} = 1 \right\}, \tag{7}$$

$$W_{L_\infty}^{co} = \left\{ \sum_{i=1}^{m} G_i f_i | f_i \in friction\ cone,\ f_{n_i} \leq 1 \right\}, \tag{8}$$

Because the calculation of the Minkowski sum is more difficult, we chose the Equation (9) to compute the grasp wrench space (GWS). We can use some properties of GWS to evaluate the grasp quality. If the interior of $W_{L_1}^{co}$ is nonempty and contains the origin of the wrench space $\mathbb{R}^6$, the grasp has force closure and is stable. Furthermore, the minimum of the largest resultant wrenches that a grasp can resist, over all wrench directions, with limited contact forces is an important quality measure of a force closure grasp. It can be formulated as the minimum distance between the origin of the wrench space $\mathbb{R}^6$ and the boundary of $W_{L_1}^{co}$, and the magnitude of the quality measure Q is defined as:

$$Q = \min_{v \in W_{L_1}^{co}} \|v\|, \tag{9}$$

where v is $\mathbb{R}^6$ vector. Figure 18 shows an example of the quality measure in the simulator. The negative score in Figure 18 means that the zero set {0} was not contained in the GWS. In other words, the grasp was unstable. For grasps with a positive score, the higher the score, the better the grasp's capability of resisting a perturbation. Usually, Q ranges from 0.01 to 0.30 for stable grasping. We considered that grasps of more than 0.01 were sufficiently stable.

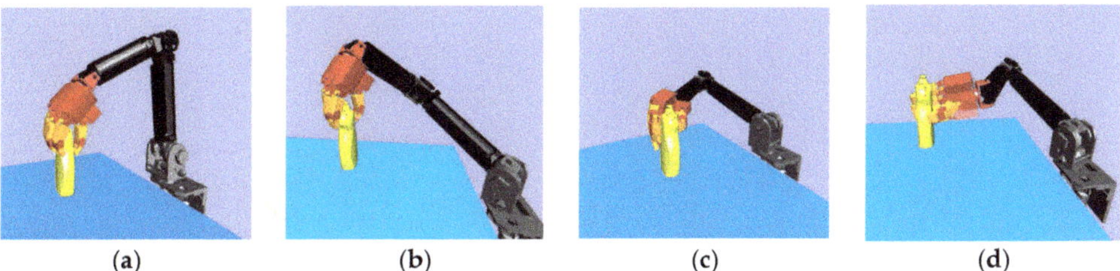

Figure 18. Quality measure example for a bottle: (**a**) Path size: 0.0135; (**b**) Path size: 0.001; (**c**) Path size: 0.017; (**d**) Path size: −0.27.

4. Real-Time Grip Force Selection and Control

The previous sections illustrated our method of estimating an object's shape and determining a robust and safe grasping configuration. However, while executing the grasping motion, the appropriate force required to immobilize an object still needs to be set. The force regulation is also very important for a precision grasp, especially when grasping is followed by manipulation.

This paper presents a methodology for choosing an appropriate grasping force when a multi-finger robot hand grasps and lifts an object without knowing its weight, coefficient of static friction, or stiffness. To fulfill these requirements, a method that measures object stiffness and detects slippage by using a multi-sensor approach is proposed. The desired grasping force can be set with an upper limit to avoid damage and a lower limit to avoid slippage according to the object's measured characteristics. Figure 19 shows the concept of a grip force selection structure.

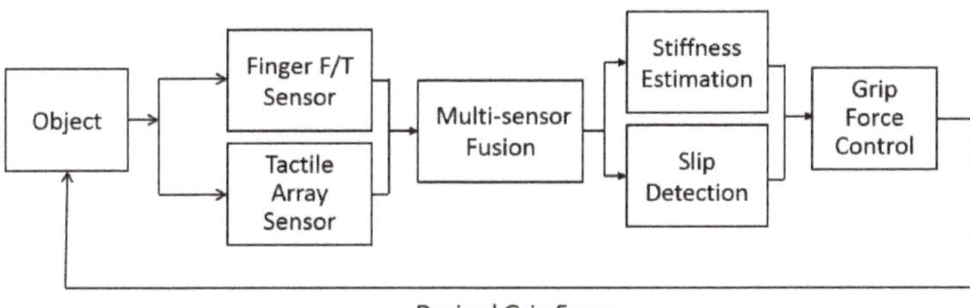

Figure 19. Grip force selection structure.

The control flowchart shown in Figure 20 consists of two stages, the pre-grasp stage and the grasping stage. In the pre-grasp stage, the robot hand is positioned, and using the position controller, the robot hand fingers are closed on the target object until the normal force feedback is above a certain threshold. The threshold is chosen according to the stiffness measurement using a very small value in order to avoid damaging the object. Once all the fingers are in contact with the object, the position controller is stopped, and the hand enters the grasping stage. In the grasping stage, the robot hand will be switched to position-based force control.

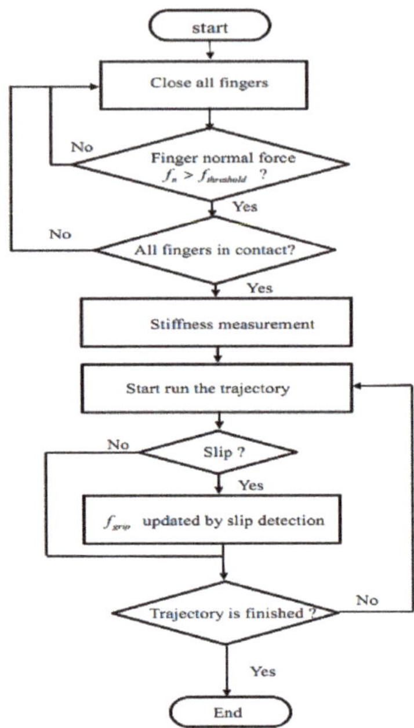

Figure 20. Control diagram of the grip controller.

4.1. Stiffness Measurement

When the robot hand is to grasp an object, the stiffness of the target object is a very important parameter that must be determined. Objects in our study were therefore divided into three categories according to their stiffness: a rigid object, a soft object, and a very soft object. Each category was defined by the upper limit grasp force.

A rigid object has high stiffness, and will not be deformed by the grasping force, such as a glass cup or metal bottle. For this category of objects, the upper limit force is defined as the maximum limit of the applied force of the robot finger.

Soft objects like plastic or paper cups have medium level stiffness and can be deformed when the robot hand increases the grasping force. For this category, the force that deforms objects is greater than the lift-off grasping force, so the appropriate upper limit force is set to have a minimal difference between the two forces.

The third category contains very soft objects such as sponges and other objects with little stiffness. This kind of object is greatly deformed by a slight grasping force. However, this deformation differs from soft objects because, in this case, it is acceptable for the robot hand to keep a stable grasp even if it causes a slight deformation of the object surface.

The concept of our method is that objects with different stiffness will have different mechanical responses to the initial contact. Thus, the force data to determine the stiffness is collected and processed. More details about our method will be discussed in the next subsection.

The stiffness measurement is taken using the following steps. First, the robot closes its fingers until it detects initial contact with the object. Once the hand contacts with the surface of the object, it continues to close for a fixed distance along the normal direction of the finger surfaces. Then the force curve is analyzed and features are selected to determine the object's stiffness. Both three-axis force/torque (F/T) sensor [21] and tactile sensor arrays [5] are employed to collect the force data, and the obtained fusion results are shown in Figure 21. As seen from the force curve, different stiffness will have different force curves. The harder the object, the faster the force increases and the greater the steady force value when the loading is finished. What's more, there are transient and steady states encountered during the stiffness measurement process. For the transient state, the hand makes contact with the object at a constant velocity. The results indicate that the force is proportional to contact speed for harder objects. Thus, the force curve slope U_d was extracted as a feature to estimate the stiffness of a grasped object as:

$$U_d = \frac{1}{N_1} \sum_{k=i}^{k=N_1+i} F_k, \tag{10}$$

where i is the time at which the hand starts to close, and N_1 is the sampling number of the transient state. F_k is the normal force in the contact point at time k. For steady state, the object has been grasped. Further, the gripping force and the reverse force due to deformation have been balanced. According to Hooke's law, $F = KX$, the steady pressure value is proportional to the stiffness of the object. Thus, the final steady pressure value of the curve can be used as a feature to estimate the stiffness of a grasped object:

$$U_s = \frac{1}{N_2} \sum_{k=j}^{k=N_2+j} F_k, \tag{11}$$

where j is the time at which the hand stops, and N_2 is the sampling number of the steady state. Based on the above analysis, the characteristics of force during contact and steady-state force characteristics at equilibrium can be obtained.

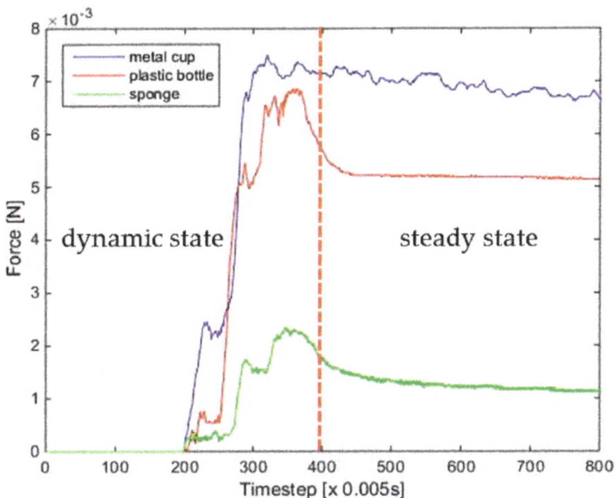

Figure 21. Force curves of different categories of objects.

After selecting the appropriate features, the k-nearest neighbors (K-NN) algorithm [22,23], was used to fuse these features and obtain more accurate stiffness recognition results. The K-NN algorithm is used to classify sample points into several distinct classes, and is summarized below:

- A positive integer k is specified along with a new sample.
- The k entries in the database that are closest to the new sample are selected.
- The most common classification of these entries is determined.
- This is the classification given to the new sample.

4.2. Slipping Detection and Avoidance

Robust slip detection ability [24] is one of the most important features needed for a grasping task. Knowledge about slip can help the robot prevent the object from falling down its hand. Furthermore, the sensation of slip is critical for a robot to grasp an object with minimal force [25].

In this study, we combined tactile sensor arrays [5] and three-axis force/torque (F/T) sensor [21] to obtain the slip signals. The framework of slip signal detection is shown in Figure 22. The slip feature based on frequency analysis and the slip feature based on motion estimation will be fused by support vector machine (SVM) algorithm.

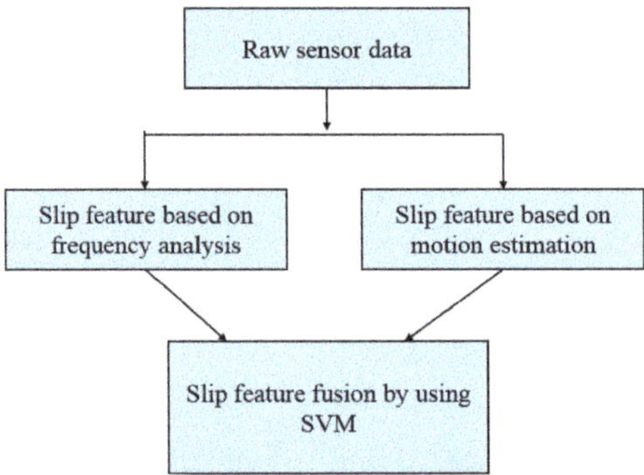

Figure 22. Slip detection and avoidance framework.

A number of studies have developed a method that observes the frequency content of the contact forces when slippage occurs [26,27]. When the relative speed of the fingers and objects are relatively low, the theory of friction and vibration shows that the slippage is an intermittent vibration that causes vibration of the gripping force. Slippage can, therefore, be detected by the high-frequency signal it generates. Thus, a wavelet transform [28,29], was used to analyze and process the finger three-axis force/torque (F/T) sensor information and extract the real-time slip feature. Considering the complexity and real-time performance of the algorithm, the simplest Haar wavelet in wavelet transforms was adapted to analyze the sensor information. The Haar scaling function is defined as:

$$\phi(x) = \begin{cases} 1, & if\, 0 \le x < 1 \\ 0, & elsewhere \end{cases}, \tag{12}$$

The Haar wavelet function looks like:

$$\psi(x) = \phi(2x) - \phi(2x-1), \tag{13}$$

The sensor force data $f\,(t)$can be represented as the linear sum of the Haar scaling function and the Haar wavelet using Haar decomposition as:

$$f_j(t) = w_{j-1} + f_{j-1}(t), \tag{14}$$

where:

$$w_{j-1} = \sum_{k \in Z} b_k^{j-1} \psi(2^{j-1}t - k), \tag{15}$$

$$f_{j-1} = \sum_{k \in Z} a_k^{j-1} \phi(2^{j-1}t - k), \tag{16}$$

$$b_k^{j-1} = \frac{a_{2k}^j - a_{2k+1}^j}{2},\ a_k^{j-1} = \frac{a_{2k}^j + a_{2k+1}^j}{2}, \tag{17}$$

and a_k^{j-1} are called the detail coefficient. The Haar wavelet is employed to transform the contact force information into the slip process and to extract the slip signal by using the detail coefficient. It is known from the definition of the detail coefficient that it can be expressed by the difference of the force signals of the adjacent two moments.

Vibration-based methods can suffer from increased sensitivity to robot motion [30]. A very simple alternative is the tracking of the center of mass (the intensity weighted centroid) [31]. But this method is quite sensitive to noise and saturation effects. In addition, a shifting of weight is interpreted as a translation even if the pressure profile is not moving at all. Thus, we considered using methods based on the convolution calculation [32].

The overall workflow is shown in Figure 23. Let the pixel intensity function of a sensor matrix T_n at time step n be denoted as $t_n(x,y)$, $T_n \in \mathbb{R}^{M \times N}$. The two-dimensional (2D) convolution of the discrete matrices, also known as complex-conjugate multiplication, is defined as:

$$t_{n-1}(x,y)^* t_n(x,y) = \sum_{i=-\infty}^{\infty} \sum_{j=-\infty}^{\infty} t_{n-1}(y,j) t_n(x-j, y-j), \tag{18}$$

This results in the convolution matrix C of size $(2M-1) \times (2N-1)$. The elements of C are therefore a measure of similarity between the two images, and the translation of the peak from the origin indicates the shift between them. In the context of slip detection, this relationship can be interpreted as a slip vector between two similar tactile sensor arrays profiles.

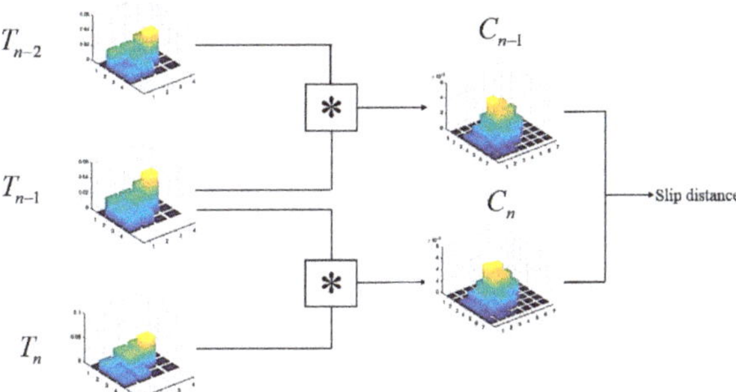

Figure 23. Slip detector based on motion estimation workflow.

The decision is made to implement a slip detection algorithm by Alcazar and Barajas [33]. First, two index matrices A and B of the same size as the convolution matrix and consisting of repeating rows and columns, respectively, are defined as:

$$A = \begin{bmatrix} 1 & 2 & \cdots & 2N-1 \\ 1 & 2 & \cdots & 2N-1 \\ \vdots & \vdots & \ddots & \vdots \\ 1 & 2 & \cdots & 2N-1 \end{bmatrix} - NB = \begin{bmatrix} 1 & 1 & \cdots & 1 \\ 2 & 2 & \cdots & 2 \\ \vdots & \vdots & \ddots & \vdots \\ 2M-1 & 2M-1 & \cdots & 2M-1 \end{bmatrix} - M, \tag{19}$$

In the slip detection loop, the convolution matrix C_{n-1} of the first pair of tactile matrices T_{n-2} and T_{n-1} is computed. A slip index along the X direction at time T is defined by:

$$\Delta x_{n-1} = E\left(\frac{A \cdot \mu_c^T}{sum(\mu_c)}\right), \tag{20}$$

Similarly, a slip index along the Y direction at time T is computed by:

$$\Delta y_{n-1} = E\left(\frac{\mu_r^T \cdot B}{sum(\mu_r)}\right), \tag{21}$$

where μ_c is a row vector containing the mean value of each column from the convolution matrix C_{n-1}. In contrast, μ_r is a column vector containing the mean value of each row from the convolution matrix C_{n-1}. $E()$ and $sum()$ denote the mean value and the sum of the elements of the vector, respectively. At the next time, the previous step was repeated. Again, the column and row means of defining the resulting convolution matrix C_n and the displacements were computed. The final slip signal was computed with:

$$X_n = \sqrt{(\Delta x_n - \Delta x_{n-1})^2 + (\Delta y_n - \Delta y_{n-1})^2}, \tag{22}$$

In this study, we fused the two slip signals mentioned above along with the SVM to improve the identification accuracy of slip detection performance. The SVM is a machine learning method based on statistical learning theory (SLT) [34] that minimizes empirical risk and can, therefore, solve linear and nonlinear classification problems.

4.3. Real-Time Grasping Control

The purpose of the grasping control is to adjust the grasping force to prevent slippage and limit deformation of unknown objects. The controller must have a slippage adaptive function to adjust the grasping force in real time.

Let F_{des} be the desired grip force (normal force) in each fingertip in a no-slip situation. Let F_{grip} be the updated desired grip force that accounts for any slip conditions. Define F_{grip} to be:

$$F_{grip}[k+1] = F_{grip}[k] + \alpha\beta, \tag{23}$$

where α is a gain factor that is derived according to the object's stiffness. A high stiffness object will have a bigger α, and this can help the hand quickly eliminate slippage. A low stiffness object will have a smaller α in order to avoid distortion of the object by increasing the grip force too quickly. In this paper, we have defined three levels of rigid categories, so there will be three corresponding α values. β is the slip signal derived from SVM results.

When slippage is detected, the regulated force must be applied almost instantaneously to ensure that the robot hand acts immediately. If the force is applied too slowly, the object will continue to slip and may even fall. Thus, a proportional-derivative (PD) position controller was used to achieve stable grasping control as follows:

$$U = \left\{ \begin{array}{cc} K_p(F_{grip} - F) + K_d(\frac{dF}{dt}), & if\ F \leq F_{\text{limit}} \\ 0, & if\ F > F_{\text{limit}} \end{array} \right\}, \tag{24}$$

where U is the output of position controller that determines each fingertip's normal direction in Cartesian space, K_p and K_d are the proportional and derivative parameters, and F is the real feedback force from the sensors in the fingertips. F_{limit} is set according to the object's stiffness.

5. Experiment Results

This section describes how the NTU 6-DOF robot arm [5,6,35–37] and the NTU five-finger robot hand were equipped with additional hardware and software to enable the resultant grasp of unknown objects. We combined the robot arm and robot hand to facilitate planning and real-time control, as shown in Figure 24.

This paper presents the on-going research of the National Taiwan University (NTU) five-finger robot hand. The purpose of the development of the NTU five-finger robot hand is for delicate dynamic grasp. To design a motor-driven, dexterous robot hand, we analyzed the human hand. Our design features a customized fingertip three-axis force/torque (F/T) sensor and joint torque sensors integrated into each finger, along with powerful super-flat brushless DC (BLDC) motors (The MAXON (Sachseln, Switzerland) motor measures only 23 mm in the outer diameter and creates a unit that is only 18.5 mm in height, with a motor weight of just 15 g. The rated speed and torque of the motor idle at 4140 rpm; the 3*W* motors are available in a 9*V* version and provide a maximum torque of 8.04 mNm. The motors include digital Hall sensors and digital magnetic Encoders. A linear per position resolution of 0.001 mm is obtained by coupling a MAXON Encoder MR (Type M 512 cpt) to the motor. and tiny harmonic drivers (HD) (Harmonic Drive™ gear AG, Limburg an der Lahn, Germany, HDUC-5-100-BLR, gear ratio 1:100). By using a steel coupling mechanism, the phalanx distal's transmission ratio is exactly 1:1 in the whole movement range. The rest of this paper presents a control strategy for the NTU five-finger robot hand. For robust grasp, we implemented classical impedance control. Our goal is to develop 21 DOFs dexterous robot hand. The hand has an independent palm and five identical modular fingers; each finger has three DOFs and four joints. We designed this robot hand by improving previously developed robot hand designs and by analyzing actual human hand motions. This NTU motor-driven, the five-finger robot hand can be equipped with a three-axis force/torque sensor [21] at each fingertip, as well as with developed distributed tactile sensor arrays with 376 detecting points on its surface [5]. The hand can communicate with external sources using technologies such as EtherCAT and controller area networks (CAN bus). To evaluate the performance of this robot hand design, we analyzed workspace, intersection volume, and manipulability, as shown in Figure 24.

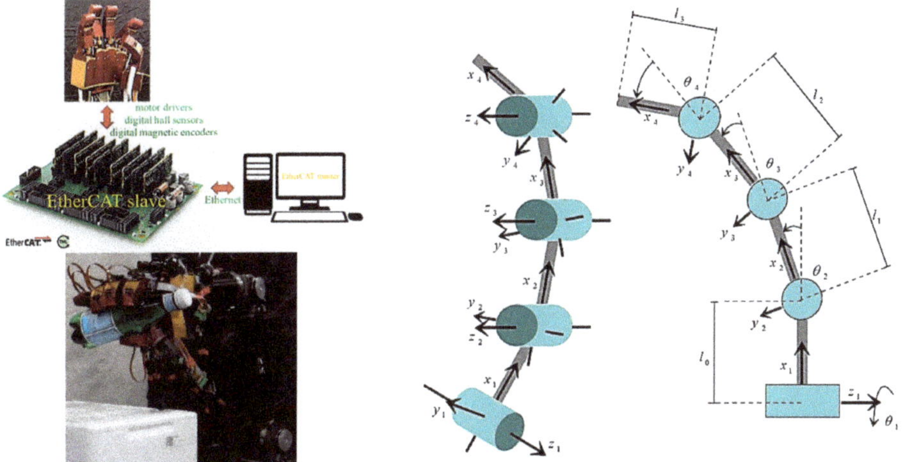

Figure 24. NTU five-finger hand with modular fingers (Four link manipulator to model one finger of the robot hand).

A dexterous robot hand needs at a minimum a set of force and position sensors to enable control schemes like position control and impedance control in autonomous operation and teleoperation. The aim of the sensory design is to integrate into the artificial hand a great number of different sensors in order to confer to the hand functionalities similar to that of a human hand. Sensor equipment for the NTU five-finger robot hand is shown in Table 4.

Table 4. Sensor equipment of one finger.

Sensor Type	Count/Finger
joint torque {current (torque) control loop in BLDC motor}	3
joint position	3
motor speed	3
distributed tactile sensor arrays (376 detecting points) (Tekscan, Inc., South Boston, MA, USA)	1
three-axis force/torque (each fingertip)	1
six-axis force/torque (the wrist joint) (Mini 40, ATI Industrial Automation, Apex, NC, USA)	1

The maximum payload of the NTU 6-DOF robot arm is over two kg. The specification of the NTU 6-DOF robot arm is shown in Table 5.

Table 5. Specification of NTU 6-DOF robot arm.

Total Weight	About 5.2 kg (exclude the shoulder)
Max. Payload	Over 2 kg
Max. Joint Speed	Exceed 110 deg./sec.
Max. Reachable distance	510 mm
Max. width	110 mm
Motor	Brushed DC X6
Reduction device	Harmonic drive
Transmission	Timing Belt, Gear

Because there are two kinds of the sensor on the fingertip, a three-axis F/T sensor and tactile sensor arrays, we calibrate and fuse the two sensors' normal force. We use weight groups (20, 50, 70, 100, 120, 150, 170 and 200 g) as the standard value and record the error between the measured value of the two sensors' normal force and weight's standard value. The result is shown in Figure 25. From the figure, in the range of 0 to 100 g, the linearity of the two sensors is different, while in the range of 100 to 200 g, the measured values of the two sensors are quite similar.

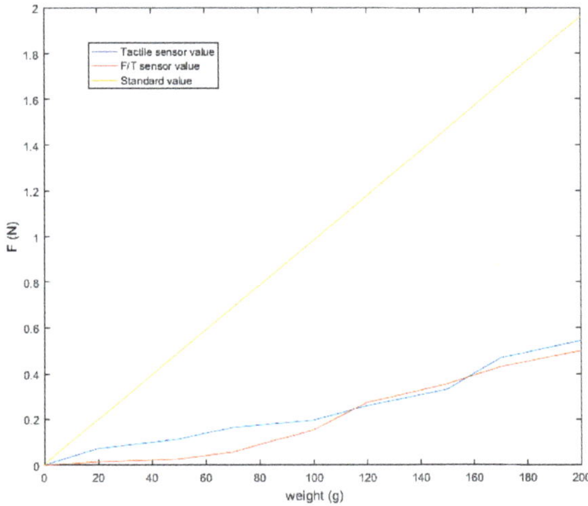

Figure 25. Force sensor calibration.

Therefore, we fuse the sensor data in the two areas separately using the least squares method. The result is shown in Figure 26. Let f_1 and f_2 denote three-axis F/T sensor and tactile sensor arrays measurements of normal force, respectively. In the 0–0.98 N range, the fusion force is $F = 1.8413f_1 + 3.6176f_2 - 0.0245$. In the 0.98–1.96 N range, the fusion force is $F = 4.9991f_1 - 1.2253f_2 + 0.1103$. The results show that the fused sensor values have good linearity with the standard value in the whole range.

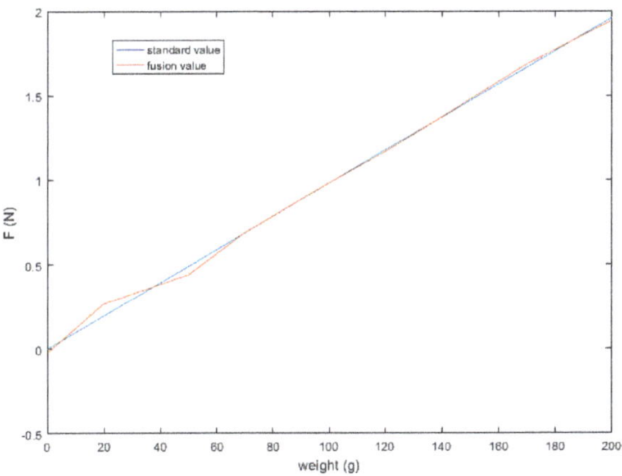

Figure 26. Sensor fusion result.

5.1. Experiment 1: Slipping Detection and Avoidance

This experiment was mainly aimed at verifying the accuracy of slip detection and the quick response of dynamic adjustment to the grasping force.

In order to collect the sensor data during the slip, we set up the experiment, as shown in Figure 27. First, we give the fixed initial grasp pose and grip force. When an initial steady grip was reached, a heavy load was added to the object (rice in the cup), as shown in Figure 28a. As the weight increased, the object slipped (Figure 28b), so the additional weight was used as a disturbance that caused relative slippage. Finally, the object slipped from the hand (Figure 28c), and we recorded the three-axis force/torque (F/T) sensor and tactile sensor arrays data.

Figure 27. Experiment setup for slip detection.

(**a**)

(**b**)

(**c**)

Figure 28. The process of slipping: (**a**) The heavy load was added to the object (rice in the cup); (**b**) Weight increased; (**c**) The object slipped from the hand.

The tangential force collected by the three-axis force/torque (F/T) sensor in the thumb is shown in the upper graph of Figure 29. From this figure, we can see that the raw data contained high-frequency random noise. In order to remove the noise, we used a low-pass filter to process the raw data. The result after filtering is shown in the bottom graph of Figure 29. The upper graph in Figure 30 presents the discrete wavelet transform from the finger three-axis force/torque (F/T) sensor. The bottom graph shows the displacement value from the tactile sensor arrays data. When t = 1.28 s, slippage occurred.

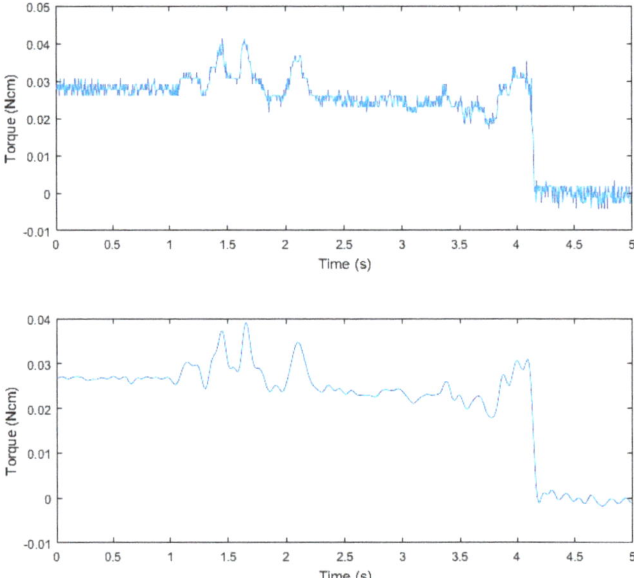

Figure 29. The tangential force of finger three-axis force/torque (F/T) sensor.

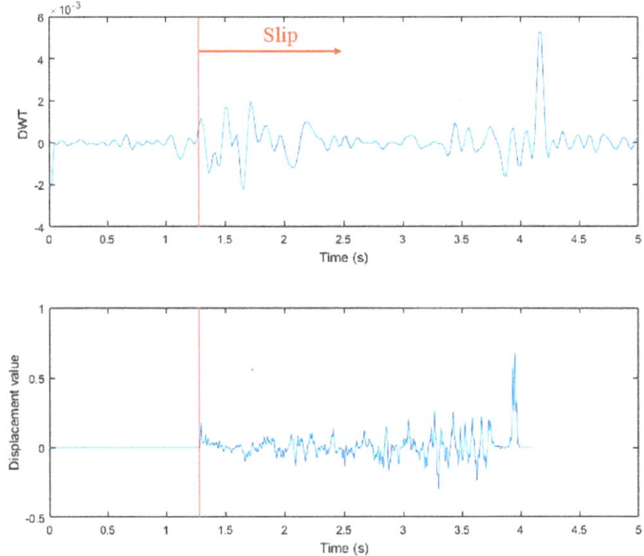

Figure 30. Results of slip detection experiment.

The slip experiment dataset is shown in Figure 31, where the red spots represent the slip situation, and the blue spots represent the no-slip situation. Then we used LIBSVM [38] to train the model. In this study, we used a grid search method to find the best parameter (c, g) for an RBF kernel. As shown in Figure 32, we found that the best parameter was (8, 32), with a cross-validation rate of 97.5%.

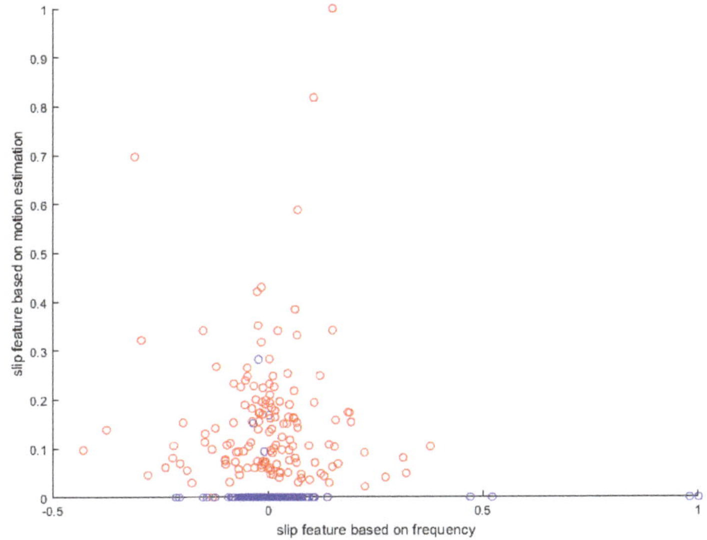

Figure 31. Dataset.

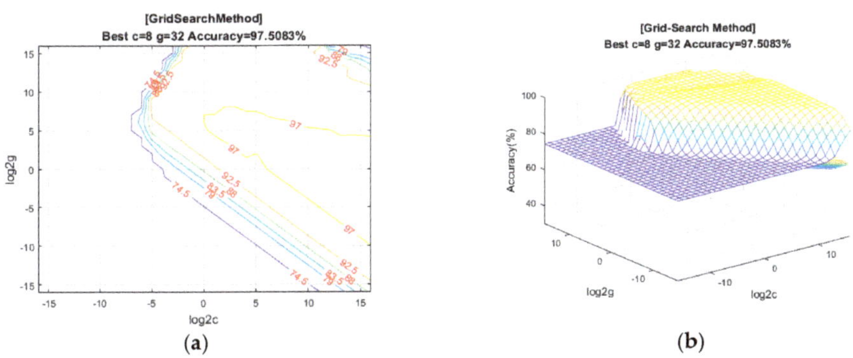

(a) **(b)**

Figure 32. The result of the grid search method: (**a**) Grid-Search Method; (**b**) The cross-validation rate.

Next, we used the SVM predict function to help the robot hand detect the slip signal in real time. Similar to the previous experiment, the task of this experiment was to grasp the plastic cup with three fingers and gradually add rice to the cup. This time, we used slip detection to adjust the grip force in time to suppress the slip. The experimental process is shown in Figure 33. In this experiment, when an initial steady grip was reached, a heavy load was added to the object (rice in the cup) (Figure 33a–c). The additional weight was used as a disturbance that caused relative slippage, we succeeded in avoiding the slippage (Figure 33d–f). However, because we had not measured the stiffness of the object, when the robot hand came into contact with the object (heavy load), it caused deformation of the plastic cup (Figure 33g–i). The slip signal and the grip force change are shown in Figures 34 and 35.

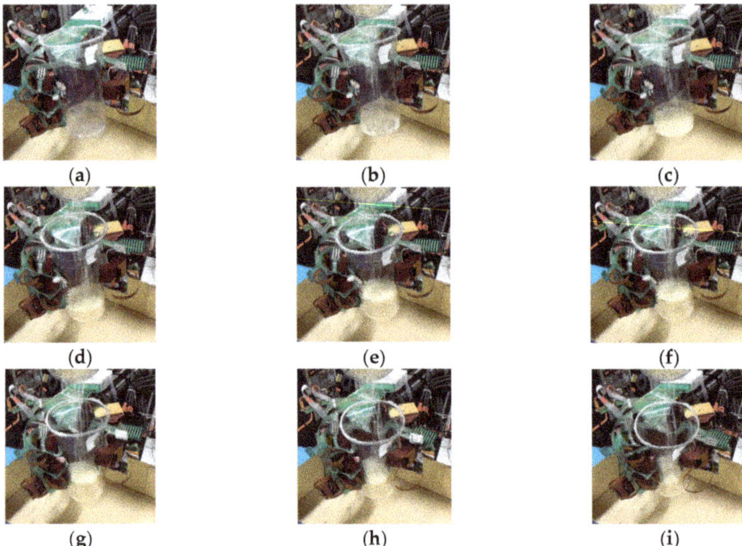

Figure 33. Slip prevention experiment: (**a**–**c**) When an initial steady grip was reached, a heavy load was added to the object (rice in the cup); (**d**–**f**) The additional weight was used as a disturbance that caused relative slippage, we succeeded in avoiding the slippage; (**g**–**i**) Finally, when the robot hand came into contact with the object (heavy load), it caused deformation of the plastic cup.

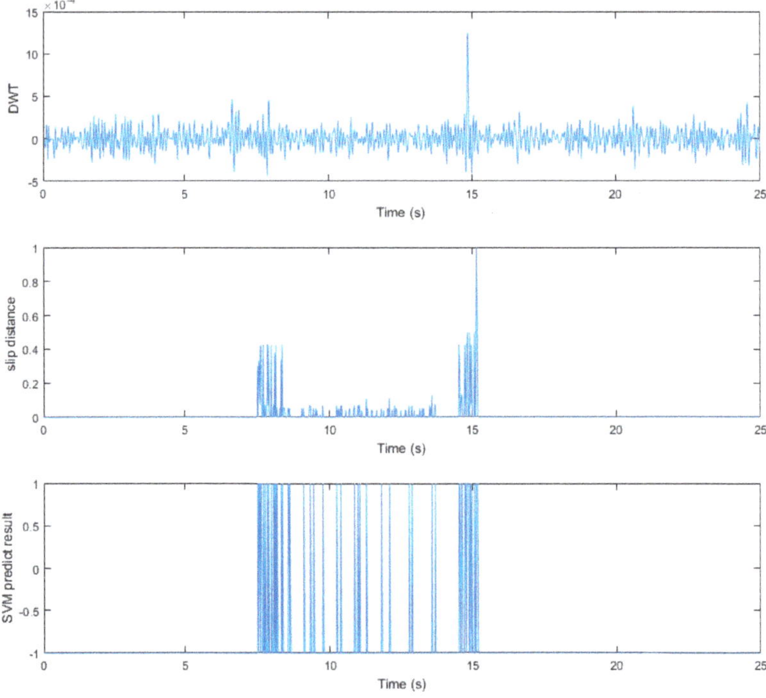

Figure 34. Slip signal.

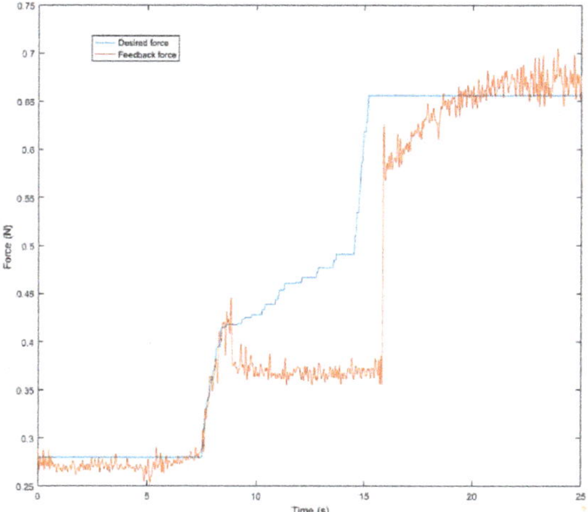

Figure 35. Grip force.

5.2. Experiment 2: Grasping in a Cluttered Scene

This experiment demonstrated grasping in a cluttered scene. The experimental environment used is shown in Figure 36. The objects were set on the blue table, and the robot hand-arm system was located at one end of the table. Two depth sensors were placed on one end of the table and on the side of the robot hand-arm system to capture the point cloud of the environment.

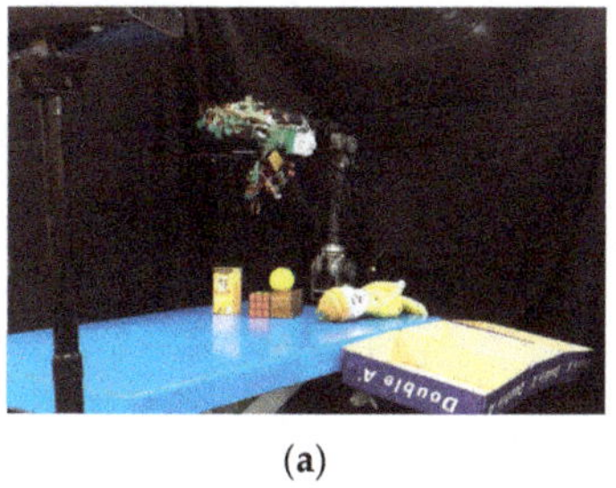

(**a**)

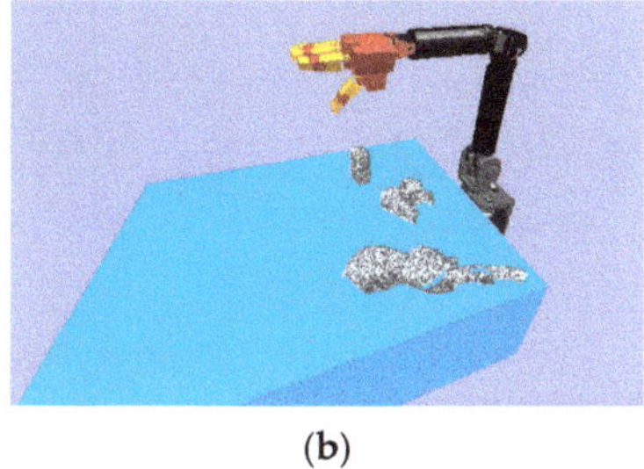

(**b**)

Figure 36. The experimental environment: (**a**) Real environment; (**b**) Objects in the simulator.

Because the NTU robot hand is a right hand, we predefined our task as being to clear the table from the right side to the left side. The first step was to grasp the milk tea. In the robot simulator block, the simulator chose the target object (milk tea). Then, the grasp strategy was used to find the grasp configuration of the robot hand-arm. The path planning used the RRT-Connect algorithm to find collision-free paths for grasping, as shown in Figure 37b.

Once the hand contacted the object, it first took a rigid measurement, as shown in Figure 38a. According to the result of K-NN (Figure 38b), the first object belonged to the low stiffness category. The planner generated a collision-free path from the current position to the target box area (Figure 37c), and then, in the grasping stage, the robot hand turned to position-based force control. According to the multi-sensor feedback as well as the slip detection and stiffness measurement, the robot hand could grasp the object successfully and place it in a specific area according to the object's stiffness. The procedure used for the real robot hand-arm system is shown in Figure 39.

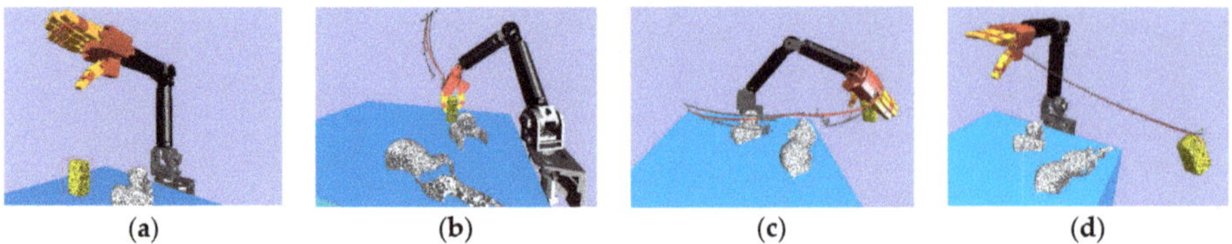

Figure 37. Experiment 2 in a simulator: (**a**) The target object (milk tea); (**b**) Collision-free paths for grasping; (**c**) RRT-Connect algorithm; (**d**) Successfully separated and placed in different boxes.

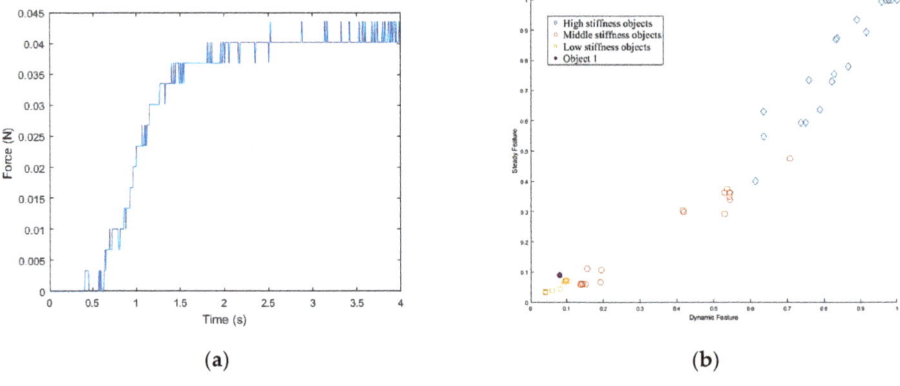

Figure 38. Object 1 stiffness measurement: (**a**) Stiffness; (**b**) K-NN.

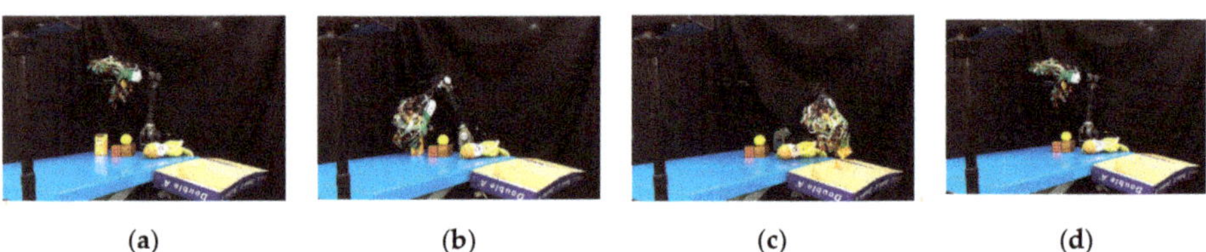

Figure 39. A procedure of grasping the first object: (**a**) The target object (milk tea); (**b**) Collision-free paths for grasping; (**c**) RRT-Connect algorithm; (**d**) Successfully separated and placed in different boxes.

Next, we found that in the middle of the table, there were three objects stacked together, but the vision system considered them to be one object only. As shown in Figure 40a, the target object was approximated by using a few bounding boxes, and the robot hand is first considered grasping the top box. The grasp poses are shown in Figure 40b. Similar to grasping object 1, the hand determined the stiffness measurement (Figure 41a), and K-NN indicated that object 2 belonged to the medium stiffness category. The procedures for the simulation and the real robot hand-arm are shown in Figures 40 and 42, respectively.

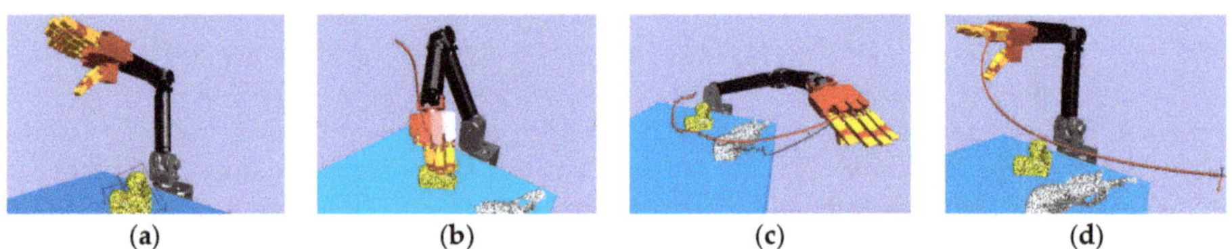

Figure 40. A procedure of grasping the second object in the simulator: (**a**) The target object (top box); (**b**) Collision-free paths for grasping; (**c**) RRT-Connect algorithm; (**d**) Successfully separated and placed in different boxes.

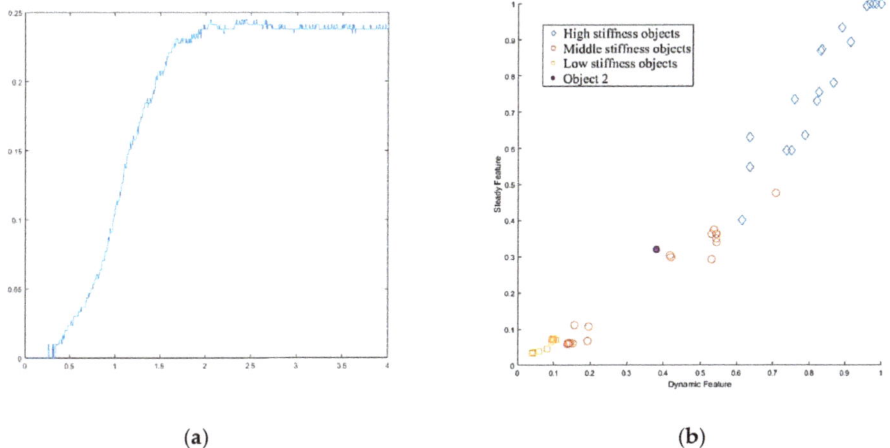

(a) (b)

Figure 41. Object 2 stiffness measurement: (**a**) Stiffness; (**b**) K-NN.

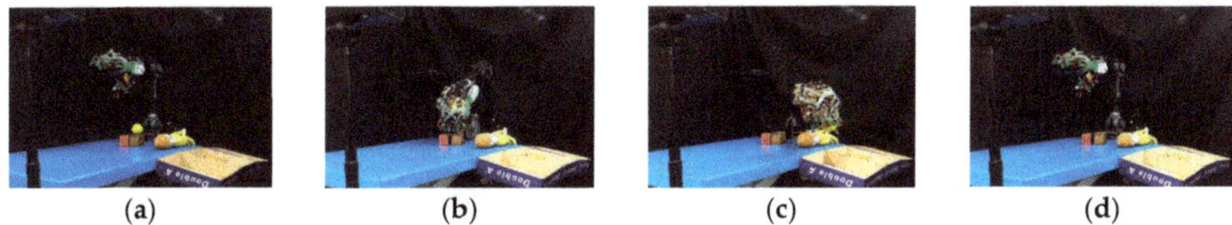

(a) (b) (c) (d)

Figure 42. A procedure of grasping second object: (**a**) The target object (top box); (**b**) Collision-free paths for grasping; (**c**) RRT-Connect algorithm; (**d**) Successfully separated and placed in different boxes.

The vision system collected the point cloud in the scene again (Figure 43a) to update the object state according to the object state inference. After grasping object 2, there was still a point cloud in the same location, meaning there were multiple objects stacked in the same place, and the robot hand considered the next object remaining on the table to be object 3. Grasp planning was performed in a similar manner as the previous objects. The stiffness measurement (Figure 44a) and K-NN result (Figure 44b) indicated that object 3 belonged to the category of high stiffness. The procedures for the simulation and the real robot hand-arm are shown in Figures 43 and 45, respectively. Finally, the last object in the original location was considered to be object 4 is shown in Figure 46. The stiffness measurement (Figure 47a) and K-NN result (Figure 47b) indicated that object 4 also belonged to the high stiffness category. As the robot interacted with them, all the real objects were gradually separated. The procedures for the real robot hand-arm system and the simulation are shown from Figures 46–48.

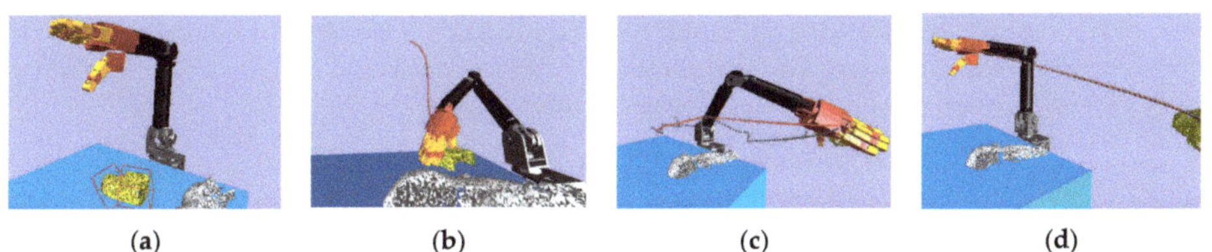

(a) (b) (c) (d)

Figure 43. A procedure of grasping the third object in the simulator: (**a**) The target object; (**b**) Collision-free paths for grasping; (**c**) RRT-Connect algorithm; (**d**) Successfully separated and placed in different boxes.

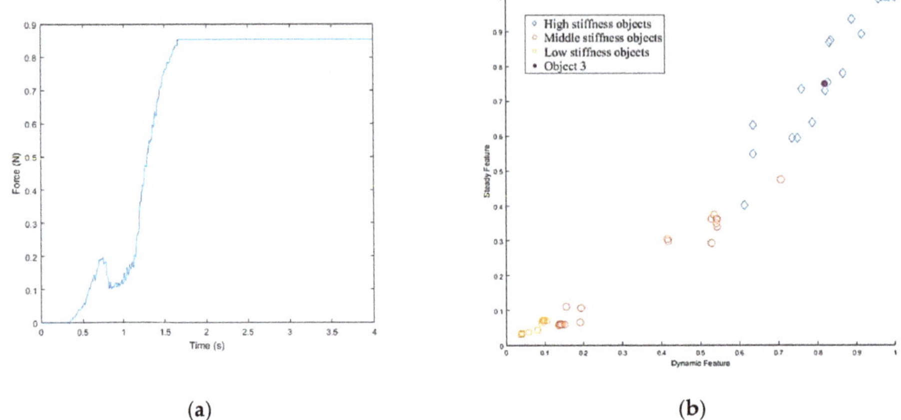

(a) (b)

Figure 44. Object 3 stiffness measurement: (**a**) Stiffness; (**b**) K-NN.

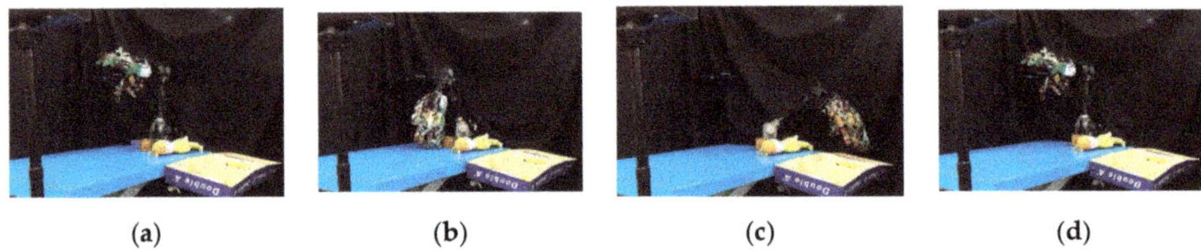

(a) (b) (c) (d)

Figure 45. A procedure of grasping the third object: (**a**) The target object; (**b**) Collision-free paths for grasping; (**c**) RRT-Connect algorithm; (**d**) Successfully separated and placed in different boxes.

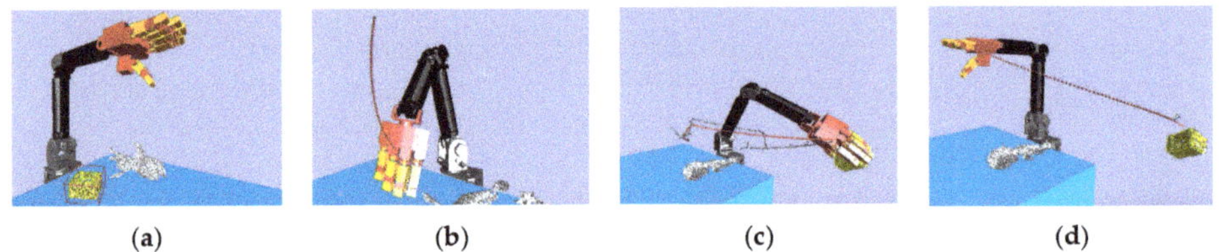

(a) (b) (c) (d)

Figure 46. A procedure of grasping the fourth object in the simulator: (**a**) The target object; (**b**) Collision-free paths for grasping; (**c**) RRT-Connect algorithm; (**d**) Successfully separated and placed in different boxes.

The last object on the table, in a different location, was a cloth puppet, for which the bounding box is shown in Figure 49a. Based on this bounding box, the grasp planner chose an appropriate grasp configuration. According to the result of K-NN (Figure 50b), object 5 belonged to the low stiffness category. The procedures for the real robot hand-arm system and the simulation are shown in Figures 49 and 51, respectively. The classification results are shown in Figure 52.

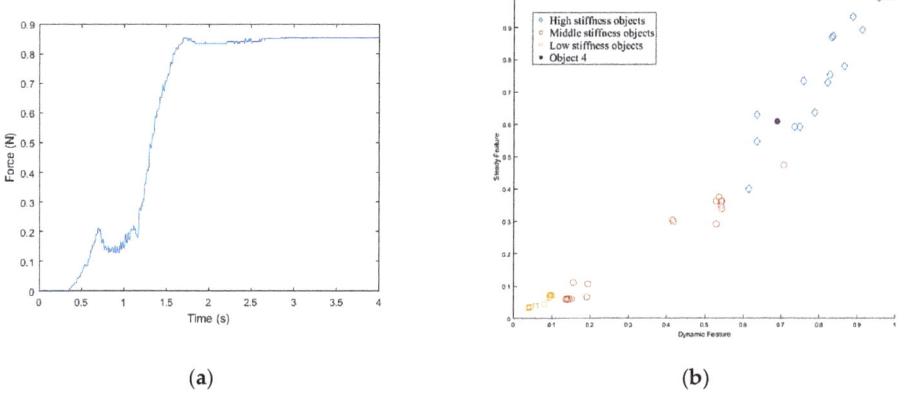

Figure 47. Object 4 stiffness measurement: (**a**) Stiffness; (**b**) K-NN.

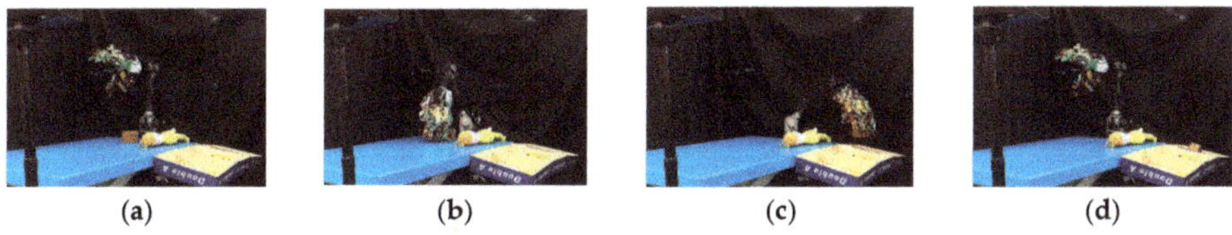

Figure 48. A procedure of grasping the fourth object: (**a**) The target object; (**b**) Collision-free paths for grasping; (**c**) RRT-Connect algorithm; (**d**) Successfully separated and placed in different boxes.

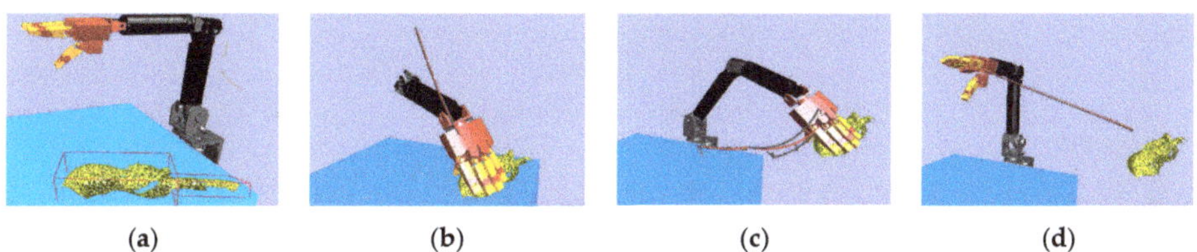

Figure 49. A procedure of grasping the fifth object in the simulator: (**a**) The target object (low stiffness category); (**b**) Collision-free paths for grasping; (**c**) RRT-Connect algorithm; (**d**) Successfully separated and placed in different boxes.

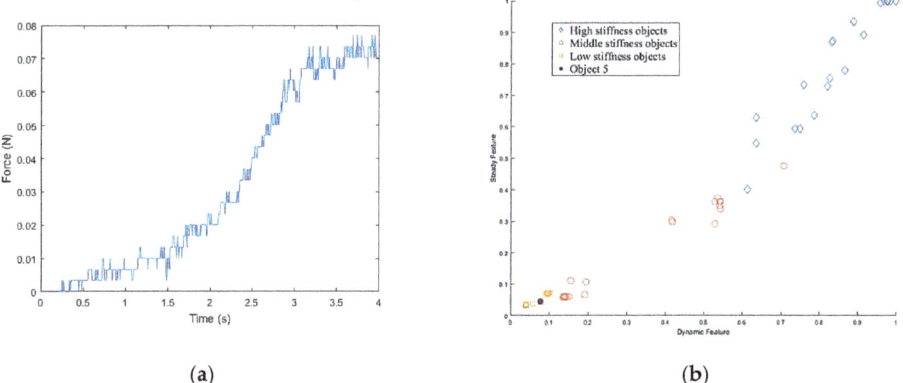

Figure 50. Object 5 stiffness measurement: (**a**) Stiffness; (**b**) K-NN.

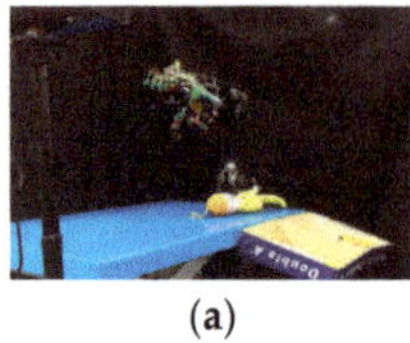
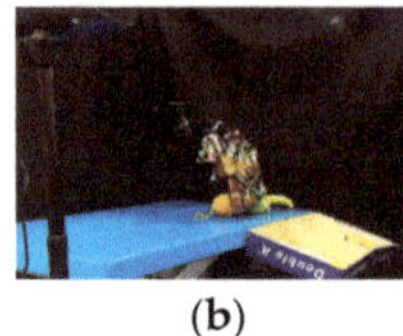
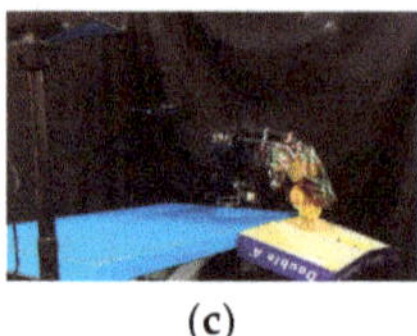
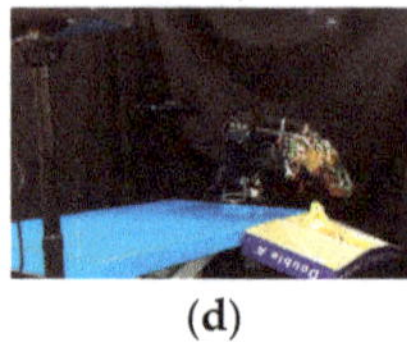

(a) (b) (c) (d)

Figure 51. A procedure of grasping the fifth object: (**a**) The target object (low stiffness category); (**b**) Collision-free paths for grasping; (**c**) RRT-Connect algorithm; (**d**) Successfully separated and placed in different boxes.

As also shown in Figure 52, all the objects on the table were successfully separated and placed in different boxes based on the stiffness of the object.

Figure 52. Classification results.

6. Conclusions

The aim of those experiments was to train a robot hand-arm system to grasp unknown objects. Grasping an object is a simple task for a human, who can grasp and manipulate any object whether he has seen it or not. We want a robot that can work in a human environment in a similar manner to grasp a variety of objects of different materials, shapes, and sizes in a cluttered scene. Therefore, we focused on two issues. The first issue was the methodology for setting the real-time grasping force of the fingers when grasping an object with unknown stiffness, weight, and friction. The second issue was on-line sorting objects in a cluttered scene.

When the task is to grasp an unknown object, it is the most concern to avoid dropping it due to slipping, which is considered a serious error because of breakage of the object or unpredictable disasters. What is more, crushing objects or grasping them with excessive force should also be avoided. In order to grasp an unknown object with an appropriate force, we used multi-sensor inputs and sensor fusion technology to detect slippage and measure the stiffness of objects. The experiments demonstrated that multi-sensor information can ensure that the robot hand grasp unknown objects safely and make the robot hand's motion more like that of a human.

Physically interacting to improve the understanding of an environment is a natural behavior for humans, and based on this, we proposed an algorithm that combines perception and manipulation to enable a robot to sort objects in a cluttered space according to one specific property (in this case, stiffness). In contrast to pure visual perceptual approaches, our method can quickly rearrange objects. Our experiments show that sorting is more viable and reliable when using this method to deal with a quantity of unknown objects.

Author Contributions: Conceptualization, S.-Q.J., M.-B.H. and H.-P.H.; methodology, S.-Q.J., M.-B.H. and H.-P.H.; software, S.-Q.J. and M.-B.H.; validation, S.-Q.J. and M.-B.H.; formal analysis, S.-Q.J. and M.-B.H.; investigation, S.-Q.J. and M.-B.H.; resources, H.-P.H.; data curation, S.-Q.J. and M.-B.H.; writing—original draft preparation, S.-Q.J. and M.-B.H.; writing—review and editing, M.-B.H. and H.-P.H.; visualization, M.-B.H. and H.-P.H.; supervision, M.-B.H. and H.-P.H.; project administration, H.-P.H.; funding acquisition, H.-P.H.

Funding: This research was funded by HIWIN Technologies Corporation, Taiwan, grant number 104-S-A13. This work was partially supported by the Ministry of Science and Technology (MOST), Taiwan, grant number 107-2221-E-002-176-MY3.

Acknowledgments: The authors would like to thank the Ministry of Science and Technology (MOST).

Conflicts of Interest: The authors declare no conflict of interest.

References

1. Huang, H.-P.; Yan, J.-L.; Huang, T.-H.; Huang, M.-B. IoT-based networking for humanoid robots. *J. Chin. Inst. Eng.* **2017**, *40*, 603–613. [CrossRef]
2. Saudabayev, A.; Varol, H.A. Sensors for robotic hands: A survey of state of the art. *IEEE Access* **2015**, *3*, 1765–1782. [CrossRef]
3. Huang, H.-P.; Yan, J.-L.; Cheng, T.-H. Development and fuzzy control of a pipe inspection robot. *IEEE Trans. Ind. Electron.* **2010**, *57*, 1088–1095. [CrossRef]
4. Huang, M.-B.; Huang, H.-P.; Cheng, C.-C.; Cheng, C.-A. Efficient grasp synthesis and control strategy for robot hand-arm system. In Proceedings of the 11th IEEE International Conference on Automation Science and Engineering (CASE), Gothenburg, Sweden, 24–28 August 2015.
5. Lee, W.-Y.; Huang, M.-B.; Huang, H.-P. Learning robot tactile sensing of object for shape recognition using multi-fingered robot hands. In Proceedings of the 26th IEEE International Symposium on Robot and Human Interactive Communication (RO-MAN), Lisbon, Portugal, 28 August–1 September 2017.
6. Liu, Y.-R.; Huang, M.-B.; Huang, H.-P. Automated grasp planning and path planning for a robot hand-arm system. In Proceedings of the 11th IEEE/SICE International Symposium on System Integration (SII), Paris, France, 14–16 January 2019.
7. Huang, M.-B.; Huang, H.-P. Innovative human-like dual robotic hand mechatronic design and its chess-playing experiment. *IEEE Access* **2019**, *7*, 7872–7888. [CrossRef]
8. Rusu, R.B. Semantic 3D object maps for everyday manipulation in human living environments. *Künstl. Intell.* **2010**, *24*, 345–348. [CrossRef]
9. Huebner, K.; Ruthotto, S.; Kragic, D. Minimum volume bounding box decomposition for shape approximation in robot grasping. In Proceedings of the 2008 IEEE International Conference on Robotics and Automation (ICRA), Pasadena, CA, USA, 19–23 May 2008.
10. Lei, Q.; Chen, G.; Wisse, M. Fast grasping of unknown objects using principal component analysis. *AIP Adv.* **2017**, *7*, 095126. [CrossRef]
11. Liu, Z.; Kamogawa, H.; Ota, J. Fast grasping of unknown objects through automatic determination of the required number of mobile robots. *Adv. Robot.* **2013**, *27*, 445–458. [CrossRef]
12. Faria, D.R.; Dias, J. 3D hand trajectory segmentation by curvatures and hand orientation for classification through a probabilistic approach. In Proceedings of the 2009 IEEE/RSJ International Conference on Intelligent Robots and Systems (IROS), St. Louis, MO, USA, 10–15 October 2009.
13. Hirano, Y.; Kitahama, K.-I.; Yoshizawa, S. Image-based object recognition and dexterous hand/arm motion planning using RRTs for grasping in cluttered scene. In Proceedings of the 2005 IEEE/RSJ International Conference on Intelligent Robots and Systems (IROS), Edmonton, AB, Canada, 2–6 August 2005.
14. Yang, K.; Sukkarieh, S. 3D smooth path planning for a UAV in cluttered natural environments. In Proceedings of the 2008 IEEE/RSJ International Conference on Intelligent Robots and Systems (IROS), Nice, France, 22–26 September 2008.
15. Bicchi, A.; Kumar, V. Robotic grasping and contact: A review. In Proceedings of the 2000 IEEE International Conference on Robotics and Automation (ICRA), San Francisco, CA, USA, 24–28 April 2000.
16. Murray, R.M.; Li, Z.; Sastry, S.S. Multifingered hand kinematics. In *A Mathematical Introduction to Robotic Manipulation*, 1st ed.; CRC Press: Boca Raton, FL, USA, 1994; pp. 211–240.
17. Roa, M.A.; Suárez, R. Grasp quality measures: Review and performance. *Auton. Robot.* **2015**, *38*, 65–88. [CrossRef] [PubMed]
18. Zheng, Y. An efficient algorithm for a grasp quality measure. *IEEE Trans. Robot.* **2013**, *29*, 579–585. [CrossRef]
19. Ferrari, C.; Canny, J. Planning optimal grasps. In Proceedings of the 1992 IEEE International Conference on Robotics and Automation (ICRA), Nice, France, 12–14 May 1992.
20. Boyd, S.P.; Wegbreit, B. Fast computation of optimal contact forces. *IEEE Trans. Robot.* **2007**, *23*, 1117–1132. [CrossRef]
21. WACOH-TECH Inc. Available online: https://wacoh-tech.com/products/mudynpick/maf-3.html (accessed on 3 September 1988).

22. Altman, N.S. An introduction to kernel and nearest-neighbor nonparametric regression. *Am. Stat.* **1992**, *46*, 175–185.
23. Kozma, L. k Nearest Neighbors algorithm (kNN). Helsinki University of Technology, Special Course in Computer and Information Science. Available online: http://www.lkozma.net/knn2.pdf (accessed on 20 February 2008).
24. Shaw-Cortez, W.; Oetomo, D.; Manzie, C.; Choong, P. Tactile-based blind grasping: A discrete-time object manipulation controller for robotic hands. *IEEE Robot. Autom. Lett.* **2018**, *3*, 1064–1071. [CrossRef]
25. Johansson, R.S.; Westling, G. Roles of glabrous skin receptors and sensorimotor memory in automatic control of precision grip when lifting rougher or more slippery objects. *Exp. Brain Res.* **1984**, *56*, 550–564. [CrossRef] [PubMed]
26. Deng, H.; Zhong, G.; Li, X.; Nie, W. Slippage and deformation preventive control of bionic prosthetic hands. *IEEE/ASME Trans. Mech.* **2017**, *22*, 888–897. [CrossRef]
27. Teshigawara, S.; Tadakuma, K.; Ming, A.; Ishikawa, M.; Shimojo, M. High sensitivity initial slip sensor for dexterous grasp. In Proceedings of the 2010 IEEE International Conference on Robotics and Automation (ICRA), Anchorage, AK, USA, 3–7 May 2010.
28. Heil, C.; Colella, D. Dilation equations and the smoothness of compactly supported wavelets. In *Wavelets: Mathematics and Applications*, 1st ed.; CRC Press: Boca Raton, FL, USA, 1993; pp. 163–202.
29. Boggess, A.; Narcowich, F.J. Other wavelet topics. In *A First Course in Wavelets with Fourier Analysis*, 2nd ed.; John Wiley & Sons: Hoboken, NJ, USA, 2009; pp. 250–272.
30. Tremblay, M.R.; Cutkosky, M.R. Estimating friction using incipient slip sensing during a manipulation task. In Proceedings of the 1993 IEEE International Conference on Robotics and Automation (ICRA), Atlanta, GA, USA, 2–6 May 1993.
31. Kappassov, Z.; Corrales, J.-A.; Perdereau, V. Tactile sensing in dexterous robot hands—Review. *Robot. Auton. Syst.* **2015**, *74*, 195–220. [CrossRef]
32. Jähne, B. Image representation. In *Digital Image Processing*, 6th ed.; Springer: Berlin, Germany, 2005; pp. 41–62.
33. Alcazar, J.A.; Barajas, L.G. Estimating object grasp sliding via pressure array sensing. In Proceedings of the 2012 IEEE International Conference on Robotics and Automation (ICRA), Saint Paul, MN, USA, 14–18 May 2012.
34. Vapnik, V.N. Constructing learning algorithms. In *The Nature of Statistical Learning Theory*, 1st ed.; Springer: New York, NY, USA, 1995; pp. 119–156.
35. Cheng, C.-A.; Huang, H.-P. Learn the Lagrangian: A vector-valued RKHS approach to identifying Lagrangian systems. *IEEE Trans. Cybern.* **2016**, *46*, 3247–3258. [CrossRef] [PubMed]
36. Cheng, C.-A.; Huang, H.-P.; Hsu, H.-K.; Lai, W.-Z.; Cheng, C.-C. Learning the inverse dynamics of robotic manipulators in structured reproducing kernel Hilbert space. *IEEE Trans. Cybern.* **2016**, *46*, 1691–1703. [CrossRef] [PubMed]
37. Lo, S.-Y.; Cheng, C.-A.; Huang, H.-P. Virtual impedance control for safe human-robot interaction. *J. Intell. Robot. Syst.* **2016**, *82*, 3–19. [CrossRef]
38. LIBSVM: A Library for Support Vector Machines. 2001. Available online: https://www.csie.ntu.edu.tw/~cjlin/libsvm (accessed on 1 September 2001).

Article

Localization of Sliding Movements Using Soft Tactile Sensing Systems with Three-axis Accelerometers

Hiep Xuan Trinh [1,*], Yuki Iwamoto [1], Van Anh Ho [2] and Koji Shibuya [1]

1 Department of Mechanical and Systems Engineering, Faculty of Science and Technology, Ryukoku University, 1-5 Yokotani, Seta Oe-cho, Otsu, Shiga 520-2194, Japan; t18m026@mail.ryukoku.ac.jp (Y.I.); koji@rins.ryukoku.ac.jp (K.S.)

2 School of Material Science, Japan Advanced Institute of Science and Technology (JAIST), 1-1 Asahidai, Nomi, Ishikawa 923-1292, Japan; van-ho@jaist.ac.jp

* Correspondence: t17d501@mail.ryukoku.ac.jp; Tel.: +81-77-543-7444

Received: 5 April 2019; Accepted: 29 April 2019; Published: 30 April 2019

Abstract: This paper presents a soft tactile sensor system for the localization of sliding movements on a large contact surface using an accelerometer. The system consists of a silicone rubber base with a chamber covered by a thin silicone skin in which a three-axis accelerometer is embedded. By pressurizing the chamber, the skin inflates, changing its sensitivity to the sliding movement on the skin's surface. Based on the output responses of the accelerometer, the sensor system localizes the sliding motion. First, we present the idea, design, fabrication process, and the operation principle of our proposed sensor. Next, we created a numerical simulation model to investigate the dynamic changes of the accelerometer's posture under sliding actions. Finally, experiments were conducted with various sliding conditions. By confirming the numerical simulation, dynamic analysis, and experimental results, we determined that the sensor system can detect the sliding movements, including the sliding directions, velocity, and localization of an object. We also point out the role of pressurization in the sensing system's sensitivity under sliding movements, implying the ideal pressurization for it. We also discuss its limitations and applicability. This paper reflects our developed research in intelligent integration and soft morphological computation for soft tactile sensing systems.

Keywords: soft tactile sensing system; localization and object detecting; three-axis accelerometer

---

1. Introduction

Tactile sensors, according to [1], are defined as devices that can detect and measure a contact's properties in a predetermined area and subsequently pre-processes the signals with the sensing elements before sending them to higher levels of perceptual interpretation. Human tactile sensing generally serves as a reference point in robotics. Human sensory psychophysics can be described by four main attributes: location, modality, intensity, and timing [2]. Many researchers focus on artificial tactile sensors that mimic these capabilities for applications in robotics [3–5]. In terms of localization and object detection, research is currently focusing on designing and fabricating artificial tactile sensor arrays that can cover large surfaces to obtain rich contact information. For instance, Damian et al. [6] presented an artificial ridged skin for detecting the position and speed of a sliding object using force sensors that were embedded in nonuniform arranged-parallel ridges. Another work [7] proposed a real-time sensor fusion algorithm for estimation object position, translation, and rotation during grasping by pressure array sensing. Although these tactile sensor arrays can detect an object's localization on a large contact surface with improved accuracy and spatial resolution, they require many sensing elements, complicating integration and manufacturing.

In addition, due to the complexity of the output data, tactile sensing array normally needs a computational program with complicated algorithms for data processing [8].

Recent research in morphological computation and embodied intelligence is introduced as a method to reduce the complexity of entire robotics systems by creating different sensing and actuating functions [9,10]. Soft robotics is an example that utilizes embodiment to facilitate control and perception [11]. One crucial element of soft robotics research is soft tactile sensors, which are normally implemented by embedding sensing elements in a soft body [12]. The sensing elements can be such conventional sensors as strain gauges [13], magnets and Hall sensors [14], or pressure sensors [15], which transduce physical interaction into electrical output signals through the soft body's deformation. By exploiting the soft, elastic properties of soft material, soft tactile sensors suggest the potential of using the characteristics of a biological body to reduce complexity [16]. Based on the idea of morphological computation to obtain rich sensory information for soft tactile sensors, we previously proposed Wrin'Tac, a soft active tactile sensor system that can change its sensitivity for different sensing tasks using a strain gauge as its only sensing element [17,18]. Although Wrin'Tac can successfully change its sensitivity and the sense sliding movements on its surface, it cannot localize them because the strain gauge's output is only one dimensional, which means that we cannot detect the position of a sliding movement if the sensor provides identical output for different movements. Inspired by how a spider detects its prey, we present in this paper a tactile sensing system that is sensitive to different sliding motions in large contact areas using just a three-axis accelerometer as a sensing element. This sensing system consists of a soft base with a chamber covered by a thin silicone skin in which a three-axis accelerometer is embedded. We integrated pneumatic actuation to generate a large inflated skin's surface that easily transforms the deformation at the contact's position to the accelerometer's position. Under a sliding motion, the accelerometer's posture changes continuously, depending on the contact's location, resulting in a variation of an accelerometer's output voltages. Thus, our sensing system can detect and localize sliding motions with different directions. Several studies have utilized accelerometers for fabrication of tactile sensors [19–21]. A previous work [19] used a miniature accelerometer that was attached to a human finger to measure the acceleration at the radial skin to estimate the surface undulation. It was attached to several pneumatic grippers to classify the hardness of different cylinders, to estimate the pneumatic pressure, and to assess the firmness of eggplants and mangoes [20]. The authors of [21] proposed a bio-mimetic fingertip on which they embedded three commercial accelerometers and force sensors to detect force and vibration modalities. Most accelerometer applications in tactile sensors focus on estimating physical quantities such as hardness and firmness and using measured vibrations for surface identification. In terms of detecting and localization sliding movements, there is a lack of interest in accelerometer applications in tactile sensing systems.

In this paper, to clarify this sensing system's ability, we constructed a numerical analysis model by 3D simulation and conducted experimental validations with dynamic analyses in different sliding actions. Our proposed sensing system has the following advances: easy fabrication, low cost, and simple integration to soft robotics. Our research is using intelligent embodiment to reduce the complexity of soft tactile sensor systems. The following are the main contributions of this paper:

(1) We integrated our proposed idea, design, and fabrication of a soft tactile sensing system a pneumatic actuator and a three-axis accelerometer as a sensing element for detecting and localization sliding movements in a large contact area.
(2) We clarified the operation principle of our sensing system and conducted a numerical simulation to study the dynamic responses of the sensing elements under a sliding motion with morphological changes.
(3) We proposed dynamic analyses and interpolated functions based on experiment results to verify the sensing system's ability.
(4) We elaborated the role of pressurization on the response of the sensing system under sliding movements.

2. Materials and Methods

2.1. Idea

Nature provides many examples of amazing touch organs with tactile sensing that have inspired robotic applications [22–24]. Our study was inspired by the prey detection of spiders, which lurk at the bottom of their webs. Although spiders do not have great eyesight, and they usually use the vibrations of the web strands to locate their prey. Spiders interpret vibrations with organs on their legs called slit sensillae, which are small, hypersensitive grooves that deform with even the slightest disturbance [25,26].

Based on the notion of prey detection by spiders, we propose an active tactile sensing system for detecting and localizing sliding objects on large contact areas. Our tactile sensing system integrates a sensing element that is embedded in soft skin and actuation that can generate a curved skin's surface on a sensing system to resemble the role of spider's web (Figure 1b). We deploy a three-axis accelerometer as the sensing element and pressurized air as the actuation. Under pressurization, a curved surface was generated, and when an object contacts and slides on it, the deformation at the contact's position is conveyed to the accelerometer's location that provides information about the acceleration of the three axes (x, y and z axes). This information is sufficient for detecting the sliding directions and localizing the sliding movement. In our sensing system, the inflated skin's surface spreads the deformation, which mimics the role of a spider's web in a vibration's transmission. Figure 1 shows schematic drawings of our soft tactile sensing system.

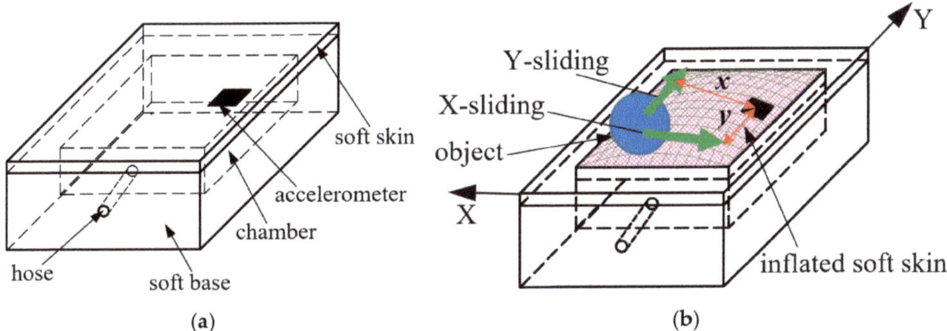

Figure 1. Schematic drawing of soft tactile sensing system: One soft skin layer with embedded accelerometer covers another soft base. In this base, a chamber is cut from its surface and has a hose for input compressed air. Under pressurization inside it, soft skin is inflated, changing the accelerometer's posture. When object slides on inflated soft skin, accelerometer is sensitive to deformation at contact positions. (**a**) Initial state: without pressurization, soft skin is flat. (**b**) With pressurization, soft skin is inflated, making sensitivity of soft tactile system to sliding action on its surface.

In this paper, we focus on the localization of sliding movements in two X- and Y- directions with different distances from a sliding line to the accelerometer (Figure 1b). For simplicity, we suppose that first a sliding object slightly contacts the soft skin (outside the inflated area) and then slides over the inflated area at a constant velocity without changing the initial contact depth. The localization of the sliding object can be characterized by the sliding directions and its position coordinate (x,y). For the X-sliding case, y is the distance from the sliding line to the accelerometer and $x = vt$ whereas for the Y-sliding case, x is the distance from the sliding line to the accelerometer and $y = vt$. Here v is the magnitude of sliding velocity. Thus, for localization, first we demonstrate that the sensing system can detect the sliding motion and its directions. Based on the accelerometer output, the sliding velocity's magnitudes and the distances from the sliding line to the accelerometer are estimated. Then the sliding object can be localized.

2.2. Design and Fabrication

Our tactile sensor system has two parts. The first is a soft, 64 × 64 × 20 mm base with a 50 × 50 × 10 mm chamber is cut from its surface. The soft base was made from KE-1603 (Shin-Etsu, Tokyo, Japan) silicone rubber from Shin-Etsu Silicone [27] and fabricated with a 3D printing mold. Then the silicone rubber in a liquid state was poured into it and cured. The second part is soft, thin 60 × 60 × 4 mm skin made from EcoFlex10 (Smooth-on, Macungie, PA, USA) that completely covers the chamber's base.

To fabricate the skin, we used a small chamber to install the accelerometer using a mold. After installing an accelerometer in the small chamber, silicone rubber in liquid state EcoFlex10 part A was mixed with EcoFlex10 part B [28] at a ratio of 1:1 with a mixer for three minutes, and then the mixed liquid was poured to entirely cover the skin's surface and cured at 70 °C for 40 min. The 3 × 5 × 1 mm accelerometer used in this paper is MMA7361L by Freescale Semiconductor (Austin, TX, USA) mounted on a 10 × 10 × 1.7 mm circuit board [29]. Next the soft skin with an embedded accelerometer was attached to the soft base with adhesive glue that is specialized for silicone rubber attachment. In the final step, a 4-mm diameter tube was put and glued for directing the compressed air in the chamber. The design and a photograph of the tactile sensing system are shown in Figure 2.

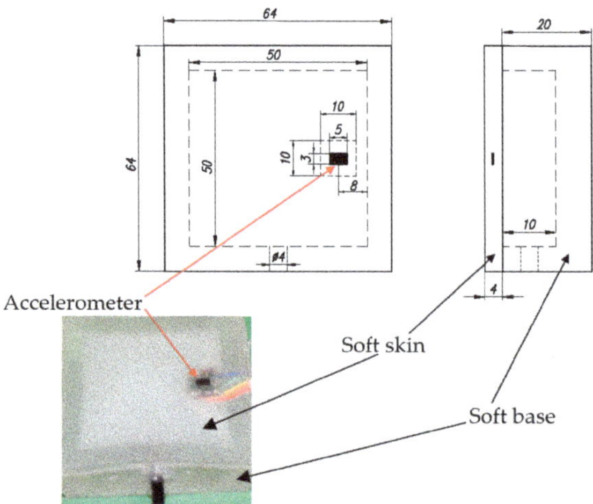

Figure 2. Dimension and photograph of fabricated tactile sensing system.

2.3. Operation Principle

The sensing element used in this research is a MMA7361L, a surface-micromachined integrated-circuit accelerometer by a Freescale Semiconductor, which can provide three-axis acceleration information. The device consists of a surface-micromachined capacitive sensing cell and a signal conditioning ASIC contained in a single package. ASIC uses switched capacitor techniques to measure the capacitors and provide a high-level output voltage that is ratiometric and proportional to the acceleration [29].

In our research, we use the ability to calculate the static acceleration based on the posture of the accelerometer. When the object contacts the inflated soft skin surface, the contact area is deformed. Due to the soft properties of the skin material (Ecoflex10) the skin area at the sensing element's position is also deformed, changing the accelerometer's posture. When the accelerometer's posture changes, its xyz axis coordinate also rotates (Figure 3a,b). With each position of the accelerometer, its output is the value of the projection of the gravitational acceleration on the Ox, Oy, Oz axis.

Let θ be the tilt angle of each axis with horizontal axis Δ (Figure 3c,d) where the positive direction of θ is clockwise (the rotational direction toward gravitational acceleration g). That means according to the smallest angle to coincide with x, y or z axis, the value of θ is positive if the rotational direction of horizontal axis Δ is clockwise and if θ has a negative value in the contrast rotational case.

Figure 3 shows that for each axis the acceleration is:

$$a = gsin\theta. \tag{1}$$

For the Ox, Oy, Oz axis, output acceleration a_x, a_y, a_z can be calculated:

$$a_x = gsin\theta_x,\ a_y = gsin\theta_y,\ a_z = gsin\theta_z\ . \tag{2}$$

Output voltage $V_x,\ V_y,\ V_z$ of the accelerometer is:

$$V_x = ka_x,\ V_y = ka_y, V_z = ka_z, \tag{3}$$

where k is the calibration factor.

When the object slides on the sensor's surface, the accelerometer's posture changes continuously, depending on the contact position. Angles $\theta_x, \theta_y, \theta_z$ are functions of time.

Finally, we get:

$$\begin{aligned} V_x &= ka_x = kgsin\theta_x(t) = V_x(t) \\ V_y &= ka_y = kgsin\theta_y(t) = V_y(t) \\ V_z &= ka_z = kgsin\theta_z(t) = V_z(t). \end{aligned} \tag{4}$$

The values of output voltages $V_x,\ V_y,\ V_z$ depend on angles $\theta_x(t),\ \theta_y(t),\ \theta_z(t)$. Thus, output voltages $V_x,\ V_y,\ V_z$ depend on the contact's position of the sliding object on the sensor's surface. Based on the values of $V_x,\ V_y,\ V_z$, the sensing system can detect and locate the sliding movements.

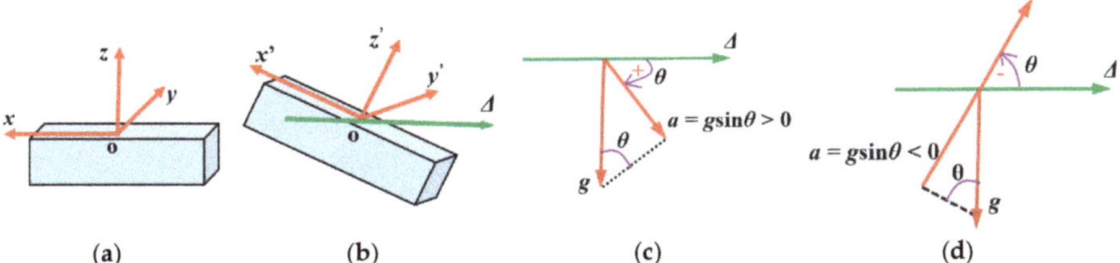

Figure 3. Schematic diagram shows change of accelerometer's posture, its coordinates, and calculation of acceleration on each axis based on gravitational acceleration and tilt angle. (**a**) Initial position of accelerometer and its coordinate Oxyz. (**b**) Coordinate Ox'y'z' when accelerometer's posture changes. (**c**) Acceleration on axis when tilt angle $\theta > 0$. (**d**) Acceleration on axis when tilt angle $\theta < 0$.

3. Numerical Simulation

As mentioned in Section 2.3, when the object slides on the inflated skin's surface, the accelerometer's posture and its coordinates change, depending on the contact's localization. Therefore, to clarify the effect of the sliding action on the accelerometer's posture, first we exploited a finite element (FE) model with commercial software (Abaqus, Dassault Systemes, Johnston, IA, USA) to conduct a dynamic simulation.

Compared to the design of the soft tactile sensor system in Figure 1, we proposed a three-dimensional (3D) model (Figure 4) that is comprised of four different parts: a soft base, a skin layer, an accelerometer inside the skin, and a hemi-spherical indenter. The sizes of the soft base, the width of the skin layer, and the accelerometer were identical to those mentioned in the sensor system's design. To reduce the computational cost, the thickness of the skin layer and the accelerometer were set smaller than their values in the design: 2-mm skin layer and 0.5-mm accelerometer in simulation compared to 4 mm and 1 mm in the sensing's design. The hemi-spherical indenter has a diameter of 6 mm. The values of the Young's modulus of the soft base and the skin were set to 1.2 and 0.07 MPa. The indenter and the accelerometer are expected to be rigid with a much larger Young's modulus (2000 MPa) than the skin and the base. The FE model is meshed with 47,656 linear hexahedral elements of type C3D8R.

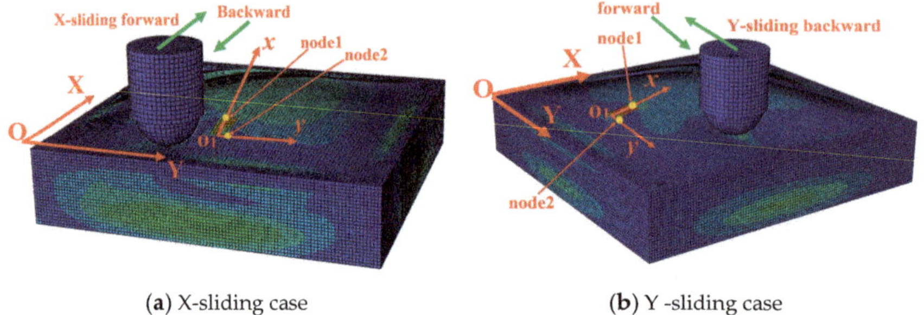

(a) X-sliding case (b) Y -sliding case

Figure 4. Dynamic simulation of soft active tactile sensing system. (**a**) Indenter is displaced 0.5 mm toward the skin, then is forced to slide along X axis with forward and backward directions. (**b**) Indenter is displaced 0.5 mm toward the skin, then is forced to slide along Y axis with forward and backward directions.

Each simulation trial was conducted in four continuous steps: pressurization of the chamber, push indenter, forward and backward slides of the indenter. In the first step, the chamber is pressurized from 0 Pa to a particular value in 0.5 s. In the second step, the indenter is displaced 0.5 mm toward the skin for 0.25 s. The interaction between the indenter and the skin surface is modeled as normal and tangential contact using the penalty formulation with a friction coefficient of 0.1. In the third step, the indenter is forced to slide horizontally right to left (forward direction) at a constant velocity of 200 mm/s in 0.25 s without changing the contact depth. In the last step, the indenter is slid back at the same velocity to its original position (backward direction). The third and last steps were conducted in both the X- and Y-sliding directions (Figure 4). Some simulation trials were also conducted with different pressurizations for comparison.

In the simulation, the accelerometer was attached to local coordinate O_1xyz, which is identical to the defined coordinate in the actual sensor system. In the O_1x, O_1y axes, nodes 1 and 2 were chosen to represent each axis ($O_1x \equiv O_1$node 1, $O_1y \equiv O_1$node 2) (notation $\equiv$ indicates that they coincide). In our tactile sensing system, we used output voltages V_x, V_y for detecting the object's localization. Thus, in the simulation, we only consider the position of two axes: O_1x and O_1y.

As discussed in Section 2.3, the accelerometer output depends on the angle between O_1x, O_1y axes and horizontal axis. Thus, to investigate the position of the O_1x, O_1y axes when the accelerometer's posture changes under sliding motions, we extracted the displacements of two nodes: nodes 1 and 2.

In the xy coordinate (Figure 5), angles θ_x, θ_y between O_1x, O_1y and horizontal axis Δ depends on displacement Dx, Dy of nodes 1 and 2.

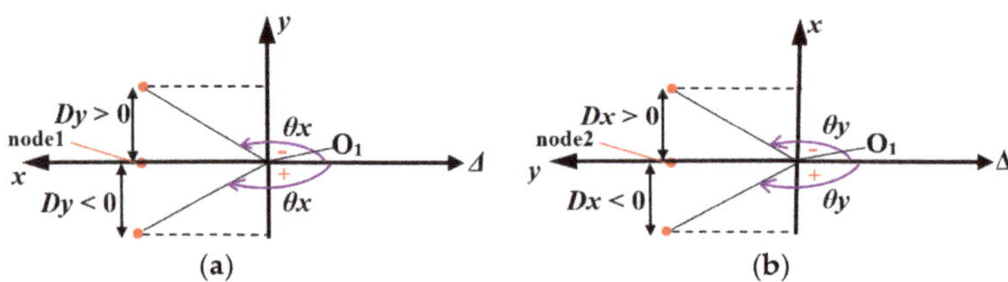

Figure 5. Relationship beween displacement of node 1 (Dy), node 2 (Dx) with value of angle θ_x, θ_y. (**a**) Displacement of node 1 Dy represents position of axis O_1x and value of angle θ_x. $Dy > 0 \rightarrow \theta_x < 0$, $Dy < 0 \rightarrow \theta_x > 0$. (**b**) Displacement of node 2 Dx represents position of axis O_1y and value of angle θ_y. $Dx > 0 \rightarrow \theta_x < 0, Dx < 0 \rightarrow \theta_x > 0$.

For node 1, $\theta_x > 0$ if the displacement $Dy_{node1} > 0$ and $\theta_x < 0$ if $Dy_{node1} < 0$ (Figure 5a).

For node 2, $\theta_y > 0$ if $Dx_{node2} > 0$, and $\theta_y < 0$ if $Dx_{node2} < 0$ (Figure 5b).

Thus, based on the change of displacement Dy, Dx of nodes 1 and 2, we can predict the dynamic output of our tactile sensing system.

The simulation results in Figure 4 show that with pressurization, the skin layer was inflated, causing an inclination of the accelerometer's posture, which is possibly more sensitive to sliding motions. From the simulation results in Figure 6, when the chamber is pressurized, under the indenter's sliding traction, coordinate axis O_1x, O_1y of the accelerometer changes correspondingly. With the X-sliding action, displacement Dy (blue line) changes from positive to negative in the forward case and negative to positive in the backward case. The value of Dx (red line) is negative for all the times in both cases. With Y-sliding, the sign of Dx changes but Dy has a constant sign. That means with X-sliding, the sign of θ_x changes and θ_y has a constant sign, and with Y-sliding, the sign of θ_x is constant and θ_y has a changed sign.

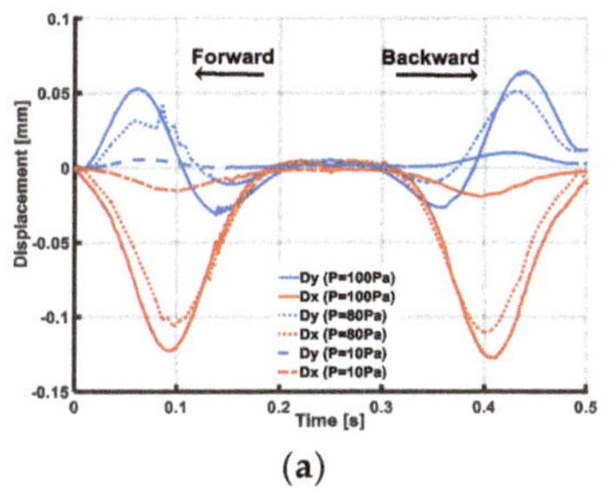

(a)

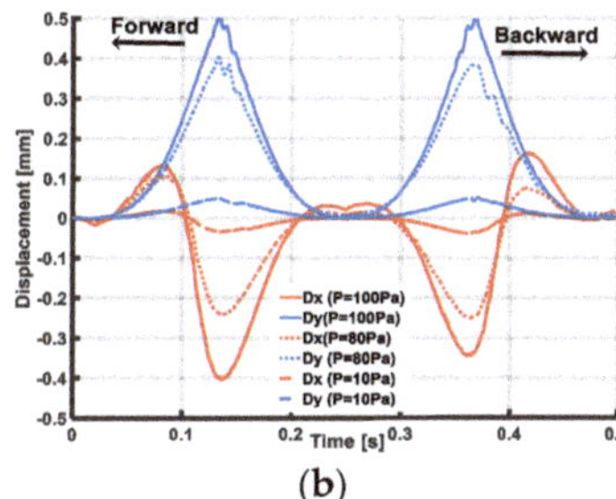

(b)

Figure 6. Simulation results show displacements Dy, Dx under various pressurizations: (**a**) X-sliding and (**b**) Y-sliding.

The simulation results imply that, by investigating the output acceleration of the accelemeter on the x, y axis, the tactile sensing system can detect the sliding action with forward and backward actions in both the X- sliding and Y- sliding cases.

To clarify the role of the inflated skin's morphology, we conducted simulations with various pressurizations (Figure 6), with higher pressure values, the Dx and Dy displacement increased, indicating that the sensitivity and output voltage of the sensing system was increased with higher pressurization.

4. Experimental Setup

We conducted experiments with our soft tactile system to assess the possibility of localizing sliding movements on the skin's surface. Figure 7 shows the experimental setup that consists of two linear stages (PG615-R05 AG-C, Suruga Seiki, Japan) and an indenter. The linear stages provide precise movements with 2-μm resolution. The indenter, which is a rigid spherical 6-mm-diameter ball, is attached vertically to the linear stage. The tactile sensor system was placed in a solid box fabricated by a 3D printer. The box was fixed and placed horizontally to the linear stage, which moves the sensing system back and forth. A laser sensor (Keyence-IL-030) measured the height of the inflated skin, and this value estimates the skin's deformation and the accelerometer's posture when the sensor system is not being probed.

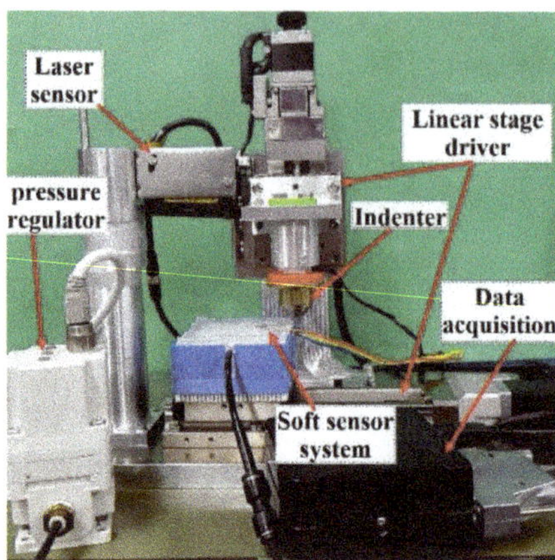

Figure 7. Experimental setup: sensing system is set on horizontal linear stage. Indenter is fixed to vertical linear stage. Skin is inflated by air regulator. Laser sensor measures inflated skin's height. Data from accelerometer are obtained through data acquisition system before being sent to computer.

We defined the two sliding directions as the X- and Y-axis directions along the x and y axes of the accelerometer. When the indenter moves toward the right relative to the sensor system, we call the direction forward and the opposite direction is backward. With both the X- and Y-sliding directions, there were six sliding cases (Figure 8a).

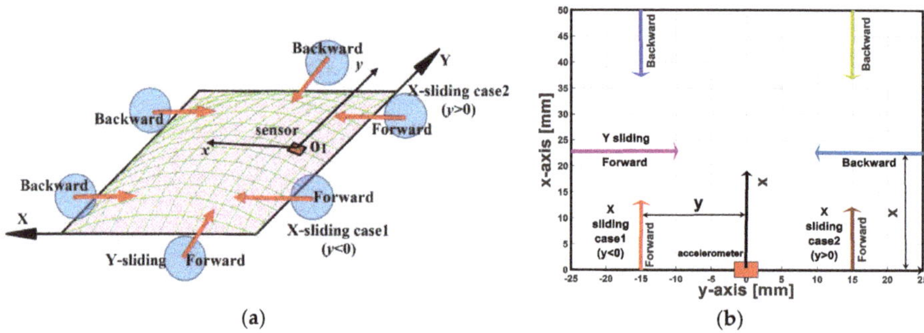

Figure 8. Schematic drawing illustrates sliding cases were conducted in experiments. (**a**) Definition of six sliding directions with both X, Y axes. (**b**) Sliding movements with various contact positions: different values of y, x for X, Y-sliding.

To investigate the sensor system's capability to detect and localize a sliding object on a large contact area, we conducted experiments with different distances from the accelerometer to the indenter (Figure 8b).

- With X-sliding case: Distance value y increased from 0 to 25 mm or from −25 to 0 mm with increased 2.5-mm steps.
- With Y-sliding case: Distance value x rose from 0 to 25 mm with increased 2.5-mm steps. At a particular distance, the sliding motions were conducted with various velocities from 0.5 to 4 mm/s with increased steps of 0.5 mm/s.

All the signals from the accelerometer were sent to ADC (24 bit, Kyowa Electronic Instruments Co., Tokyo, Japan) before being sent to a computer where the data were recorded by commercial software. The ADC converter's sampling rate was set to 1000 Hz. The skin was inflated by pressurization with a pneumatic actuation system with a compressor (JUNair) and a regulator. The air pressure was regulated within a range of 0.01 to 0.06 MPa.

5. Results and Discussion

5.1. Output Voltage of Accelerometer Under Static Pressurization Without Sliding Motion

As discussion in Section 2.3 and the simulation results in Section 3, the skin layer was inflated under pressurization, resulting in an inclination of the accelerometer's posture. Thus, the accelerometer's output voltage has initial values that are caused by static pressure at the no-sliding cases. To investigate the output response of the sensor system just under the effect of sliding traction, this initial output voltage of the accelerometer was measured for each value of pressurization and subtracted from the acquired output of the accelerometer during sliding operations. In addition, based on the accelerometer's output, we also estimated the height of the inflated skin and compared it to the measured height from the laser sensor.

Without pressurization, since the skin is flat, the accelerometer lies horizontally (Figure 9a). Under pressurization, the accelerometer inclined, and the angle between the accelerometer and horizontal plane was β . θ_x, θ_y is the angle between axis O_1x, O_1y and horizontal axis Δ (Figure 9b). Output voltages V_x, V_y are shown in Figure 9c.

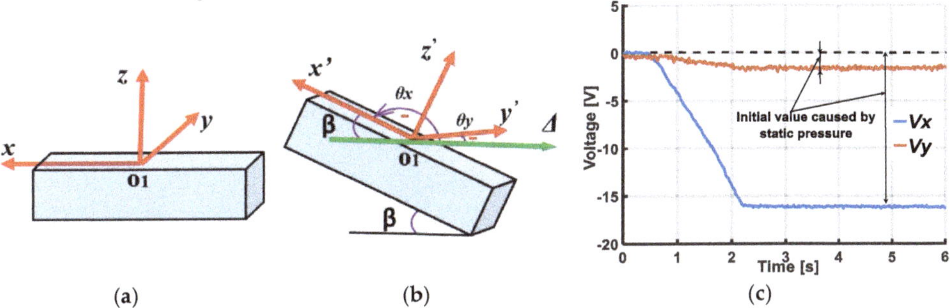

Figure 9. Posture of accelerometer changed and its output voltages under static pressurization. (**a**) Initial posture of accelerometer without pressurization. (**b**) Posture of accelerometer under static pressurization. (c) Output voltages of accelerometer under static pressurization.

As in Section 2.3, output voltage V_x can be calculated:

$$V_x = ka_x = kgsin\theta_x. \tag{5}$$

In Figure 9b, since $\beta + \theta_x = \pi$, $sin\theta_x = sin\,\beta$, we get the following:

$$\beta = acsin\left(\frac{V_x}{gk}\right). \tag{6}$$

The posture of the accelerometer can be evaluated from value β.

For an estimation of the shape of the inflated skin from output voltage V_x of the accelerometer, the skin's surface shape can be approximated by the following quadratic curve (Figure 10a):

$$y = -ax^2 + h, \quad (a > 0), \tag{7}$$

where h is the skin's highest height. The absolute value of the slope of the curve at x_0 is $2ax_0$, where x_0 represents the accelerometer's position. The slope equals $\tan\beta$, and we obtain

$$a = \frac{\tan\beta}{2x_0}. \tag{8}$$

The distance from the y axis to the position at which the curve intersects the x axis is $\sqrt{h/a}$. If this distance is represented by D, the height of the inflated skin is computed:

$$h = \frac{D^2 \tan\beta}{2x_0}. \tag{9}$$

Since the dimension of the skin is 50 × 50 × 4 mm, D = 25 mm. β is calculated from Equation (6). The computed value of the inflated skin's height is then compared to the measured values using a laser sensor (Figure 10b) and plotted in Figure 11 with error less than 12%. Thus, based on the accelerometer's output voltage we can accurately estimate the accelerometer's posture as well as the inflated skin's shape.

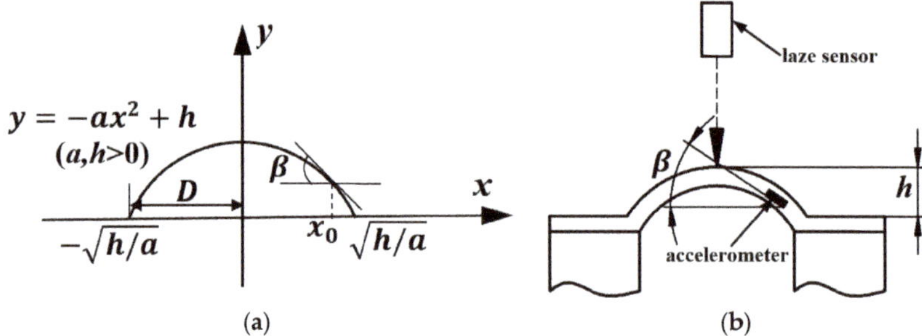

Figure 10. Schematic diagrams to calculate output voltage V_x under static pressurization (**a**) Skin's surface approximated by quadratic curve. (**b**) Schematic drawing of experimental setup for measuring inflated skin's height.

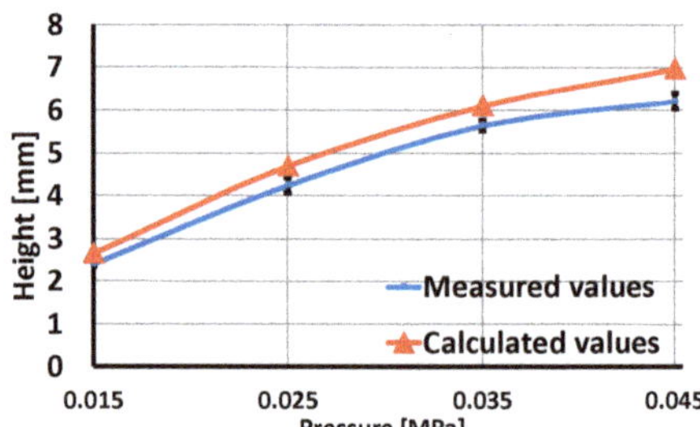

Figure 11. Comparison of calculated and measured values of inflated skin's height.

5.2. Output Voltages of Accelerometer Under Sliding Action

5.2.1. Dynamic Analysis for Detecting Sliding Directions

To localize the sliding movements, first the sensing system needs to recognize the motion's directions. In this section, we propose dynamic analysis based on the accelerometer's output for detecting the sliding directions. We conducted experiments with various sliding tractions (Figure 8a) and divided them into three main cases (Figure 8b):

(1) X-sliding case 1: Indenter slides along the X axis with contact position y < 0.
(2) X-sliding case 2: Indenter slides along the X axis with contact position y > 0.
(3) Y-sliding: Indenter slides along the Y axis.

When the indenter moves on the sensing system's surface, the accelerometer's output and its posture depend on the contact's position. For analysing the accelerometer's posture and its output, we divided the contact surface into four contact zones (Figure 12a).

When the indenter contacts the sensing system's surface in contact zone 1, the accelerometer tends to incline toward point 1 (Figure 12b), and its posture with the corresponding xy coordinate is illustrated as in Figure 13a. Similarly, with contact zones 2, 3, and 4, the accelerometer inclines toward points 2, 3, and 4, and Figure 13b–d show the accelerometer's posture and its xy coordinate, respectively. The analysis shown in Figure 13 suggests the following:

- With contact zone 1: The values of angles θ_x, θ_y between O1x and O1y and horizontal axis Δ are negative, and thus output voltages V_x and V_y are negative: $\theta_x < 0,\ \theta_y < 0 \rightarrow V_x < 0,\ V_y < 0$.
- With contact zone 2: $\theta_x > 0,\ \theta_y < 0 \rightarrow V_x > 0,\ V_y < 0$.
- With contact zone 3: $\theta_x < 0,\ \theta_y > 0 \rightarrow V_x < 0,\ V_y > 0$.
- With contact zone 4: $\theta_x > 0,\ \theta_y > 0 \rightarrow V_x > 0,\ V_y > 0$.

The output voltages of the sensing system with each sliding case are shown in Figure 14. With X-sliding case 1, when the indenter starts to make contact and slides over the inflated skin's surface in contact zone 1, measured output voltages V_x, V_y are negative, by keeping moving, the indenter slides to the other side of the skin's surface in contact zone 2, output V_x increases to a positive peak value before it returns to the initial state when the indenter moves off of the skin's surface in contact zone 2. While V_y increases from a negative value to an initial value. When the indenter slides backward, its contact with the skin's surface changes inversely from contact zone 2 to contact zone 1, and thus V_x changes from a positive to a negative value and V_y keeps its negative value (Figure 14a).

With X-sliding case 2, the contact's position changes from zone 3 to 4 with forward sliding and from zone 4 to 3 with backward sliding. The changed behaviors of V_x and V_y can be analysed as in the X-sliding case 1. V_x increases from negative to positive, and V_y remains positive for forward sliding. With backward sliding, V_x decreases from positive to negative, and the value of V_y remains positive (Figure 14b).

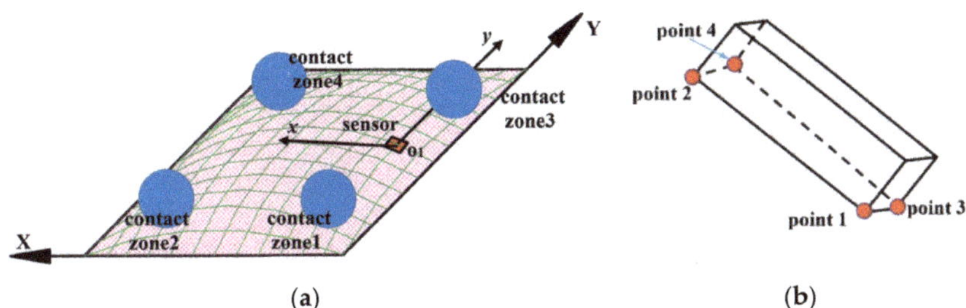

Figure 12. Divided contact zones on skin's surface and intial posture of accelerometer. (**a**) Depending on location of indenter on skin's surface, contact area can be categorized into four zones. (**b**) Posture of accelerometer under static pressurization without sliding actions.

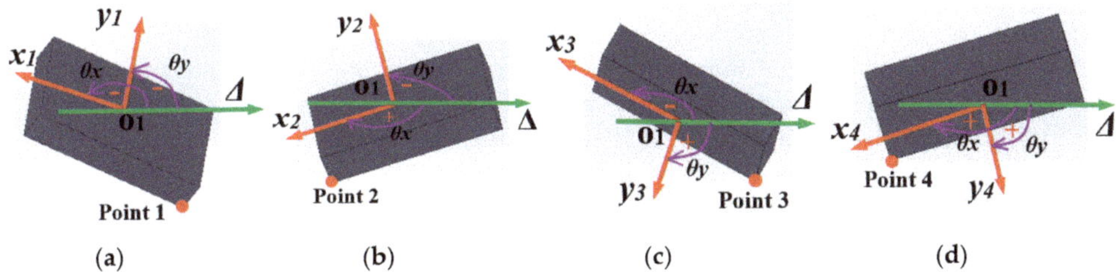

Figure 13. Accelerometer's posture and its outputs with different contact zones. (**a**) Contact zone 1: Accelerometer is inclined toward point 1 $\theta_x < 0 \rightarrow V_x < 0,\ \theta_y < 0 \rightarrow V_y < 0$. (**b**) Contact zone 2: Accelerometer is inclined toward point 2 $\theta_x > 0 \rightarrow V_x > 0,\ \theta_y < 0 \rightarrow V_y < 0$. (**c**) Contact zone 3: Accelerometer is inclined toward point 3 $\theta_x < 0 \rightarrow V_x < 0,\ \theta_y > 0 \rightarrow V_y > 0$. (**d**) Contact zone 3: Accelerometer is inclined toward point 4 $\theta_x > 0 \rightarrow V_x > 0,\ \theta_y > 0 \rightarrow V_y > 0$.

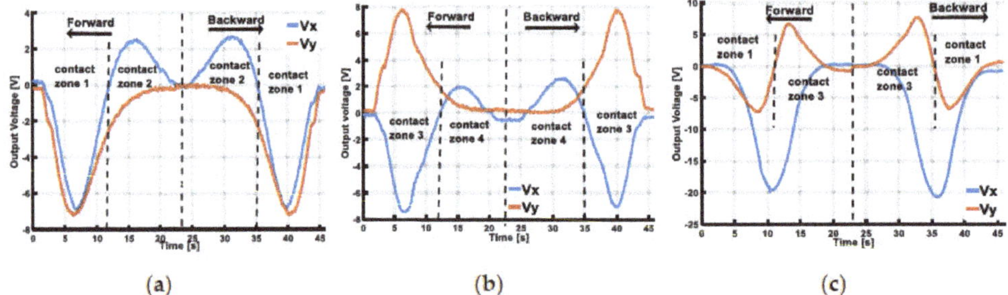

Figure 14. Output voltage V_x, V_y of accelerometer with different sliding cases. V_x, V_y values depend on contact zone between indenter and skin's surface. Based on changed behaviour of V_x, V_y, our sensing system can temporally detect sliding directions: forward and backward in both X- and Y-sliding cases. (**a**) X-sliding case 1. (**b**) X-sliding case 2. (**c**) Y-sliding.

In terms of the Y-sliding case, with the forward sliding direction, the contact zone changes from 1 to 3 and $V_y < 0$ rises to $V_y > 0$. With backward sliding, V_y changes from positive to negative. The value of V_x is negative in both the forward and backward directions (Figure 14c). In other words, with X-sliding, the sign of V_x changes but V_y has a constant sign. Whereas with Y-sliding, the sign of V_x is constant and V_y has a changed sign. This resemble the discussion of the displacement's change (Dx, Dy) in the simulation results, plotted in Figure 6, Section 3. A dynamic analysis of the experimental results agrees well with the simulation results. Based on output voltages V_x, V_y of the accelerometer, we conclude that our sensing system can temporally detect the sliding action with two forward and backward directions in the X- and Y-sliding cases.

5.2.2. Estimation of Sliding's Localization.

We presented the idea of sliding movement localization in Section 2.1 and in Section 5.2.1 demonstrated that the sensing system can temporally detect a sliding motion and its directions. Based on the accelerometer's output voltage, in this section we confirm that the sliding object can be localized through the distance from the sliding line to the accelerometer and the magnitude of the sliding velocity.

Because the deformation of the soft skin at the accelerometer's position depends on the distance from the object to the accelerometer, the accelerometer's outputs are a function of the object's contact point. Therefore, based on the accelerometer's output voltages, a soft tactile sensing system can estimate distances x, y between the accelerometer and the object's sliding line (Figure 8b). To clarify this capacity, we experimented with various distances x, y for both the X- and Y-sliding cases (Figure 8b).

Figures 15a and 16a show the accelerometer's output voltages V_x, V_y with different distances y, x of the X- and Y-sliding cases, when the soft skin is inflated by pressurization $P = 0.035$ MPa. These figures suggest that the values of V_x, V_y generally depend on distances y and x. When distances y, x increase, the output voltage decreases.

In Figure 15a, $V_x min$ and $V_x max$ are the minimum and maximum values of output voltage V_x with corresponding moments t_{min} and t_{max}. In Figure 16a, $(V_y min, t_{min})$ and $(V_y max, t_{max})$ are similarly minimum and maximum magnitudes with corresponding moments of V_y.

Figures 15a and 16a show that $V_x min$, $V_x max$ and $V_y min$, $V_y max$ depend on the y and x values. Thus, based on the magnitude of the minimum and maximum voltage, the distance from the accelerometer to the object's sliding line can be estimated and expressed as the following interpolated functions:

$$\begin{aligned} y &= f_1(V_x min);\ y = f_2(V_x max) \text{ for X-sliding} \\ x &= f_3\left(V_y min\right);\ x = f_4\left(V_y max\right) \text{ for Y-sliding.} \end{aligned} \tag{10}$$

Figures 15a and 16a also indicate that the t_{min} and t_{max} values are not dependent on distances y or x. t_{min}, t_{max} only depend on the magnitude of velocity v (Figures 15b and 16b).

The magnitude of the sliding velocity can be expressed by:

$$v = \chi_1(t_{min});\ v = \chi_2(t_{max}) \quad (11)$$

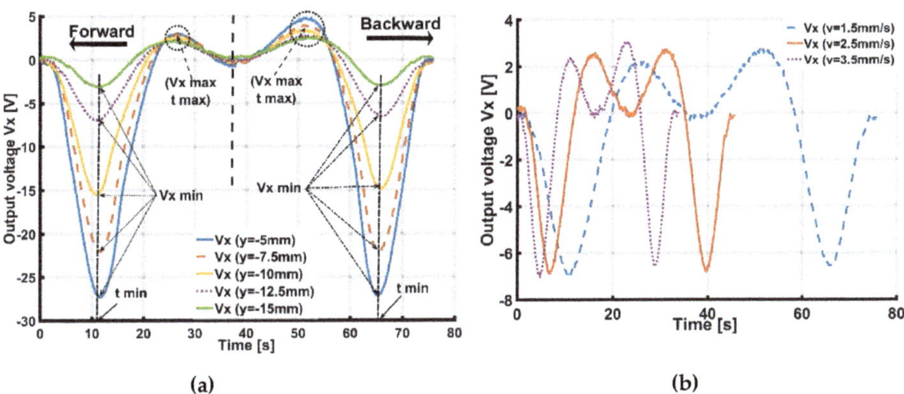

Figure 15. Output voltage V_x of accelerometer with X-sliding case 1 when distance y increases from 5 to 15 mm. Value of $V_x min$, $V_x max$ only depends on distance y and t_{min}, t_{max} depends on magnitude of velocity v. Based on accelerometer's output voltage, distance y from accelerometer to sliding line and magnitude of velocity v can be estimated. (**a**) Output voltage V_x with different distances y (mm). (**b**) Output voltage V_x with different magnitudes of velocity v (mm/s).

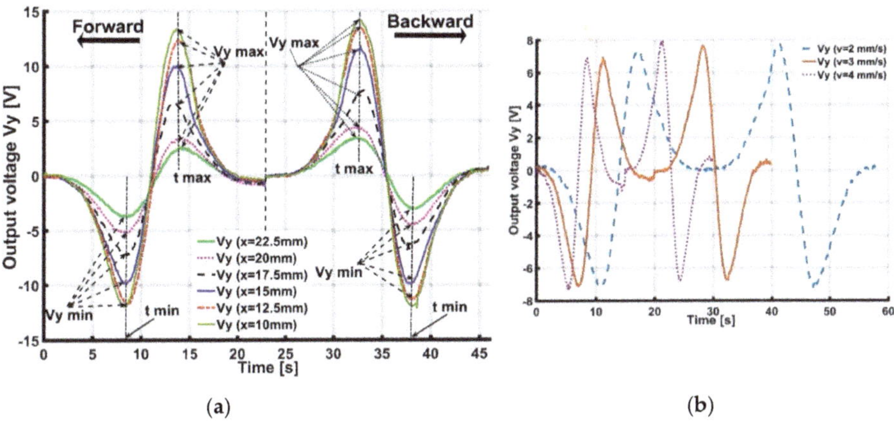

Figure 16. Output voltage V_y of accelerometer with Y-sliding when distance x increases from 10 to 22.5 mm. Values of $V_y min$, $V_y max$ only depend on distance x and t_{min}, t_{max} depends on magnitude of velocity v. Based on accelerometer's output voltage, distance x from accelerometer to sliding line and magnitude of velocity v can be estimated. (**a**) Output voltage V_y (V) with different distances x (mm). (**b**) Output voltage V_y (V) with different velocities $v(mm/s)$.

From our experimental results with various values of distances and velocity magnitudes, we extracted the values of the minimum, maximum output voltage, and corresponding moment t_{min}, t_{max} to assess the interpolated functions (Figures 17–20). These functions estimate the contact's distances x or y and velocity's magnitude v. The dependences of the velocity's magnitude on the values of t_{min}, t_{max} and distances y, x on $V_{x\ min}$, $V_{y\ min}$ are not linear. Thus, we chose quadratic equations and cubic functions for the interpolation of the velocities and the distances. These functions create the best fit experimental data without requiring complicated computations.

Based on the output voltages of the accelerometer with the interpolated functions, our tactile sensing system can estimate the magnitude of the velocity and the distance from the accelerometer to sliding lines x or y. Figures 21 and 22 present the calculated and actual values of the velocity and the distance from the accelerometer to the object's sliding line with an error bar.

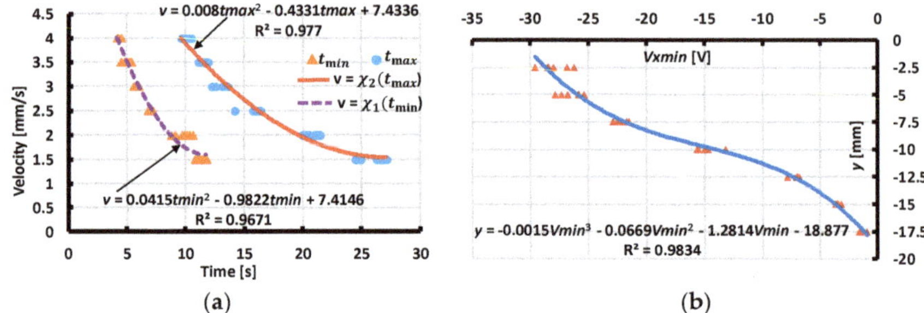

(a) (b)

Figure 17. Interpolating velocity v and distance y for case of X-sliding forward. (**a**) Interpolating velocity v from t_{min}, t_{max}. (**b**) Interpolating distance y from output voltage $V_x min$.

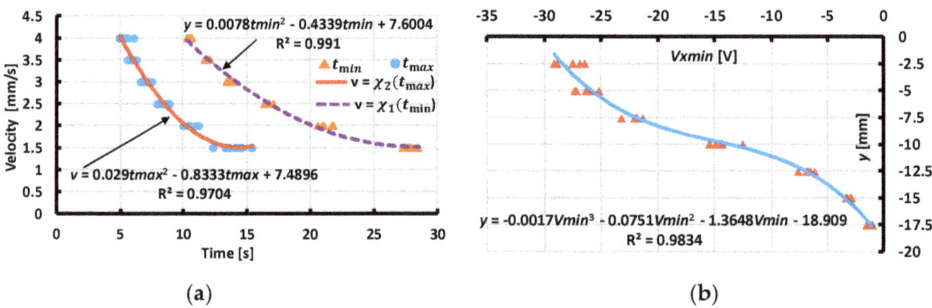

(a) (b)

Figure 18. Interpolating velocity v and distance y for case of X-sliding backward (**a**) Interpolating velocity v from t_{min}, t_{max}. (**b**) Interpolating distance y from output voltage $V_x min$.

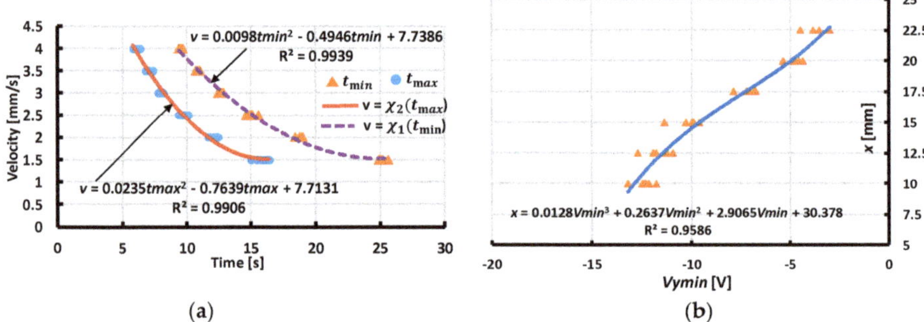

(a) (b)

Figure 19. Interpolating velocity v and distance x for case of Y-sliding forward. (**a**) Interpolating velocity v from t_{min}, t_{max}. (**b**) Interpolating distance x from output voltage $V_y min$.

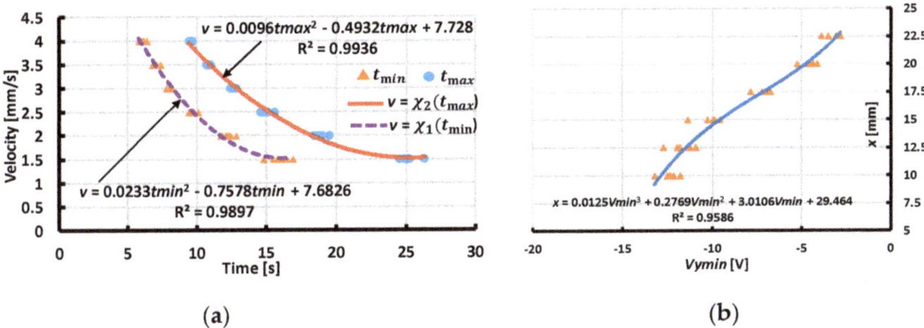

(**a**) (**b**)

Figure 20. Interpolating velocity v (mm/s) and distance x (mm) for case of Y-sliding backward. (**a**) Interpolating velocity v from t_{min}, t_{max}. (**b**) Interpolating distance x from output voltage $V_y min$.

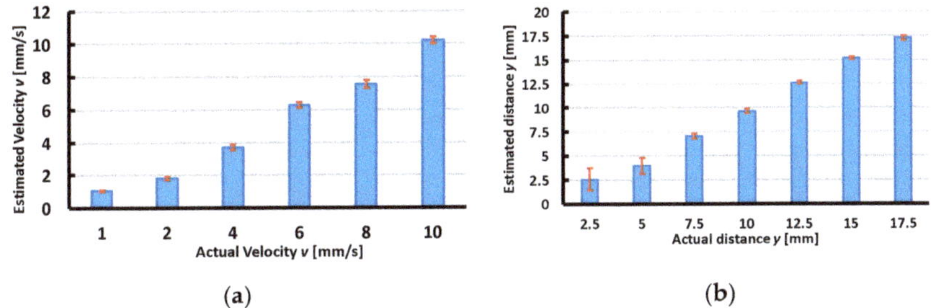

(**a**) (**b**)

Figure 21. Calculated results compare to actual values with X-sliding case. (**a**) Comparison of calculated and actual velocity v (mm/s). (**b**) Comparison of calculated and actual distance y (mm).

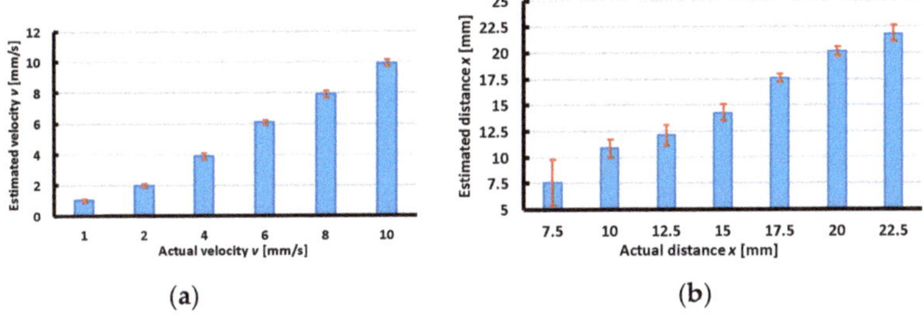

(**a**) (**b**)

Figure 22. Calculated results compared to actual values with Y-sliding case. (**a**) Comparison of calculated and actual velocity. (**b**) Comparison of calculated and actual distance x (mm).

To estimate the velocity, the maximum errors between the calculated and actual magnitude were less than 8%. This indicates the sensing system's accuracy for the velocity's estimation. Note that when the velocity's magnitude increases, the values of t_{min}, t_{max} decrease, and thus based on experiments we acknowledged the sensing system can estimate a maximum velocity of about 13 mm/s.

To estimate the distance from the accelerometer to the object's sliding line, for the X-sliding direction, Figure 21a shows that the maximum error is less than 15% with distance $y \geq 5$ mm. Nonetheless, with distance $y = 2.5$ mm, the error is about 40% due to the effect of the nonlinearity of the large deformation at the accelerometer's position when the indenter is close to the accelerometer. The experimental results presented in Figures 15a and 16a also show that the output voltage decreases when the distance increases. If the minimum and maximum output voltages approach zero, the sensing system cannot detect the sliding actions. Therefore, for the X-sliding case, the tactile sensing system has a detectable area with distance $-17.5 \leq y \leq 17.5$ mm (Figure 23a), and in this area the sensing system can detect the sliding direction and estimate the velocity's magnitude where the maximum error is less than 8%.

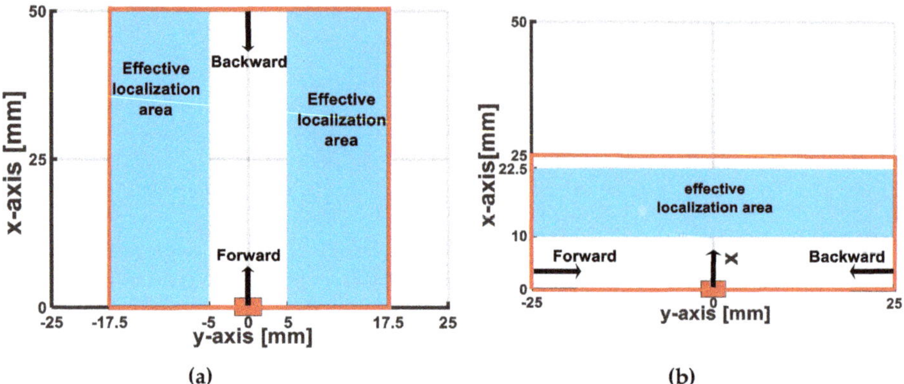

Figure 23. Schematic illustration of detectable area of tactile sensing system. In detectable area (inside red lines), sensing system can detect sliding direction and estimate velocity where maximum errors are less than 8%, and in effective localization area (blue areas), distance from accelerometer to sliding objects can be estimated where maximum errors are less than 15%. (**a**) X-sliding case. (**b**) Y-sliding case.

Inside the detectable area, there are two effective localization areas with distances $-17.5 \leq y \leq -5$ mm and $5 \leq y \leq 17.5$ mm (blue area in Figure 23a), where the sensing system can estimate the distance from the accelerometer to the object's sliding line where the maximum error is less than 15%. Similarly, in the Y-sliding direction, the detectable area is evaluated with distance $0 \leq x \leq 25$ mm and the effective localization area is $10 \leq x \leq 22.5$ mm (Figure 23b).

In the effective localization area, based on the interpolated values of velocity magnitude v and the distance between the accelerometer and the sliding line, the sliding object can be localized through the values of its coordinates x and y. For the X-sliding case, $(x, y) = (vt,\ y)$, and for the Y-sliding case, $(x, y) = (x,\ vt)$. Figures 24 and 25 illustrate the comparisons of the actual object's position and its estimated localization with various sliding cases.

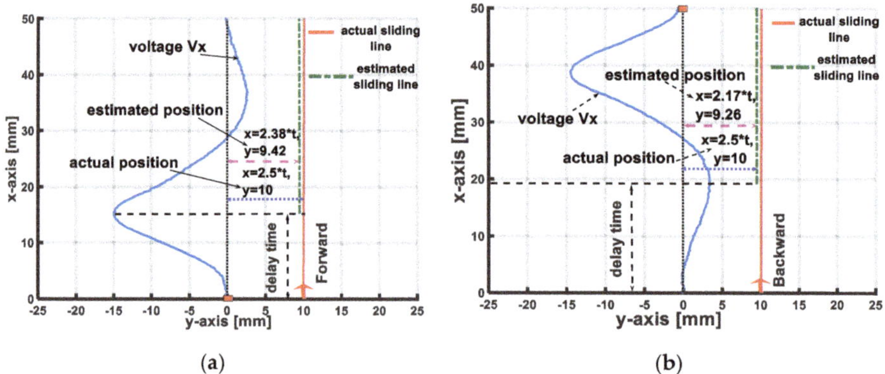

Figure 24. Comparison of actual sliding position and estimated position for X-sliding cases. Actual sliding trajectory is red line, after delay time (when accelerometer's output voltage reaches extreme values), sensing system can localize sliding object's positions with errors less than 15% (localized sliding trajectory is blue line). (**a**) X-sliding case: forward direction. (**b**) X-sliding case: backward direction.

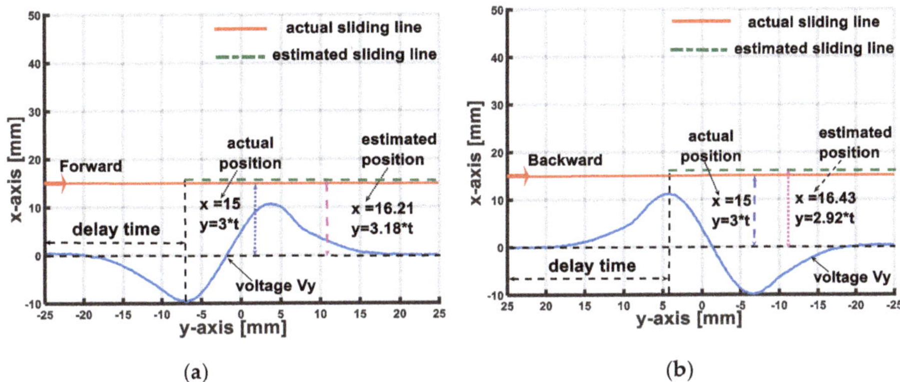

Figure 25. Comparison of actual sliding position and estimated position for Y-sliding cases. (**a**) Y-sliding case: forward direction. (**b**) Y-sliding case: backward direction.

Because we use interpolated functions to estimate the values of sliding velocity v and distances x, y, for the sliding object's localization, the sensing system needs a delayed period for the accelerometer's output voltage to reach extreme values. The value of the delayed time depends on the magnitude's velocity and can be estimated from interpolated function $v = \chi_1(t_{min})$; $v = \chi_2(t_{max})$.

In the X-sliding case with forward direction, the delayed time is about $t_{delay} = \frac{18}{v}$ (s), and with backward direction, it is $t_{delay} = \frac{20.5}{v}$ (s). For the Y-sliding case with forward direction, $t_{delay} = \frac{17.5}{v}$ (s) and with backward direction: $t_{delay} = \frac{21.5}{v}$ (s), compared to the total time of one sliding action on the inflated skin's surface $t_{sliding} = \frac{50}{v}$ (s). This result denotes that in a sliding action, after the indenter slides about $\frac{2}{5}$ of the distance, the sensing system can localize its position with the maximum errors are less than 15%.

5.3. Role of Pressurization

In this proposed tactile sensing system, inflated soft skin is generated by the activation of pressurization. Pressurization plays the actuation role that creates the continued change in the accelerometer's position under the object's sliding action. Figure 11 in Section 5.1 shows that with higher pressurization, the inflated skin's height increases, leading to higher deformation of the skin's surface under identical sliding motions. Resulting in increased output voltage of the accelerometer. Thus, the pressure's value greatly affects the sensitivity of the sensing system with sliding motions.

Figure 26 shows the output voltage responses of the accelerometer under sliding actions with various pressurization's conditions. With higher pressure value, the output voltage increases. Thus, sensitivity to the sliding action increases with extra pressurization. This result resembles the simulation results in Figure 6 of Section 3 and suggests the potential of numerical simulations for predicting the responses of our sensing system under various sliding actions, although in the numerical simulations, the pressurization was setup at a much smaller value than in experiments to save computational cost. Thus, by confirming the simulation and experimental results, we conclude that with higher pressure values, the sensing system's sensitivity increases with sliding motions.

Nonetheless, at higher pressure ($P = 0.045$ MPa) the output voltage is not much bigger than at pressure $P = 0.035$ MPa due to the nonlinearity of the skin's deformation with high pressure. In addition, high pressurization might prevent sliding movements. With the presented configuration of the soft active tactile sensor system in this paper, the ideal pressurization is 0.035 MPa, which was used in the experimental results in Section 5.2. This result also suggests that for generating and maintaining the inflated skin's surface, the required pressurization is small (maximum, 0.035 MPa).

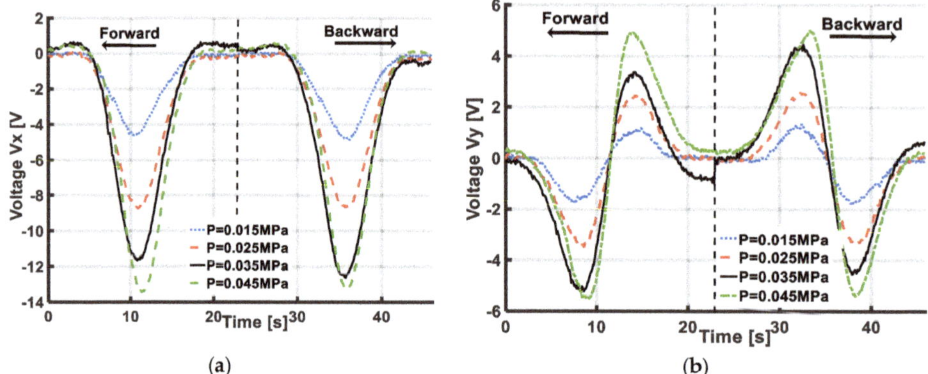

Figure 26. Accelerometer's output voltage with different pressurizations in Y-sliding case. With higher pressure value, sensing system is more sensitive to sliding action. Nonetheless, high pressurization is relatively ineffective. (**a**) Output voltage V_x (V) with different pressurizations. (**b**) Output voltage V_y (V) with different pressurizations.

5.4. Limitations and Applicability of Sensing System

We analyzed our sensing system's ability on the detecting and localization of sliding movements and validated it by simulations and experiments. Nonetheless, it continues to suffer from fundamental issues and limitations.

The accelerometer's output depends on the contact location and the contact force. We assume that first the sliding object slightly contacts the soft skin outside the inflated area and slides without changing the initial contact depth. Thus, under specific pressurization, the contact force at each contact point is identical for all sliding movements. We can ignore the effect of the contact force on the accelerometer's output voltage. In the case of changing the contact depth during sliding movements, the effect of the force at each contact point must be considered. To solve this issue, in the next step we will propose a finite element model to estimate the deformation of the inflated soft skin's surface under various contact forces. This model could also be used to locate the indentations on soft skin and estimate the normal contact force. In addition, the effect of acceleration due to the friction force between the sliding object and the skin's surface was not mentioned in this paper. This effect could be reduced by filtering, such as with a Kalman filter.

A sensing system can be used for specific devices and an accelerometer's output may change if the entire device changes its posture. Thus, this effect must be investigated and eliminated. To solve this issue, it may require an algorithm that calculates the accelerometer's output based on the whole device's posture. These calculated values will be subtracted from the accelerometer's acquired output for sensing.

Another current limitation of our sensing system is that it can detect and localize sliding motions with two X- and Y- directions because we only deploy the accelerometer as a sensing element. In the future we will use more accelerometers, each of which will be placed at a suitable initial posture and position. Based on the output responses of all the accelerometers, we will build a method using machine learning to localize the sliding motions with diagonal directions. To accomplish complicated tasks, a sensing system must be trained with training data in various conditions of sliding motions. After training, the acquired program can be used as the sensing system's brain that should be attached to the main processing system. The processing system can help a sensing system be localized temporally with improved revolutions. We must tackle this critical issue in the future to fulfill this sensing system. In addition, we will use another accelerometer with a flexible circuit to replace the current accelerometer with a solid circuit board. For example, the previously proposed flexible one [21] would be easier to embed in soft skin.

Our sensing system is anticipated for applications in soft robotics. For instance, it could be attached to a soft robotics hand's palm or other body parts of robots that frequently make contact with sliding objects. Normally, without pressurization, a soft skin surface plays the role of robotic skin and is inflated for detecting and localization sliding tasks, which are crucial for robots to make subsequent decisions. Because the pressurization for activating the inflated skin is small (maximum, 0.035 MPa), a sensing system can be activated with a small pump and a simple pneumatic controller without affecting the portability of the whole device. In addition, our sensor system is a promising device for detecting sliding objects and localizing their position in an environment with limited vision. In this paper we avoided applying our sensing system for specific applications. Instead, we set a prototype design, fabrication, and experiment with analysis to validate its ability. Thus, other researchers can exploit it to design specifically purposed structures.

6. Conclusions

We introduce a soft tactile sensing system with a three-axis accelerometer for detecting and localizing sliding motions on large contact surfaces. Our numerical simulation and experimental results show that it can detect temporally sliding directions in both X- and Y-direction sliding cases. In addition, interpolated functions were given to estimate the magnitude's sliding velocity and the distance from the accelerometer to the sliding line to localize the sliding object. The detection area also validated that the sensing system can detect and localize sliding tractions on large contact areas. Future work will investigate the sensing system's ability with various sliding cases, including diagonal directions and changing the contact depth by proposing finite element model and machine learning methods. Then we will apply our sensing system to soft robotics.

Author Contributions: H.X.T., Y.I. conceived, designed, performed the experiments. Data analysis, writing the first draft of the manuscript by H.X.T. and it was revised and substantially improved by V.A.H. and K.S. Methodology, project administration was made by K.S. and V.A.H.

Funding: This research was supported by the Joint Research Center for Science and Technology of Ryukoku University and JSPS Grant-in-aid for Scientific Research (KAKENHI) grant number 18H01406.

Conflicts of Interest: The authors declare no conflict of interest.

References

1. Ravinder, D.; Maurizio, V. Tactile Sensing for Robotic Applications. In *Sensors: Focus on Tactile Force and Stress Sensors*; Jose, R., Senentxu, M., Eds.; I-Tech: Vienna, Austria, 2008; pp. 289–304.
2. Eric, K. *Principles of Neural Science*, 5th ed.; McGrraw-Hill: New York, NY, USA, 2000.
3. Ravinder, D.; Giorgio, M.; Maurizio, V.; Guilio, S. Tactile sensing—From Humans to Humanoids. *IEEE Trans. Rob.* **2010**, *26*, 1–20.
4. Kwonsik, S.; Minkyung, S.; Eunmin, C.; Hyunchul, P.; Ji, W.C.; Yuljae, C.; Jung, I.S.; Seung, N.C.; Jae, E.J. Artificial Tactile Sensor Structure for Surface Topography Through Sliding. *IEEE/ASME Trans. Mechatron.* **2018**, *23*, 2638–2649.
5. Huang, Y.; Yuan, H.; Kan, W.; Guo, X.; Liu, C.; Liu, P. A flexible three axial capacitive tactile sensor with multilayered dielectric for artificial skin applications. *Microsys. Technol.* **2007**, *23*, 1847–1852. [CrossRef]
6. Dana, D.; Taylor, N.; Rolf, P.; Allison, O. Artificial Tactile Sensing of Position and Slip Speed by Exploiting Geometrical Features. *IEEE/ASME Trans. Mechatron.* **2015**, *20*, 263–274.
7. Javier, A.; Leandro, B. Estimating Object Grasping Sliding via Pressure Array Sensing. In Proceedings of the IEEE International Conference on Robotics and Automation, Saint Paul, MN, USA, 14–18 May 2012; pp. 1740–1746.
8. Paolo, G.; Luigi, P.; Lucia, S.; Maurizio, V.; Rodolfo, Z. Computational Intelligence Techniques for Tactile Sensing Systems. *Sensors* **2014**, *14*, 10952–10976.
9. Pfeifer, R.; Lungarella, M.; Idia, F. Self-organization, embodiment and biologically inspired robotics. *Science* **2007**, *318*, 1088–1093. [CrossRef] [PubMed]

10. Hauser, H.; Rudolf, M.; Kohei, N. Morphological computation-The physical body as a computational resource. Self-published. **2014**, *20*, 226–244.
11. Muller, V.C.; Hoffman, M. What is morphological computation? On how the body contributes to cognition and control. *Artif. Life* **2017**, *23*, 1–24. [CrossRef] [PubMed]
12. Ahanat, K.; Juan, A.C.R.; Veronique, P. Tactile sensing in dexterous robot hands-Review. *Rob. Auton. Syst.* **2015**, *74*, 195–220.
13. Trinh, H.X.; Shibuya, K. Tactile sensing system with wrinkle's morphological change: Modeling. In Proceedings of the IEEE International Conference on Soft Robotics, Livorno, Italy, 24–28 April 2018; pp. 362–368.
14. Chathuranga, D.S.; Wang, Z.; Noh, Y.; Thrishantha, N.; Shinichi, H. Magnetic and mechanical modelling of a soft three-axis force sensor. *IEEE Sens. J.* **2016**, *16*, 5298–5307. [CrossRef]
15. Mannsfeld, S.C.; Tee, B.C.; Stoltenberg, R.M.; Chen, C.V.H.; Barman, S.; Muir, B.V.; Sokolov, A.N.; Reese, C.; Bao, Z. Highly sensitive flexible pressure sensors with microstructured rubber dielectric layers. *Nat. Mater.* **2010**, *9*, 859–864. [CrossRef] [PubMed]
16. Julius, E.B.; Van, A.H.; Hongbin, L. Morphological computation in haptic sensation and interaction: from nature to robotics. *Adv. Rob.* **2018**, *32*, 340–362.
17. Van, A.H.; Yamashita, H.; Wang, Z.; Shinichi, H.; Koji, S. Wrin'Tac: Tactile sensing system with wrinkle's morphological change. *IEEE Trans. Inf.* **2017**, *13*, 2496–2506.
18. Trinh, H.X.; Ho, V.A.; Shibuya, K. Computational model for tactile sensing system with wrinkle's morphological change. *Adv. Rob.* **2018**, *32*, 1135–1150.
19. Shunsuka, S.; Shogo, O.; Yoichiro, M.; Yoji, Y. Wearable finger pad deformation sensor for tactile textures in frequency domain by using accelerometer on finger side. *ROBOMECH J.* **2017**, *4*, 1–11.
20. Carlos, B.; Martin, M.; Pablo, B. Tactile sensing with accelerometers in prehensile grippers for robots. *Mechatronics* **2016**, *33*, 1–12.
21. Chathuranga, K.V.D.S.; Hirai, S. A bio-mimetic fingertip that detect force and vibration modalities and its application to surface identification. In Proceedings of the IEEE International Conference on Robotics and Biomimetics, Guangzhou, China, 11–14 December 2012; pp. 575–581.
22. Vallbo, A.B.; Johansson, R.S. Properties of cutaneous mechanoreceptors in the human hand related to touch sensation. *Hum. Neurobiol.* **1984**, *3*, 3–14.
23. Jackson, K.; Brooks, D.R. Do crocodiles co-opt their sense of touch to taste? a possible new type of vertebrate sensory organ. *Amphib. Reptil.* **2009**, *28*, 277–285. [CrossRef]
24. Leitch, D.B.; Catania, K.C. Structure, innervation and response properties of integumentary sensory organs in crocodilians. *J. Exp. Biol.* **2012**, *215*, 4217–4230. [CrossRef] [PubMed]
25. Fritz, V.; Paul, A.S. The role of behavior in the evolution of spiders, silks, and webs. *Annu. Rev. Ecol. Evol. Syst.* **2007**, *38*, 819–846.
26. Discovery. Spiders 'Listen' to Web Tones to Detect Prey. Available online: http://blogs.discovermagazine.com/d-brief/2014/06/04/spiders-listen-prey/ (accessed on 25 September 2018).
27. Shin-Estu Sillione. Rubber compounds. Available online: https://www.shinetsusilicone-global.com/products/type/rubb_comp/index.shtml (accessed on 27 October 2018).
28. Smooth-On, Inc. Ecoflex™ 00-10. Available online: https://www.smooth-on.com/products/ecoflex-00-10/ (accessed on 25 September 2018).
29. NXP. ±1.5g, ±6g Three Axis Low-g Micromachined Accelerometer. Available online: https://www.nxp.com/files-static/sensors/doc/data_sheet/MMA7361L.pdf (accessed on 16 October 2018).

Article

Estimating the Orientation of Objects from Tactile Sensing Data Using Machine Learning Methods and Visual Frames of Reference

Vinicius Prado da Fonseca *, Thiago Eustaquio Alves de Oliveira and Emil M. Petriu

School of Electrical Engineering and Computer Science, University of Ottawa, Ottawa, ON K1N 6N5, Canada; talvesde@uottawa.ca (T.E.A.d.O.); petriu@uottawa.ca (E.M.P.)

* Correspondence: vfons006@uottawa.ca; Tel.: +1-613-562-5800 (ext. 2146)

Received: 10 April 2019; Accepted: 10 May 2019; Published: 17 May 2019

Abstract: Underactuated hands are useful tools for robotic in-hand manipulation tasks due to their capability to seamlessly adapt to unknown objects. To enable robots using such hands to achieve and maintain stable grasping conditions even under external disturbances while keeping track of an in-hand object's state requires learning object-tactile sensing data relationships. The human somatosensory system combines visual and tactile sensing information in their "What and Where" subsystem to achieve high levels of manipulation skills. The present paper proposes an approach for estimating the pose of in-hand objects combining tactile sensing data and visual frames of reference like the human "What and Where" subsystem. The system proposed here uses machine learning methods to estimate the orientation of in-hand objects from the data gathered by tactile sensors mounted on the phalanges of underactuated fingers. While tactile sensing provides local information about objects during in-hand manipulation, a vision system generates egocentric and allocentric frames of reference. A dual fuzzy logic controller was developed to achieve and sustain stable grasping conditions autonomously while forces were applied to in-hand objects to expose the system to different object configurations. Two sets of experiments were used to explore the system capabilities. On the first set, external forces changed the orientation of objects while the fuzzy controller kept objects in-hand for tactile and visual data collection for five machine learning estimators. Among these estimators, the ridge regressor achieved an average mean squared error of 0.077°. On the second set of experiments, one of the underactuated fingers performed open-loop object rotations and data recorded were supplied to the same set of estimators. In this scenario, the Multilayer perceptron (MLP) neural network achieved the lowest mean squared error of 0.067°.

Keywords: in-hand manipulation; pose estimation; machine learning; fuzzy control; underactuated robotic hands

1. Introduction

Widespread in controlled environments such as industries, robots are moving towards unstructured settings like homes, schools, and hospitals performing high-level, complex, and fast reasoning; nevertheless, several challenges remain unsolved for robot skills achieving human-level capabilities [1]. Although robots can accurately perform several tasks such as walk, pick and place objects, understand, and communicate with people, they still present a lack of hand dexterity. Improvements on tactile sensing for in-hand manipulation and an increased understanding of human perception to action inspire and could advance robot possibilities in the scenarios mentioned above [2].

Manipulation skills developed in the human's hand and the brain have a level of ability rarely seen in other animals. Grasping and manipulating objects is a distinctive part of the human-being skill set. It is an ability evolved from the erect posture that freed our upper limbs, turning our hands into two sophisticated sets of tools [3]. Not surprisingly, humans' hand dexterity and reasoning are the holy grail of bio-inspired robotics control and actuation. Hand dexterity is the ability to interact in a useful way with objects in the real world. During robotic manipulation, a robot changes an object's state from an initial configuration to a final pose. For instance, during pick and place tasks, the goal of robotic platforms is to change the position and orientation of an object inside the manipulator's workspace. Comparatively, in-hand manipulation is the ability to change the pose of objects, from initial orientation to a given one, within one hand. Implementing robots that change objects' orientation while maintaining a stable grasp has the potential to amplify even more robot area of activity [4]. Robotic manipulator literature comprises a long list of examples of robotic hands with many levels of dexterity. From there, underactuated hands arise as an option that can achieve a reasonable level of dexterity with simplicity.

Research on haptic perception over the past decades has developed an in-depth knowledge of the psychological aspects of human touch employed to manipulate objects and perceive their characteristics. Lederman et al. [5] presented the somatosensory system divided into two subsystems: The "What" system that carries out perception and memory; and the "Where" system that deals with perception to action. In humans, the "What" system performs the recognition of surfaces and objects through their tactile properties [6]. Robots that implement similar recognition systems demonstrate their efficacy when similar objects are identified promptly even without visual feedback [7]. From this perspective, Reference [8] is one example in recent research, where authors employed camera-based sensing and deformable material as input to an object recognition system integrating grasping and object recognition. On the other hand, the "Where" system, which has a counterpart in vision, produces a description of points, surfaces, and reference frames in the world. Different from vision, touch refers to a location in the sensory organ, the skin itself, and localization in the environment. The human sensorial loop combines tactile perception in the presence of vision had attracted research interest due to increased reliability when both modalities [9]. Recent literature presents several approaches to this issue with a significant contribution on the tactile feedback topic were inspired by the concept of the tactile flow in humans, and improvements have been studied on the analysis of tactile feedback in robots using computational tactile flow [10]. Although our work is also inspired by the human somatosensory system, we took different approaches for a similar goal. Instead of using the human tactile flow as a base for our research, we developed pose estimation based on the human visuotactile system, also called the "Where" system. This intersensory interaction extends to situations combining vision and touch to include different object information, for interobject interaction or allocation of attention [5].

During visuotactile interaction, multiple frames of reference are simultaneously available on human haptic spatial localization [5]. In the "Where" system, two types of touch spatial localization are considered, one being a position on the body where the stimulus is applied, or otherwise, in the external world from where the stimulus comes. During a task, humans can use a single frame or combine multiple frames of reference, the origins of which may be visible landmarks or a body part from the individual. Even though the aforementioned haptic frame describes the contact between the skin and an external object, one could use landmark axes to specify a frame of reference external to the body (an "allocentric" frame of reference). Similarly, a local frame of reference, such as a fingertip axis, is used for localization on the body (an "egocentric" frame of reference). In summary, haptic spatial localization could be described either by a coordinate system centered on the body, egocentric or by external features in the environment such as the edges of a table, an allocentric frame of reference.

Investigations on tactile sensing have the potential to improve robotic in-hand manipulation, including, but not limited to, object characteristics extraction and feedback control. Tactile sensing provides essential information about object manipulation, solving problems such as object occlusion and object's pose estimation under stable grasping. Form, shape, and functionality of the human skin also inspired research on the tactile sensing field. A bio-inspired approach to tactile sensing has significant results in the literature, including successful works on texture classification and control feedback. The present work uses a visuotactile approach to pose estimation using data collected from bio-inspired multimodal tactile modules in conjunction with camera feedback.

Successful robotic manipulation starts with a stable object grasp; therefore, robots are expected to have robust grasping skills. Considering grasp as a control problem, in contrast to decomposing a grasping procedure into planning and execution, they do not require any specific hand–object relative pose and are more robust under pose uncertainty [11]. A model-free solution in addition to computationally inexpensive control laws uses simpler hand designs, still ensuring a stable grasping solution. Fuzzy logic control for grasp stability is present in the literature as a useful tool for in-hand manipulation [12–15]. Multifingered hands achieve stable gasping when no resultant forces act on a fully restrained object [16]. This work was developed using a fuzzy controller that was able to perform stable grasp tasks with controlled fingertip force single and dual actuated versions.

The main contributions of this paper are: (1) Tactile information to estimate the pose of unknown objects under autonomous, stable grasp; (2) Integration of bio-inspired multimodal tactile sensing modules and visual information to describe in-hand object pose; (3) Analysis of five different machine learning algorithms for tactile pose estimation. The system uses visual information similar to how humans allocate visual attention to determine frames of reference for objects of interest. Tactile information is used to learn vision-defined reference frames so that the vision system can be freed to perform other tasks [5]. Concretely, we used vision to extract an allocentric frame of reference where the object pose is located in the environment. Further, five machine learning methods using data from the tactile sensing system inferred the relation between egocentric and allocentric reference frames during haptic spatial localization. Post-grasp object rotations were performed to collect tactile information, exposing the learning system to object angles outside of the finger's actuation workspace. For this purpose, the object received external forces in two different axes, including an open-loop finger actuation experiment. A closed-loop stable grasp fuzzy controller also used the same sensory feedback signals. From all six machine learning algorithms, results using ridge regressor achieved an average mean squared error for all sizes of 1.82°.

In this paper, Section 2 presents a literature review of human haptic perception, in-hand robotic manipulation, tactile sensing, fuzzy control, and regression algorithms. Section 3 presents our prototype description, a system overview, and experimental setup. Section 4 shows experimental results for two in-hand manipulation tasks performed, followed by conclusions in Section 5.

2. Literature Review

Several papers in recent years have approached the robotic grasping and manipulation in its different aspects. Some of them are design, control, and sensing. The present work approaches in-hand manipulation aspects of control and sensing using the human "Where" sensory system as an inspiration to in-hand object pose estimation.

2.1. The Human "Where" System

Studies on the human somatosensory divided it into "What" and "Where" systems related to the functional distinction between them, the former being the system that processes surfaces, objects, and their characteristics, while the latter being responsible for describing points, surfaces, and objects in the environment. With a counterpart in vision, the "Where" system is investigated in the present paper as an approach to solve in-hand manipulation pose estimation of objects. A notable difference from vision, however, is that the localization can be related to the sensory organ, the human skin, or to the world. Accordingly, investigations focus on two types of visual–spatial localization on the human tactile field, one determining where on the body a stimulus is being activated, and another wherein the external world a body tactile sensing is being touched [5].

In order to study human touch as a spatial localization, researchers require investigation of how humans specify frames of reference [9]. A frame of reference defines Cartesian or polar coordinate systems, where its origins may be an individual's body parts or distinct points to build environment landmarks. Humans can use a single frame of reference to perform a given task, but multiple frames of reference are available. An example from [5] defines points of the body, such as fingertip axes, that act as local frames of reference. When facing the task of localizing points in the external space, humans usually refers to an "egocentric" frame of reference, which specifies distances and directions relative to the individual. By contrast a person uses an "allocentric" frame of reference when using landmarks and external axes as guidelines. The ability to use multiples frames of reference is also essential to the allocation of attention that human dexterity is known. Higher-level visual-touch interactions are a great example of how the human somatosensory system uses those frames of reference to collect information about different objects, during interobject interaction or allocation of attention.

2.2. Manipulation, Dexterity, and Underactuated Hands

Biological inspiration in robotics has been growing over the years with significant interaction between robotics and neuroscience, even though the neuroscience background is usually an endless inspiration with small impact on the final application [3]. Recent research on visuotactile grasping and manipulation has shown that robots have improved manipulation skills when simulating human tactile and visual perception to action. Bimbo et al. [17] combined vision and touch for the estimation of an object's pose during in-hand manipulation where the camera suffers from finger occlusion. A visuotactile control was developed by Li et al. [18], achieving robust in-hand manipulation and exploration of unknown objects performing robust manipulation even in the presence of accidental slippage or rolling. [19] presents another example of reliable grasps using underactuated hands, where flexible hands and visual clues were used to perform after grasping pose estimation for high-precision assembling.

Adaptive hands, such as those used in the works mentioned above, have become an exciting topic of research over the last decade, with recent development in several areas, such as new designs [20,21], reliable grasping [12,22], and dexterous in-hand manipulation [23,24]. Due to its unique ability to grasp and adapt to unknown objects, special attention has been devoted to the research of underactuated hands in unstructured environments. This characteristic is because underactuated hands have n Degrees of Freedom (DOF), and fewer than n actuators, while kinematic constraint of the finger is promoted by passive elements, e.g., springs of flexible joints, establishing the shape adaptation of the finger to the object grasped. Dynamics of underactuated hands also have outstanding results with simpler solutions, using tactile sensing with force control on top of slippage detection [25]. Odhner et al. [21] described a design where power and precise grasping are made possible by flexible joints and two pivoting fingers. Reasonable dexterity was achieved by having opposed fingers with individual actuators and tactile feedback provided by its force sensors. Even though little space for customization is observed in the solution as mentioned above, Zisimatos et al. [20] designed a single actuated robotic gripper that supported up to six modular fingers pulled by a differential disk. Later, we redesigned the base and finger to enable dual actuation for opposed fingers which, in conjunction to modified phalanges, made mounted multimodal tactile sensing modules possible with similar dexterity. Shortly, the present paper uses a prototype that falls in between the examples as mentioned earlier with a modified version of [20]'s fingers and base, using two actuators, borrowing ideas from [21].

2.3. Tactile Sensing

Although manipulation tasks could use several types of cameras and noncontact sensors for feedback, tactile sensing plays a vital role in providing accurate local data for intelligent control and manipulation [26]. With tactile data and feedback, the object's pose can be inferred using statistical and intelligent techniques [27,28]. Due to its importance in manipulation, tactile sensors have had massive development in recent years. In the literature, tactile sensors could use different technologies, such as piezoresistive, capacitive, piezoelectric, optical barometric-based, sound, and multimodal sensing [29]. Among them, multimodal techniques provide a capability that closely matches the human skin, itself a multimodal sensing organ.

Alves de Oliveira et al. [30] developed a bio-inspired multimodal sensor module observing the functionality and placement of mechanoreceptors in the human skin. Figure 1 presents the sensing module organization where a compliant cone-like structure (2) supports a magnetic, angular rate, and gravity (MARG) system (1) and guides forces acting on the module's surface to the pressure sensor localized on the bottom of the structure. In addition to force, the inertial measurement provides information about the deformation of the artificial skin when pressure is applied.

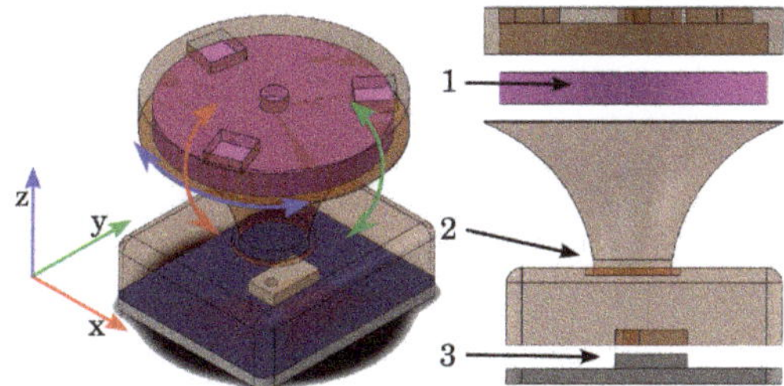

Figure 1. Components of the tactile sensing module: 1—MARG (magnetic, angular rate, and gravity) system; 2—compliant structure; 3—barometer [30].

This paper presents an application of bio-inspired tactile sensing modules and a visuotactile approach used in addition to machine learning algorithms for in-hand object pose estimation. Similar to the human somatosensory system, information about the skin itself provides an 'egocentric' frame of reference where the object's interaction to the artificial skin occurs. Real-time information obtained by the tactile modules was also used to maintain a stable grasp during experiments with an underactuated robotic hand.

2.4. Fuzzy Stable Grasping

Using fuzzy logic to address the grasping problem is one way to reduce the gap between conventional and intelligent control due to the uncertainties and complex mathematical solutions involved in the task [31]. In a recent edition of the Amazon picking challenge, the authors of [32] described problems with an open-loop control solution during pick and place tasks; therefore, autonomous, stable grasping is an essential aspect of in-hand manipulation. The authors of [33] presented a closed loop control using tactile sensing for a task of removing a book from a bookshelf when maximization of the contact surface is required.

Another aspects investigated by [12] are relations between microvibrations and grasp stability. The authors developed a dual motor fuzzy control using BioTac© sensor feedback. Grasping status and stability values were estimated using pressure and microvibration data. A second fuzzy controller provides finger directives such as pull, push, or hold using the previous stability values. In multifingered hands, when no resultant force acts on a fully restrained object, equilibrium is achieved, which is a requirement for stable grasp [16]. The present work implemented a fuzzy controller that was able to perform grasping tasks with minimum fingertip force. Our controller used microvibrations, detected from the sensor's gyroscope and force from fingertips sensor to control each finger separately. A previous version also studied used an Force-sensing resistor (FSR) sensor providing input to a single motor fuzzy controller.

The next section presents an experimental setup for our visuotactile experiment in addition to details about the prototype used, material, and methods approach.

3. Materials and Methods

This section presents materials, methods, and experimental setup for visuotactile object pose estimation. The open source design from [20] is the basis for the present implementation. It is a modular 3D printed gripper in ABS plastic providing an easily customizable platform. Our robotic hand design is a two independently controlled fingers version mounted on top of a table. For that goal, we kept Zisimatos et al.'s [20] top plate, but a modified base accommodates two motors needed to pull each finger separately. When compared to the human hand, these underactuated robotic fingers have intermediate and distal flexible joints made of Vitaflex© 30 (Smooth-On, Easton, PA, USA). Based on Zisimatos et al. [20], this top plate, including the fingers, shows a maximum force applied (and retained) of 8 N per fingertip during tests with a standard servo. Strings are attached to tip phalanges, and two fingers are pulled independently by two Dynamixel© motors (Robotis, Lake Forest, CA, USA) mounted on a modified base. From base to fingertip, the gripper is about 20 cm long. Figure 2 shows the opened gripper during experiments before a grasp attempt. At the top, the left picture shows the four tactile sensors mounted on the finger phalanges, motors, and pulleys. This viewpoint is used to place the camera, as shown in Section 3.3. The red arrows indicate the direction of force applied by the motors. On the top right of Figure 2, a side view shows details of tendons and flex joints. A detail of the pulley one appears with a red circular arrow indicating the motor actuation direction during the pulling phase. The bottom row of Figure 2 shows steps of a single finger actuation: (a) Rest position, no motor rotation; (b) initial movement with motor pulling; (c) continuous motion brings the finger to closer to the palm; and (d) around the maximum safe finger curvature.

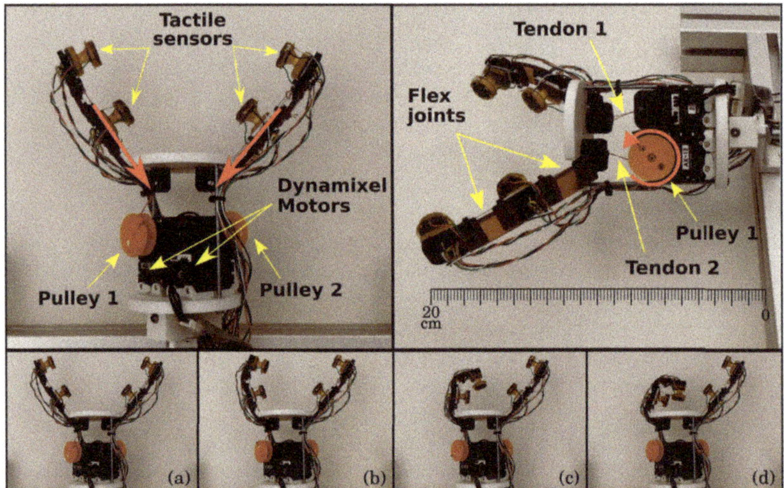

Figure 2. Top and side view of the modified gripper design. (**a**) rest position; (**b**) initial movement; (**c**) inger actuation; (**d**) maximum safe curvature.

Each finger phalange has a tactile sensor module mounted on it. Figure 2's top row shows the tactile module multimodal sensors and its placement on each finger. The module's compliant structure and material add flexibility for the fingers' functionality. All tactile modules send data to microcontrollers acting as nodes to the central computer. The software developed for this prototype is a distributed system that uses Robot Operating System (ROS) [34].

Figure 3 presents all primary ROS nodes developed for this implementation in yellow boxes inside the "Computer (ROS)" gray box. The central node runs in a laptop that concentrates control, data collection, and pose estimations. All data were collected to *ROS bags* for post-processing and pose orientation estimation. Figure 3 also shows the fuzzy control node that also receives sensor data and updates motor controllers. MCU 0 and MCU 1 are microcontrollers attached via serial connection to a main computer acting as ROS nodes. These microcontrollers receive data from magnetic, angular rate, and gravity (MARG) and barometer components of via I^2C protocol that is represented by arrows from the "Tactile sensing" module in Figure 3, which also shows I^2C communication from tactile modules multiplexed to via MUX 0 and MUX 1. There is also a USB camera represented in a blue box and a USB Serial connection from the "Motor Manager" yellow box used for computer control of Dynamixel motors represented in an orange box at the bottom of Figure 3. In the following sections, we present tactile sensing module components and organization as well as the fuzzy control system used in this experimental setup.

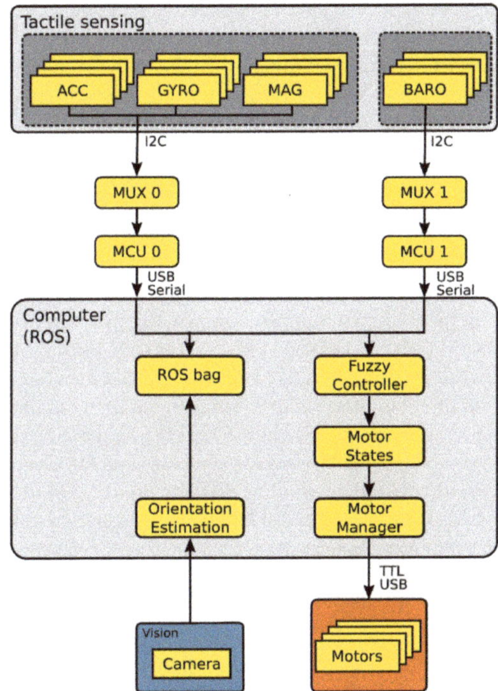

Figure 3. System overview.

3.1. Tactile Sensors

Bio-inspired multimodal tactile sensing modules which are inspired by the type, functionality, and organization of cutaneous tactile elements provide tactile information during visuotactile perception experiments. Each module contains a 9 DOF, magnetic, angular rate, and gravity (MARG) sensor, a flexible, compliant structure, and a pressure sensor placed in a structured way similar to human skin [30]. The components integrated in the tactile module are previously discussed in Section 2 and shown in Figure 1. A total of four tactile modules, one for each phalange, had data collected during experiments. Figure 3 shows four MARGs connected to microcontroller MCU 0 by a multiplexer MUX 0. In a similar manner, four deep pressure sensors are connected to microcontroller MCU 1 by a multiplexer MUX 1. The master ROS node running at the central computer demultiplexes the data and stores tactile information represented in Figure 3 by arrows from the USB serial connection to a yellow box labelled "ROS bag". The experiments and results section presents data provided by the tactile modules during the execution of rotation using both external forces and open-loop execution of in-hand manipulation.

3.2. Fuzzy Controllers

Before any manipulation took place, two fuzzy controllers based on tactile sensing information such as microvibrations and pressure maintained a stable grasp by sending motor signals to each actuator responsible for pulling the fingers [12]. The autonomous fuzzy grasping controller system on this setup provided consistent grasp force while handling different object sizes.

The present implementation used a dual fuzzy controller based on pressure and microvibrations. The pressure is measured from the deep pressure sensor readings, while gyroscope information provides microvibration values for a fuzzy feedback controller about each finger. The indicator that the object under grasp has movement is angular velocity used here to detect microvibrations. The pressure sensor shows the degree of contact with the object. In conjunction, angular velocity and pressure provided tactile feedback about stability and status to a second fuzzy grasp controller. Two gray boxes named *Tactile module 0* and *Tactile module 1* in Figure 4 describe barometer and MARG components with separated I^2C signals forwarded to independent finger fuzzy feedback controllers.

Figure 4 describes a complete system flow during the experiments. Data from tactile sensing modules (top left) provided information for the fuzzy controller (bottom left) with details described in the following sections. With a grasp controller decision based on stability and status, finger actions modified the actuators' status (blue box middle). Vision (middle blue box) provides an allocentric reference frame to the machine learning pose estimation. This allocentric reference frame used data from all four tactile modules to provide an egocentric frame of reference, which was input to the pose estimation module (right gray box). The last step was to apply five machine learning techniques and return angle estimation (green box).

Each *finger fuzzy feedback controller* provided status and stability values to a *grasp fuzzy controller*, which is indicated by a gray box inside the *fuzzy controller* box receiving status and stability inputs in Figure 4. The *grasp fuzzy grasp controller* produced actions for each finger (go forward, go backward or hold) based on status and stability information from both fingers.

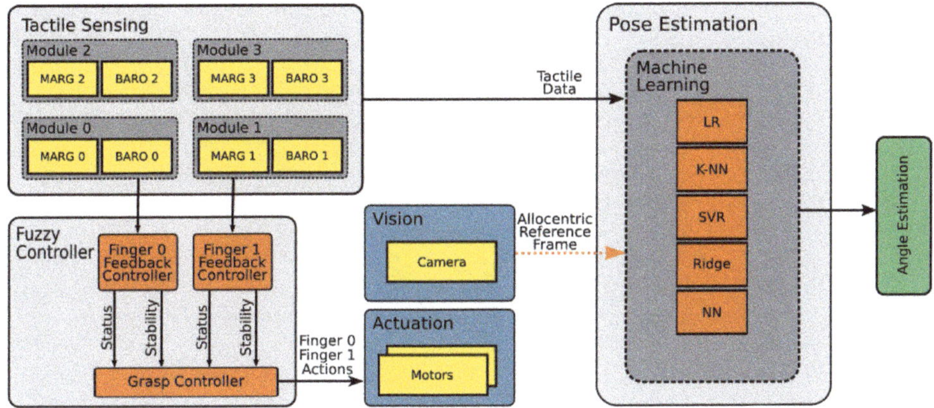

Figure 4. System flow chart.

3.2.1. Finger Fuzzy Feedback Controller

The *finger fuzzy feedback controller* used real sensor data to provide pressure status and grasp stability information for input to a *grasp fuzzy controller*. In order to describe finger fuzzy feedback controller inputs, Low and High fuzzy sets were defined for microvibrations input and Nopressure, Lowpressure, Normalpressure, and Highpressure fuzzy sets for pressure input. Stable and Notstable fuzzy sets describe finger fuzzy feedback stability output while Nottouching, touching, Holding, and Pushing define status output.

Figures 5 and 6 present input values applied for microvibrations and pressure fuzzy sets, respectively. The tactile information used here was normalized data from the barometer part of the tactile module as pressure and raw gyroscope data measuring variations on the angular velocity of the respective module. An example of possible output, Figures 7 and 8 show stability and status output form its fuzzy sets, respectively. The inference system used was Mandani based on the center of gravity. The second fuzzy controller used status and stability from this both finger's controller to produce finger actions.

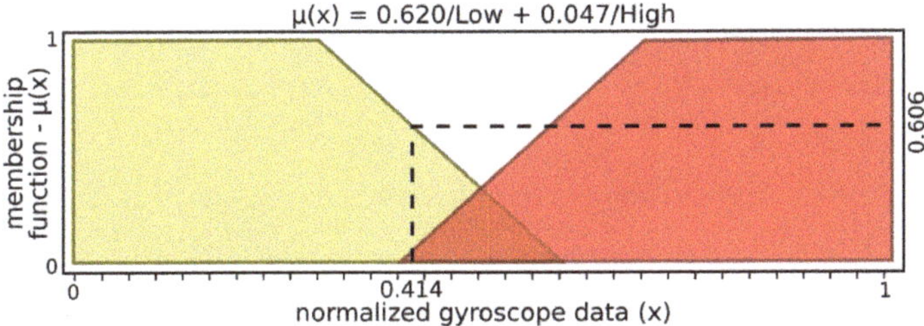

Figure 5. Input sets of microvibrations.

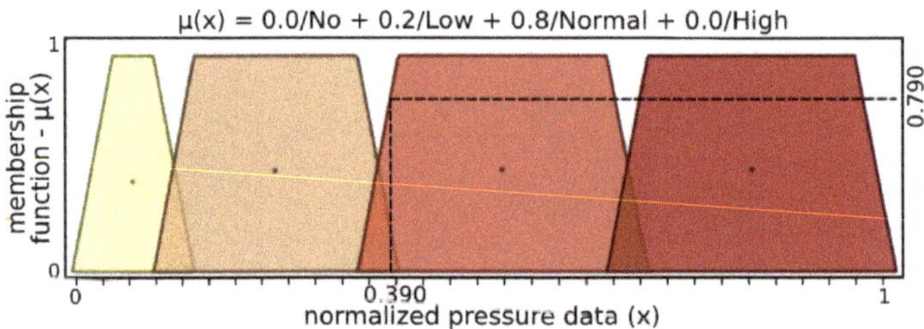

Figure 6. Input sets of pressure.

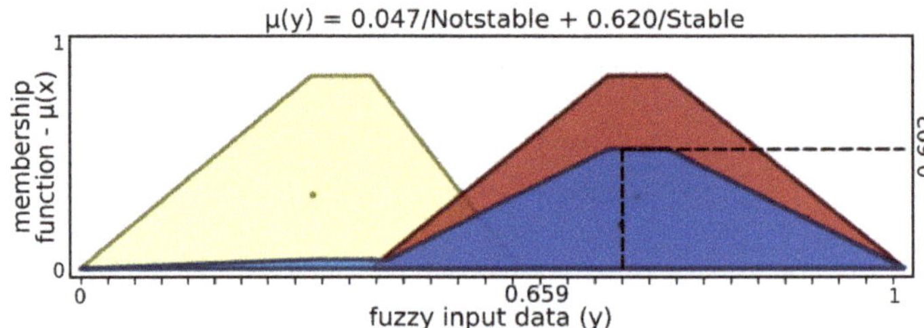

Figure 7. Output sets of stability.

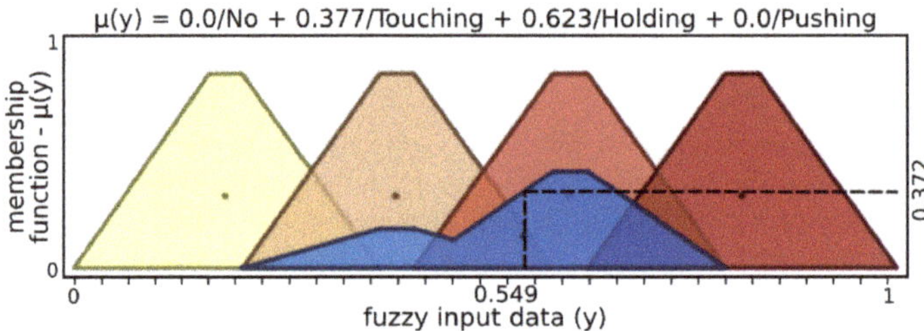

Figure 8. Output sets of status.

A rule book based on [12] was implemented for this tactile in-hand manipulation fuzzy feedback controller, and Table 1 summarizes the rules. Possible outputs for status are NT, not touching; T, touching; H, holding; and P, pushing. Stability has two possible outputs, S, stable and NS, not stable.

Table 1. Finger fuzzy feedback controller rulebook.

Presssure / Microvibrations	No	Low	Normal	High
Low	NT/S	T/S	H/S	P/S
High	NT/NS	T/NS	NS	P/NS

From the example above, microvibration data activate 0.620 of "Low" and 0.047 of "High" sets, resulting in $\mu(x) = 0.606$, where $\mu(x)$ is a membership function. In a similar way, the above example of pressure has $\mu(x) = 0.790$ with 0.2 of "Low" and 0.8 "Normal" pressure sets, while "No" and "High" sets have no participation. After inference, stability presents $\mu(y) = 0.602$, activating 0.620 of "Stable" and 0.047 of "Nonstable", while status has $\mu(y) = 0.372$, with 0.377 of "Touching" and 0.623 of "Holding".

3.2.2. Grasp Fuzzy Controller

Based on both finger status and stability, a second fuzzy controller is responsible for publishing motor values in order to maintain a stable grasp. The system overview in Figure 4 shows a box labeled grasp fuzzy controller, where stability and status data from each finger are inputs for this fuzzy controller, while its outputs update motor velocities. Above, *finger fuzzy feedback controller* provided status and stability from each finger. It is important to observe that input from both fingers is essential during the inference phase of the *grasp fuzzy controller*. Although both fingers are necessary for inference, for simplicity, Figures 9 and 10 only present one finger input example. The inference system used was also Mandani based on the center of gravity. Both outputs change finger motor velocity, updating the Dynamixel controller manager node such as presented on Figure 11, where Figures 9 and 10 show sets that define inputs of finger stability and status, respectively.

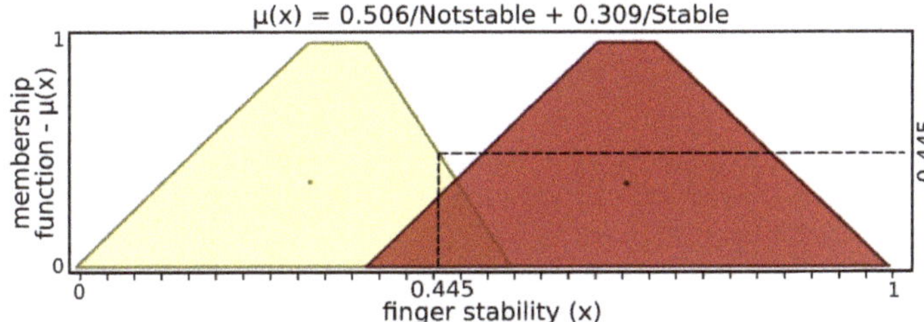

Figure 9. Input sets of stability.

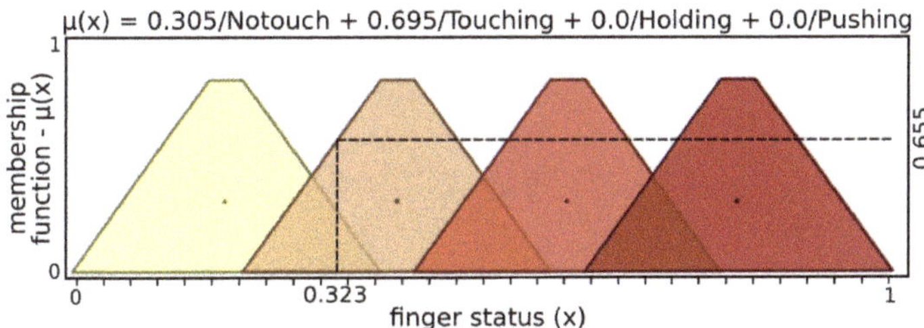

Figure 10. Input sets of status.

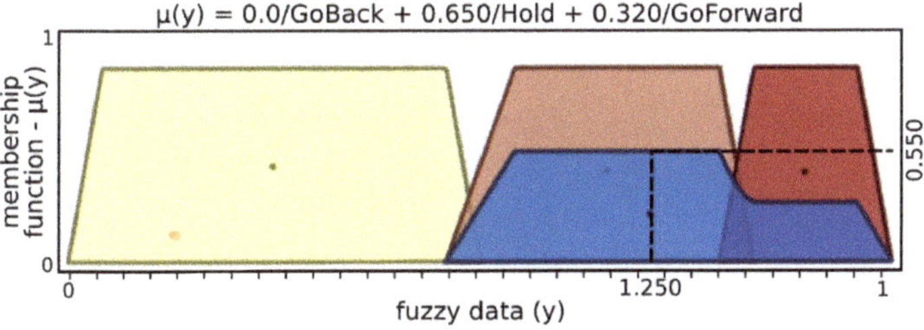

Figure 11. Action output sets.

Another rule book also based on [12] was implemented for the grasp fuzzy controller and is summarized in Table 2. Possible outputs for updating motor velocities are GF1, finger 1 go forward; GF2, finger 2 go forward; H1, finger 1 hold; H2, finger 2 hold; and GB1, finger 1 go back; GB2, finger 2 go back.

Table 2. Grasp fuzzy controller rulebook.

Finger 1 \ Finger 2	NT	T	H	HS	HNS	PS
NT	GF1/GF2	H1	GF1	GF1	GF1	H2
T	H1	GF1/GF2	H2			H2
H	GF2	H1			GF1/GF2	H2
HS	GF2			H1/H2		H2
HNS	GF2		GF1/GF2			H2
PS	H1	H1	H1	H1	H1	GB1/GB1

From the example above, stability inputs provide $\mu(x) = 0.445$, with 0.506 of "NotStable" and 0.309 of "Stable". Similarly, status input results in $\mu(x) = 0.655$, with sets Notouch of 0.305 and Touching of 0.695, while Holding and Pushing are not activated. After inference, the grasp fuzzy controller produces $\mu(y) = 0.550$, with contributions of 0.650 from "Hold" and 0.320 from "GoForward".

Figure 12 shows sensor data, motor positions, and velocities during an autonomous grasp attempt. Start and end arrows point the beginning and final phases of a grasping attempt. Motor velocities and tactile sensing data are fast changing during those two points. Straight lines after the end arrow denote that sensor data and motor position achieve stability, and the fuzzy controller is maintaining a stable grasp between a reasonable error margin.

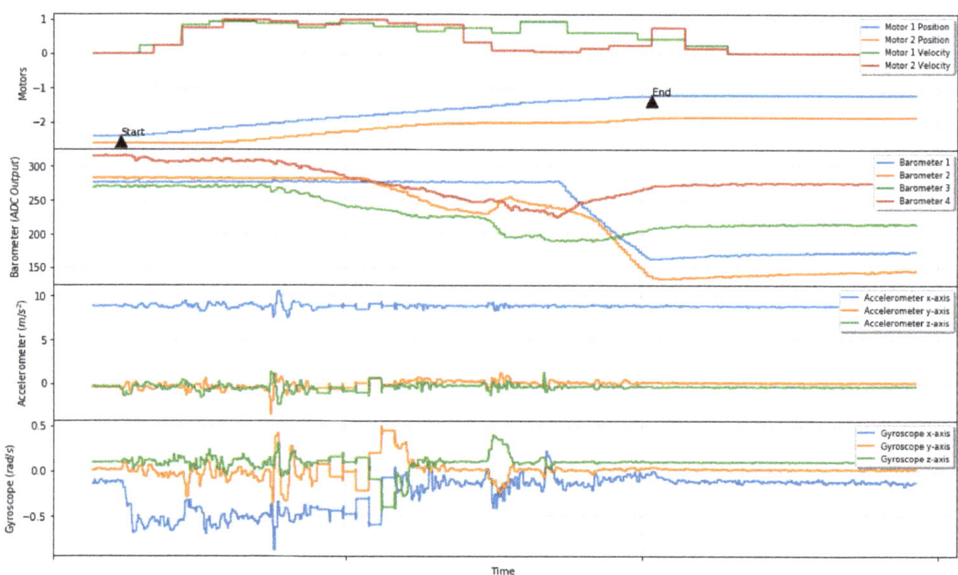

Figure 12. Actuator velocities during a stable grasping. Barometers, accelerometers, and gyroscopes values varying until stable grasp, followed by stabilization of sensor values.

3.3. Camera Setup

The top view of gripper during manipulation was used to estimate object pose variations. The middle of the USB camera image established a frame of reference fixed at the top of this setup. Inside this working space, the difference of pixels determined the angle between two color papers fixed on the object. Figure 13 presents the object angle based on a python script using OpenCV library [35] to capture in-hand angle changes between an object and the camera's visual frame of reference.

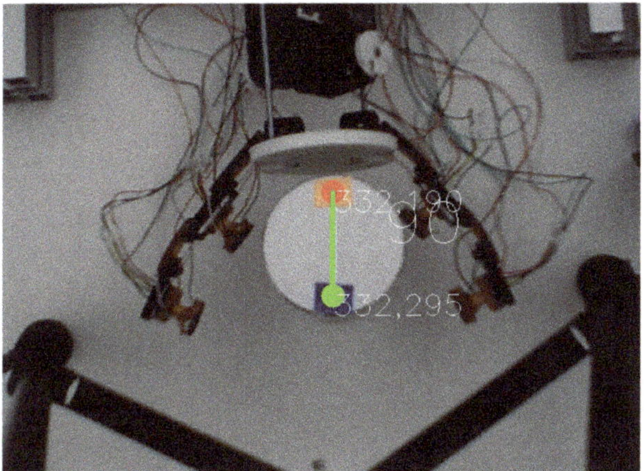

Figure 13. Angle measured from the center of color blobs in the diameter of the object.

Using the setup presented in this section, stable autonomous fuzzy grasping of objects allowed angle estimations based on initial visual inference with tactile excitation to further pose estimation. The next section presents the results of estimations using external and internal stimuli.

4. Results

This section presents the results of two visuotactile experiments on pose estimation using machine learning algorithms. The first experiment used external forces, while the second performs estimation during an open-loop finger self-actuation to rotate an object. In both experiments, images were used to train machine learning algorithms for the dynamic changes in the object's pose. During all operations, a dual fuzzy controller performs an autonomously maintained stable grasp.

4.1. Rotation Using External Forces

During data collection for this first experiment, a human operator executed rotations on a previously grasped object. Figure 14 shows three objects used during this first experiment and its respective diameter information.

Different diameter sizes were used to test the force consistency provided by the autonomous fuzzy controller even with different objects, while still allowing smooth rotation. A piece of paper with a color scheme on object top plate extremities provides a visual cue of the angle to be estimated.

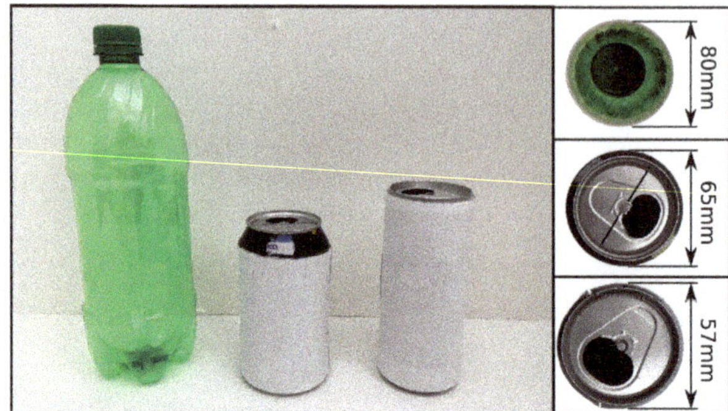

Figure 14. Object used during the first experiment.

Figure 15 shows an example of sensor data collection during rotation using external forces. The angle captured by the camera shows on the first row the execution of clockwise and counterclockwise rotations, at least three times each way, which made sure that sensors were excited with all workspace possibilities for this limited setup. A set of pictures on the second row shows the human operator rotating an object, trying to promote tactile sensor exploration of the gripper workspace. Figure 15 also presents pressure information on the third row showing sensor changes while the object pose is changing. The fourth row presents the x-axis of accelerometer data, which has less amplitude but is essential to define the direction of action, information that pressure alone is not capable of providing.

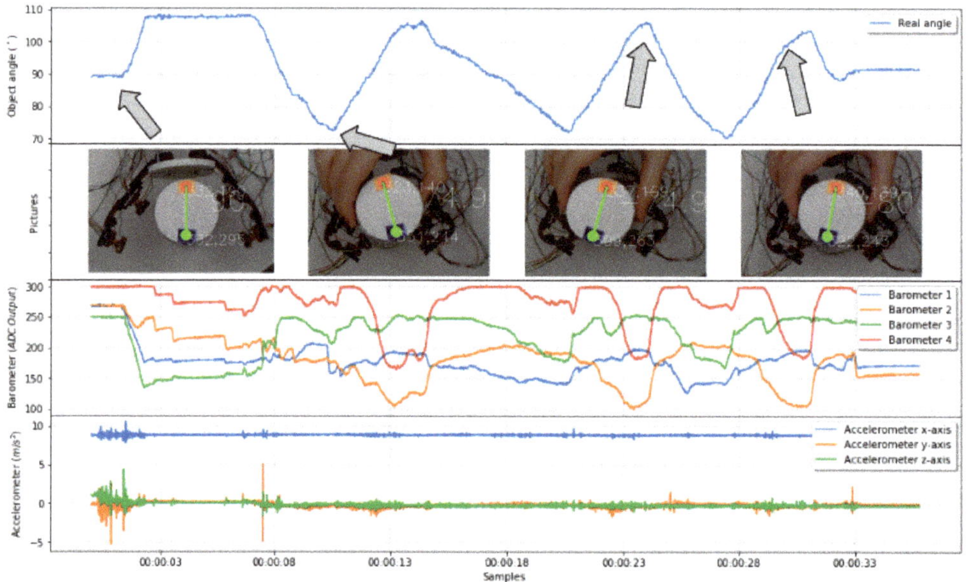

Figure 15. Object being rotated with sensors varying in accordance with the angle.

4.2. Machine Learning Results

In our visuotactile approach, this stage is where the robot has its camera pointed to the object, learning an egocentric frame of reference. Later, it could free the camera to find an allocentric frame of reference, reallocating its attention, and use only tactile information to feel any object pose changes. From grasping to rotations, we used five executions for each object with 70% of the randomized data for training. Table 3 displays the results, with the ridge regressor outperforming all other algorithms in general for the normalized angle. Denormalizing the angle, the root of the average mean squared error for all sizes for the Ridge Regressor was 1.82°.

Table 3. Angle estimation using external forces with blue indicating best results.

Regressor	Large			Medium			Small		
	MSE	R^2	EXP	MSE	R^2	EXP	MSE	R^2	EXP
Linear Regression	0.173 (0.13)	0.825 (0.12)	0.903 (0.02)	0.159 (0.12)	0.836 (0.11)	0.898 (0.04)	0.077 (0.05)	0.898 (0.05)	0.946 (0.002)
K-Nearest Neighbors	0.289 (0.15)	0.691 (0.16)	0.760 (0.09)	0.515 (0.54)	0.501 (0.41)	0.683 (0.18)	0.102 (0.02)	0.859 (0.03)	0.899 (0.02)
Support Vector Regression	0.199 (0.12)	0.794 (0.11)	0.874 (0.04)	0.228 (0.15)	0.762 (0.11)	0.829 (0.11)	0.116 (0.03)	0.836 (0.06)	0.892 (0.01)
Ridge Regression	0.173 (0.13)	0.825 (0.12)	0.903 (0.02)	0.159 (0.12)	0.836 (0.11)	0.898 (0.04)	0.077 (0.05)	0.898 (0.05)	0.946 (0.002)
Neural Network	0.171 (0.10)	0.821 (0.09)	0.890 (0.05)	0.240 (0.13)	0.743 (0.09)	0.864 (0.07)	0.085 (0.02)	0.880 (0.04)	0.925 (0.007)

Figure 16 shows extracts of real data and angle estimations using ridge regression for each of the three objects used in this experiment. Large, medium, and small diameter objects estimated data with the real angle on each row are shown from top to bottom, respectively. On the second and third lines, angle estimations for medium and small diameter objects show high accuracy for all test data presented to the ridge regressor.

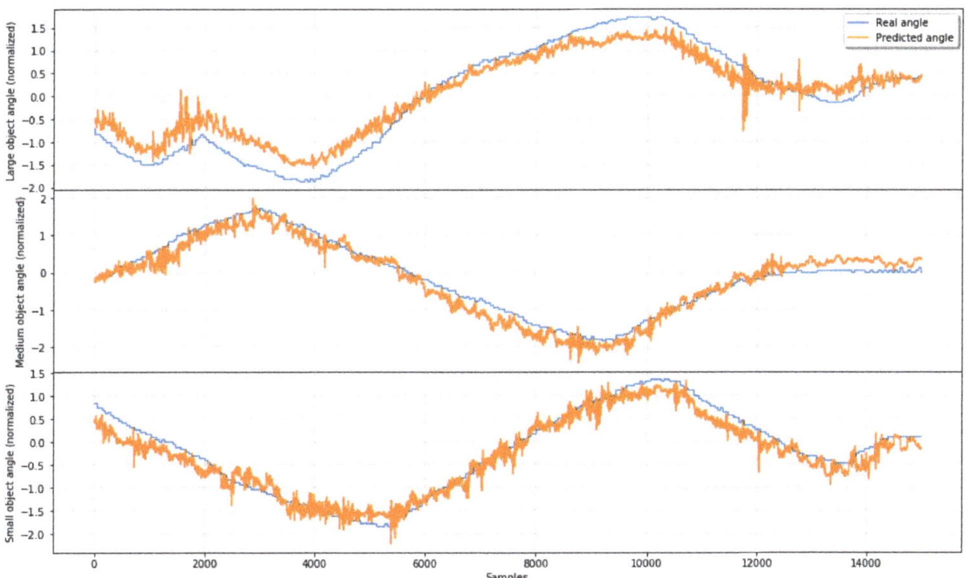

Figure 16. Angle prediction compared to real angle using ridge regression.

It is possible to observe that the large diameter object results on the first row show slightly less precision, with the predicted line not following the real angle line on some occasions. That observation is compatible with Table 3, where this object shows a mean squared error (MSE) of 0.173, while medium and small objects show 0.159 and 0.077, respectively, for the ridge regressor, while achieving a better result using a neural network.

Log files (rosbags) contained angle information at 30 frames per second, with sensor data collected at its maximum frame rate. Table 3 presents results from five methods, namely: Linear regression, K-nearest neighbors regression (KNN), support vector regression (SVR), ridge regressor, and a neural network. The KNN used five neighbours and the Minkowski metric; SVR trained with the RBF kernel and degree 3; the ridge regressor was trained using $\alpha = 0.5$; moreover, the MLP neural network contained a single hidden layer with one hundred neurons. All method predictions used the same data features: Accelerometer, barometer, and magnetometer. Method accuracy used three metrics: Mean squared error (MSE), coefficient of determination (R^2), and explained variance regression score (EXP) [36–39].

4.3. Results from Open-Loop In-Hand Manipulation Angle Estimation

During the second experiment, one finger performed in-hand manipulations, rotating the object while it was under a stable grasping. Figure 17 presents data collected during an open-loop self-rotation experiment with the object's angle in the first graph, followed by a data sensor on subsequent lines.

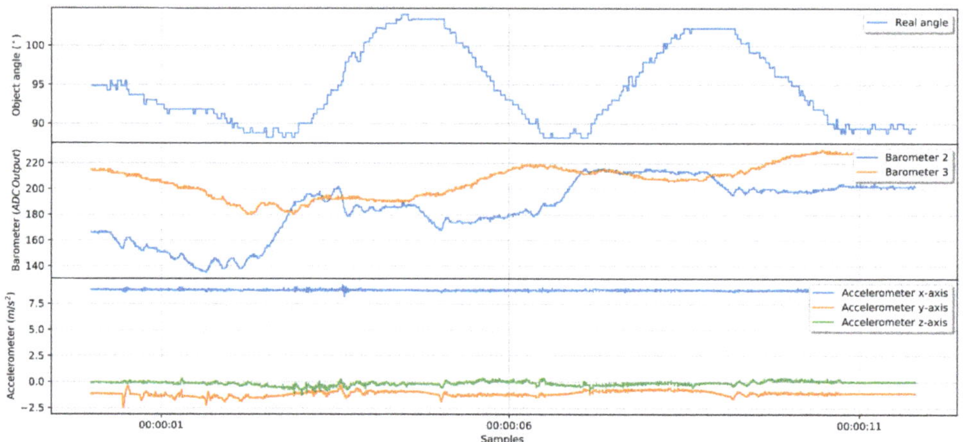

Figure 17. Object being rotated with sensors varying in accordance with the angle.

On the first row, it is possible to observe in-hand manipulations promoted by the finger changes the object angle four times. Compatible pressure data are in the second row, with both barometer measurements changing accordingly to object angle as expected. One could also observe changes in accelerometer data presented in the third row.

Table 4 presents the results of three experiments where MLP outperforms all other methods during three in-hand self-rotations of a middle-size diameter object. Figure 18 shows extracts of real data and the predicted angle using MLP during self-rotation experiments. All the machine learning techniques mentioned on Section 4.2 after retraining show similar results for the middle-size container used in the three experiments.

Table 4. Self-rotation angle estimation with blue indicating best results.

Regressor	Exp 1			Exp 2			Exp 3		
	MSE	R^2	EXP	MSE	R^2	EXP	MSE	R^2	EXP
LR	0.175	0.858	0.958	0.524	−0.155	0.781	0.531	0.492	0.743
KNN	0.389	0.686	0.686	0.203	0.552	0.553	0.193	0.814	0.818
SVR	0.218	0.824	0.830	0.107	0.762	0.812	0.268	0.743	0.770
RIDGE	0.175	0.858	0.958	0.524	−0.154	0.78	0.531	0.493	0.744
MLPR	0.077	0.937	0.976	0.051	0.885	0.898	0.067	0.936	0.946

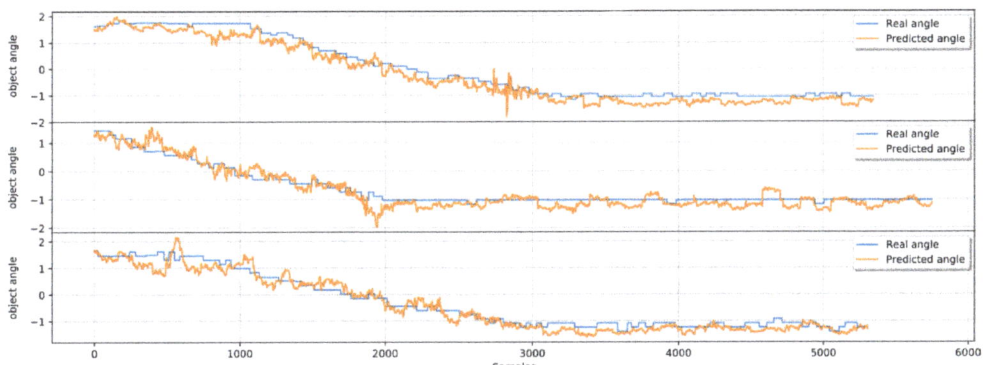

Figure 18. Prediction of angle compared to real angle using Multilayer perceptron (MLP) during self object rotation.

It is possible to see that estimations for angle during all three experiments shown in the three rows are following not only the tendency of the real angle measurement but achieving reasonable precision for all test data presented to the neural network. Observation is compatible with data from Table 4, which shows MLP results with MSE of 0.77, 0.051, and 0.067 for the first, second, and third experiments, respectively. In the next section, we draw some conclusions from the results of the experiments presented here.

5. Discussion

Underactuated hands are a useful tool for in-hand manipulation tasks due to their capability to seamlessly adapt to the contour surfaces of unknown objects and maintain such objects under grasp even when disturbed by external stimuli. These hands are incredibly versatile while grasping unknown objects, but after grasping, estimating the pose of in-hand objects becomes a challenge due to the flexibility of the fingers in such devices. To reduce the pose uncertainty of in-hand objects, we developed a visuotactile approach inspired by the human somatosensory system.

The human manipulation system operates with two subsystems with the "Where" system dealing with perception to action building egocentric and allocentric frames of reference [5]. The present paper introduced a visuotactile approach for robotic in-hand pose estimation of objects. In this research, it was used as inspiration for after grasping object pose estimation combining visual and tactile sensing information. During the experiments, autonomous, stable grasping using a dual fuzzy controller provides consistent force even while handling different objects. Two sets of experiments achieved accurate angle estimation using tactile data.

The system proposed in this paper is an attempt to emulate the human "Where" subsystem. It explores machine learning methods to find the relationship between the data collected by multimodal tactile sensing modules and the orientation of an object under grasp given by intermittent allocentric reference frames. The main advantage of such a data-driven approach is that it could extend to different gripper configurations with more sensing modules and objects with different shapes.

In the first experiment, external forces were applied to the object during a stable grasping, forcing it to change its orientation, while during the second experiment, autonomous open-loop promoted object rotation. External stimuli promote learning of some representations between tactile sensing inputs an object poses, which sometimes are not achieved by a robots finger workspace. Controlled pose disturbances during in-hand manipulation also promote robot learning of some representations between tactile sensing inputs and object poses a robot may need to be exposed to, which can be achieved through the workspace of its fingers. After an initial visual exploration, angle change estimation uses tactile data from bio-inspired sensor modules. Post-processing with ridge regressor achieved an average mean squared error of 0.077 during experiments using external forces. Compatible results using an MLP neural network achieved a mean squared error of 0.067 during autonomous open-loop rotation in the second experiment. As robots become more complex and present in unstructured environments, they have to deal with unexpected object manipulation scenarios. Being able to use its camera in conjunction with tactile sensing is essential to develop a sustainable robot presence in the modern world.

Future research could focus on integration with the somatosensory system in real-time object recognition and pose estimation. Robots that achieve a reasonable level of dexterity using the allocation of attention will improve the efficiency of use of its resources to perform a broader range of tasks in unstructured environments. Since the solution is model-free, the selected machine learning algorithms will improve its accuracy as more information is available, which will lead the investigation of this approach using a more significant number of fingers. A simple pick and place task where a table is modified from its initial position would be improved in this scenario. An "egocentric" frame of reference related to tactile sensing modules is created after exploring the object with the visuotactile system. This robot could build an "egocentric" frame of reference centered tactile control system, while a now free vision system will find a "allocentric" frame of reference (e.g., the axis of a table) for interobject interaction.

Author Contributions: V.P.d.F., T.E.A.d.O. and E.M.P. conceived and designed the experiments; V.P.d.F. analyzed the data and performed the experiments; and V.P.d.F. and T.E.A.d.O. wrote the paper.

Funding: This work was supported in part by the Discovery Grants program of the Natural Sciences and Engineering Research Council of Canada (NSERC) and in part by the Coordenação de Aperfeiçoamento de Pessoal de Nível Superior from the Brazilian Ministry of Education.

Acknowledgments: The authors would like to thank Luiz Sampaio and Onias Silveira for the help in collecting part of the test data.

Conflicts of Interest: The authors declare no conflict of interest.

References

1. Kemp, C.C.; Edsinger, A.; Torres-Jara, E. Challenges for robot manipulation in human environments [Grand challenges of robotics]. *IEEE Robot. Autom. Mag.* **2007**, *14*, 20–29. [CrossRef]
2. De Oliveira, T.E.A.; Cretu, A.M.; da Fonseca, V.P.; Petriu, E.M. Touch sensing for humanoid robots. *IEEE Instrum. Meas. Mag.* **2015**, *18*, 13–19. [CrossRef]
3. Chinellato, E.; del Pobil, A.P. *The Visual Neuroscience of Robotic Grasping*; Cognitive Systems Monographs; Springer International Publishing: Cham, Switzerland, 2016; Volume 28. [CrossRef]
4. He, J.; Zhang, J. In-hand haptic perception in dexterous manipulations. *Sci. China Inf. Sci.* **2014**, *57*, 1–11. [CrossRef]

5. Lederman, S.J.; Klatzky, R.L. Haptic perception: A tutorial. *Attent. Percept. Psychophys.* **2009**, *71*, 1439–1459. [CrossRef]
6. Rouhafzay, G.; Cretu, A.M. An Application of Deep Learning to Tactile Data for Object Recognition under Visual Guidance. *Sensors* **2019**, *19*, 1534. [CrossRef]
7. Sommer, N.; Li, M.; Billard, A. Bimanual compliant tactile exploration for grasping unknown objects. In Proceedings of the IEEE International Conference on Robotics and Automation, Hong Kong, China, 31 May–7 June 2014; pp. 6400–6407. [CrossRef]
8. Xiang, C.; Guo, J.; Rossiter, J. Soft-smart robotic end effectors with sensing, actuation, and gripping capabilities. *Smart Mater. Struct.* **2019**, *28*, 055034. [CrossRef]
9. Klatzky, R.L.; Lederman, S.J.; Matula, D.E. Haptic Exploration in the Presence of Vision. *J. Exp. Psychol. Hum. Percept. Perform.* **1993**, *19*, 726–743. [CrossRef]
10. Ganguly, K.; Sadrfaridpour, B.; Fermüller, C.; Aloimonos, Y. Computational Tactile Flow for Anthropomorphic Grippers. *arXiv* **2019**, arXiv:1903.08248.
11. Dang, H.; Allen, P.K. Stable grasping under pose uncertainty using tactile feedback. *Auton. Robots* **2014**, *36*, 309–330. [CrossRef]
12. Ciobanu, V.; Popescu, N. Tactile controller using fuzzy logic for robot inhand manipulation. In Proceedings of the 2015 19th International Conference on System Theory, Control and Computing, ICSTCC 2015—Joint Conference SINTES 19, SACCS 15, SIMSIS 19, Cheile Gradistei, Romania, 14–16 October 2015; pp. 435–440. [CrossRef]
13. Rovetta, A.; Wen, X. Fuzzy logic in robot grasping control. In Proceedings of the IROS '91:IEEE/RSJ International Workshop on Intelligent Robots and Systems '91, Osaka, Japan, 3–5 November 1991; Volume 1, pp. 1632–1637. [CrossRef]
14. Dubey, V.N.; Crowder, R.M.; Chappell, P.H. Optimal object grasp using tactile sensors and fuzzy logic. *Robotica* **1999**, *17*, 685–693. [CrossRef]
15. Petković, D.; Issa, M.; Pavlović, N.D.; Zentner, L.; Ćojbašić, Ž. Adaptive neuro fuzzy controller for adaptive compliant robotic gripper. *Expert Syst. Appl.* **2012**, *39*, 13295–13304. [CrossRef]
16. Dubey, V.N.; Crowder, R.M. Grasping and control issues in adaptive end effectors. In Proceedings of the DETC'04 ASME 2004 Design Engineering Technical Conference and Computers and Information in Engineering Conference, Salt Lake City, UT, USA, 28 September–2 October 2004; pp. 1–9.
17. Bimbo, J.; Seneviratne, L.D.; Althoefer, K.; Liu, H. Combining touch and vision for the estimation of an object's pose during manipulation. In Proceedings of the IEEE International Conference on Intelligent Robots and Systems, Tokyo, Japan, 3–7 November 2013; pp. 4021–4026. [CrossRef]
18. Li, Q.; Elbrechter, C.; Haschke, R.; Ritter, H. Integrating vision, haptics and proprioception into a feedback controller for in-hand manipulation of unknown objects. In Proceedings of the IEEE International Conference on Intelligent Robots and Systems, Tokyo, Japan, 3–7 November 2013; Volume 1, pp. 2466–2471. [CrossRef]
19. Choi, C.; Del Preto, J.; Rus, D. Using Vision for Pre- and Post-grasping Object Localization for Soft Hands. In *2016 International Symposium on Experimental Robotics*; Kulić, D., Nakamura, Y., Khatib, O., Venture, G., Eds.; Springer International Publishing: Cham, Switzerland, 2017; pp. 601–612.
20. Zisimatos, A.G.; Liarokapis, M.V.; Mavrogiannis, C.I.; Kyriakopoulos, K.J. Open-source, affordable, modular, light-weight, underactuated robot hands. In Proceedings of the 2014 IEEE/RSJ International Conference on Intelligent Robots and Systems, Chicago, IL, USA, 14–18 September 2014; pp. 3207–3212. [CrossRef]
21. Odhner, L.U.; Jentoft, L.P.; Claffee, M.R.; Corson, N.; Tenzer, Y.; Ma, R.R.; Buehler, M.; Kohout, R.; Howe, R.D.; Dollar, A.M. A compliant, underactuated hand for robust manipulation. *Int. J. Robot. Res.* **2014**, *33*, 736–752.[CrossRef]
22. Dang, H.; Allen, P.K. Semantic grasping: Planning task-specific stable robotic grasps. *Auton. Robots* **2014**, *37*, 301–316. [CrossRef]
23. Liarokapis, M.V.; Dollar, A.M. Post-Contact, In-Hand Object Motion Compensation with Adaptive Hands. *IEEE Trans. Autom. Sci. Eng.* **2018**, *15*, 456–467. [CrossRef]

24. Li, Q.; Haschke, R.; Ritter, H. A visuo-tactile control framework for manipulation and exploration of unknown objects. In Proceedings of the IEEE-RAS International Conference on Humanoid Robots, Seoul, Korea, 3–5 November 2015; pp. 610–615. [CrossRef]
25. Birglen, L.; Laliberté, T.; Gosselin, C. *Underactuated Robotic Hands*; Springer Tracts in Advanced Robotics; Springer: Berlin/Heidelberg, Germany, 2008; Volume 40. [CrossRef]
26. Dahiya, R.S.; Mittendorfer, P.; Valle, M.; Cheng, G.; Lumelsky, V.J. Directions toward effective utilization of tactile skin: A review. *IEEE Sens. J.* **2013**, *13*, 4121–4138. [CrossRef]
27. Petrovskaya, A.; Khatib, O.; Thrun, S.; Ng, A.Y. Bayesian estimation for autonomous object manipulation based on tactile sensors. In Proceedings of the IEEE International Conference on Robotics and Automation, Orlando, FL, USA, 15–19 May 2006; pp. 707–714. [CrossRef]
28. Molchanov, A.; Kroemer, O.; Su, Z.; Sukhatme, G.S. Contact localization on grasped objects using tactile sensing. In Proceedings of the IEEE International Conference on Intelligent Robots and Systems, Daejeon, Korea, 9–14 October 2016; pp. 216–222. [CrossRef]
29. Kappassov, Z.; Corrales, J.A.; Perdereau, V. Tactile sensing in dexterous robot hands—Review. *Robot. Auton. Syst.* **2015**, *74*, 195–220. [CrossRef]
30. Alves de Oliveira, T.E.; Petriu, E.; Cretu, A.M. Multimodal bio-inspired tactile sensing module. *IEEE Sens. J.* **2017**, *17*, 3231–3243. [CrossRef]
31. Passino, K. Bridging the gap between conventional and intelligent control. *IEEE Control Syst.* **1993**, *13*, 12–18. [CrossRef]
32. Eppner, C.; Höfer, S.; Jonschkowski, R.; Martín-Martín, R.; Sieverling, A.; Wall, V.; Brock, O. Lessons from the Amazon Picking Challenge: Four Aspects of Building Robotic Systems. In Proceedings of the Robotics: Scienceand Systems 2016, AnnArbor, MI, USA, 18–22 June 2016. [CrossRef]
33. Morales, A.; Prats, M.; Felip, J. *Grasping in Robotics*; Springer: London, UK, 2013; Volume 10. [CrossRef]
34. Quigley, M.; Conley, K.; Gerkey, B.; FAust, J.; Foote, T.; Leibs, J.; Berger, E.; Wheeler, R.; Mg, A. ROS: An Open-Source Robot Operating System. ICRA 2009; Volume 3, p. 5. Available online: http://www.willowgarage.com/papers/ros-open-source-robot-operating-system (accessed on 16 May 2019).
35. Bradski, G. The OpenCV Library. *Dr. Dobbs J. Softw. Tools* **2000**. Available online: http://www.drdobbs.com/open-source/the-opencv-library/184404319 (accessed on 16 May 2019).
36. Hoerl, A.E.; Kennard, R.W. Ridge Regression: Biased Estimation for Nonorthogonal Problems. *Technometrics* **2000**, *42*, 80–86. [CrossRef]
37. Cortes, C.; Vapnik, V. Support-Vector Networks. *Mach. Learn.* **1995**, *20*, 273–297. [CrossRef]
38. Altman, N.S. An Introduction to Kernel and Nearest-Neighbor Nonparametric Regression. *Am. Stat.* **1992**, *46*, 175–185.
39. Haykin, S. *Neural Networks: A Comprehensive Foundation*, 2nd ed.; Prentice Hall PTR: Upper Saddle River, NJ, USA, 1998.

Article

Soft Magnetic Powdery Sensor for Tactile Sensing

Shunsuke Nagahama [1,*], Kayo Migita [2] and Shigeki Sugano [2]

1 Waseda Research Institute for Science and Engineering, Waseda University, 3-4-1 Okubo, Shinjuku-ku, Tokyo 169-8555, Japan
2 School of Creative Science and Engineering, Waseda University, 3-4-1 Okubo, Shinjuku-ku, Tokyo 169-8555, Japan; k-migita@moegi.waseda.jp (K.M.); sugano@waseda.jp (S.S.)
* Correspondence: nagahama@aoni.waseda.jp; Tel.: +81-352-850-996

Received: 28 March 2019; Accepted: 4 June 2019; Published: 13 June 2019

Abstract: Soft resistive tactile sensors are versatile devices with applications in next-generation flexible electronics. We developed a novel type of soft resistive tactile sensor called a soft magnetic powdery sensor (soft-MPS) and evaluated its response characteristics. The soft-MPS comprises ferromagnetic powder that is immobilized in a liquid resin such as polydimethylsiloxane (PDMS) after orienting in a magnetic field. On applying an external force to the sensor, the relative distance between particles changes, thereby affecting its resistance. Since the ferromagnetic powders are in contact from the initial state, they have the ability to detect small contact forces compared to conventional resistive sensors in which the conductive powder is dispersed in a flexible material. The sensor unit can be made in any shape by controlling the layout of the magnetic field. Soft-MPSs with different hardnesses that could detect small forces were fabricated. The soft-MPS could be applied to detect collisions in robot hands/arms or in ultra-sensitive touchscreen devices.

Keywords: ferromagnetic powder; soft material; tactile sensor; magnetic field orientation

---

1. Introduction

Tactile sensors are important components in robots as they enable them to detect objects. In recent years, soft sensors have been developed for a multitude of applications, with the aim of enhancing interactions between human beings and machines. There are various types of sensors such as magnetic, piezoelectric, resistive, capacitive, and optical [1–3]. Research is also being conducted on technology that can estimate the force of an end effector or load on a joint without using a force sensor [4–6]. Capacitive touch sensors measure applied force from changes in their capacitance [7–11]. In these sensors, two sets of electrodes are placed opposite each other with a dielectric layer sandwiched in between. On applying a force, the distance between the electrodes changes; the resulting change in electrostatic capacity is a measure of the force applied. Due to the nature of the detection principle, these capacitive sensors are susceptible to environmental noise. In contrast, magnetic tactile sensors measure applied force from changes in the magnetic field [12–15]. In this method, displacement and force are estimated by measuring the change in magnetic flux density caused by the change in position of the magnetic particles embedded in the material. This is carried out by measuring the magnetic Hall voltage and the voltage change generated by self-electromotive force on the inducer. The main drawback of detecting changes in magnetic properties is that they are susceptible to the effects of geomagnetism and stray fields caused by magnetic materials in the environment. Optical tactile sensors, in contrast, measure displacement or force by observing the optical properties due to deformation of the material. Changes in light reflection due to the change in dot-patterns or deformation are observed using charge-coupled device (CCD) cameras [16–19]. However, since it is necessary to observe such changes optically, a certain distance is required between the material and the sensor, imposing restrictions on the size and geometry of the sensor. Piezoelectric tactile sensors measure applied force by evaluating the voltage generated when pressure is applied to a piezoelectric material [20,21]. However, since the piezoelectric effect arises from distortion of the crystal symmetry, it is difficult to construct flexible sensors out of these materials.

In this study, we focused on resistive sensors that are less susceptible to disturbances. Resistive tactile sensors include liquid type sensors, strain gauge sensors, and sensors using conductive threads [22–24]. Resistance-type sensors have also been fabricated from conductive powder dispersed in soft material such as Inastomer [25,26]. In this method, when the amount of conductive powder is small, conducting channels are created only upon the application of pressure; when no pressure is applied, no conduction occurs. Hence, the sensitivity is poor. However, the conductive powder is often a hard substance such as metal or carbon, so as its concentration increases, the flexibility of the sensor is lost. Magnetic compound fluid (MCF) rubber sensors are fabricated by confining a fluid in a magnetic field, which is then crosslinked by heat or electrolytic polymerization [27–30]. The hardness of this sensor depends on the hardness after polymerization.

We previously developed a type of tactile sensor that uses ferromagnetic powder as a sensing element [31]. In these "magnetic powdery sensors" (MPSs), ferromagnetic powder is confined within magnetic field lines from permanent magnets mounted on parallel plates. As the distance between the plates decreases, the spatial aggregation pattern of the ferromagnetic powder changes, which is reflected in the electrical resistance. Hence, the displacement can be measured by observing the change in resistance. One of the distinguishing features of this measurement technique is its linearity. However, since the powder is merely trapped in the magnetic field, it can be displaced upon vibration or impact. Furthermore, a mechanism such as a spring is required to restore the initial distance between the plates, which increases the size of the device.

Based on these methodologies, we devised a sensor comprising a small amount of conductive powder embedded in a soft gel matrix, wherein the powder is confined using a magnetic field to form conductive channels. Unlike in MPSs, where the powder is held in place only by the magnetic field, in these novel sensors, once the powder is trapped within the magnetic field lines, it is held in place with a silicone gel to prevent displacement. Moreover, the elasticity of the silicone rubber is used as a restoring force, negating the need for a spring-type mechanism. By utilizing this technique with a suitable powder and soft material, a highly sensitive and flexible tactile sensor can be realized. Hereafter, this sensor is referred to as a soft-MPS.

2. Materials and Methods

In a magnetic tactile sensor, the magnetic field can be oriented in different ways. General resistive sensors typically use the arrangement shown in Figure 1a. However, in this arrangement, it is necessary to place an electrode on the surface to which force is applied, which reduces the durability of the sensor. In the proposed soft-MPS, the electrodes are placed on the bottom plate alone by devising an arrangement as shown in Figure 1b.

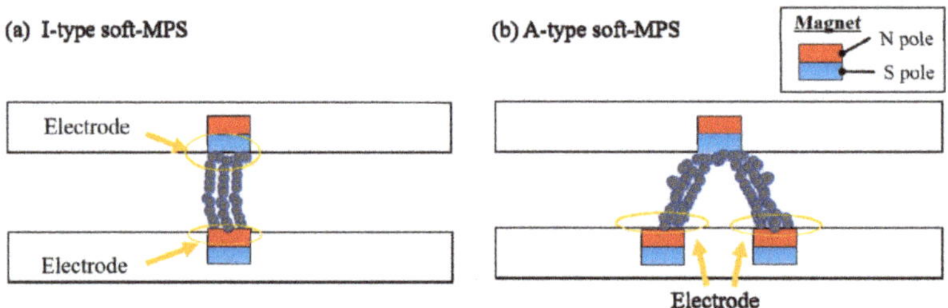

Figure 1. Magnetic field orientation of the ferromagnetic powder for two different types of magnetic resistive tactile sensors: (**a**) I-type soft-magnetic powdery sensors (MPS). One of the electrodes is set on the plate to which the external force is applied. (**b**) A-type soft-MPS. All electrodes are set on the bottom side.

Figure 2 schematically shows the manufacturing process of the soft-MPS. The soft-MPS consists of a case made of acrylonitrile butadiene styrene (ABS) and a 3D-printed magnet. First, the sensor unit was constructed by arranging iron powder along the field lines produced by a magnetic field. Following this, a sensor sheet was manufactured by fixing this iron powder in a matrix of silicone rubber. The electrode was subsequently fixed on the top of the sensor sheet using a silicone primer (PPX primer, CEMEDINE Co. Ltd., Tokyo, Japan) and a silver nano-adhesive. The primer improves the adhesion between the silicone and silver adhesive. This arrangement was left standing at room temperature for 24 h, and the soft-MPS was thus fabricated.

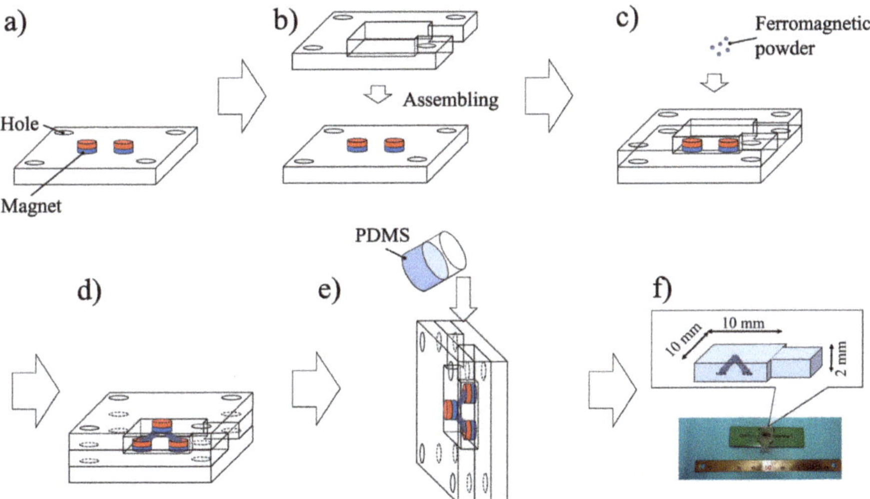

Figure 2. Fabrication process of the soft-MPS: (**a**) magnets (∅1 × 5 mm, 198 mT) are installed inside a 3D printed mold; (**b**) a spacer (thickness: 2 mm) is set for pouring the polydimethylsiloxane (PDMS) solution; (**c**) ferromagnetic powder (200 mesh) is added into the mold with the magnets; (**d**) the top lid is attached to bridge the ferromagnetic powder; (**e**) PDMS is poured into the mold; (**f**) the soft-MPS (area: 10 mm^2, thickness: 2 mm) is obtained and the sensor is fixed on a printed circuit board with a conductive adhesive.

Figure 3 shows the resistance measurement principle for the soft-MPS. On applying an external force, the intergranular distance of the ferromagnetic powder changes. As a result, the resistance changes, and the displacement as well as the force can be measured. By trapping the ferromagnetic powder in a magnetic field during the fabrication process, it becomes conductive even with a small amount of metal. Therefore, unlike a pressure-sensitive material such as Inastomer (developed by Inaba Rubber Co. Ltd., Osaka, Japan), in which a conductive substance such as carbon is dispersed in rubber, the soft-MPS has a finite electrical resistance even in the absence of an external force. When a force is applied, the ferromagnetic powder that has been oriented in the magnetic field changes its position. It is hypothesized that if the ferromagnetic powder is aligned along the magnetic field, the change in resistance due to the applied force will be small. In previous MPS experiments, the change in the resistance was linear with the change in the amount of ferromagnetic powder trapped by the magnetic field lines [31]. On the contrary, in the sensor used here, the powder is oriented by the magnetic field only at the time of manufacturing. After fabrication, the resistance changes due to the change in the positional relationship of the powder, since there is no confinement by the magnetic field. Moreover, the MPS measures the changes in resistance over a relatively large gap (about 0 to 30 mm). Since there are not many real-world scenarios where such a force is applied to a tactile sensor, it is necessary to investigate the characteristics for smaller areas and displacements, which is the case for the soft-MPS used in this study.

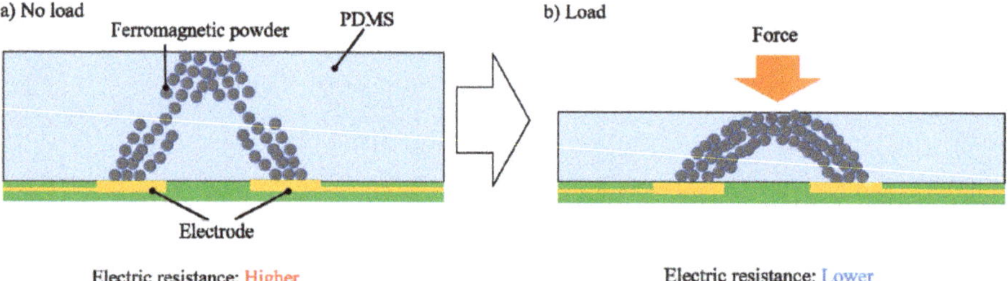

Figure 3. The principle of resistance measurement on soft-MPS for: (**a**) initial condition; (**b**) load condition.

3. Experiments and Results

We measured the change of voltage with respect to the force applied to the soft-MPS. The resistance of the soft-MPS changes upon applying an external force, and as a result, the voltage through the soft-MPS changes. A jig capable of pressing a hemisphere of diameter 10 mm on the sensor was attached to a universal testing machine (Autograph AGS-X, Shimadzu Corporation, Kyoto, Japan), and the applied force and change in voltage were measured as a result of the displacement upon pressing the sensor (Figure 4b). Since the pressure was measured, the sensitivity changes upon changing the value of resistance (R). In this study, we measured in advance the change in voltage of the sensor when R was changed, and adopted 10 kΩ for all subsequent measurements since the output of the sensor was stable at that value. The amount of ferromagnetic powder (200 mesh iron powder, Kyowa Pure Chemical Co. Ltd., Tokyo, Japan) in the soft-MPS was adjusted as shown in Figure 4c. All experiments were conducted at room temperature (23 °C).

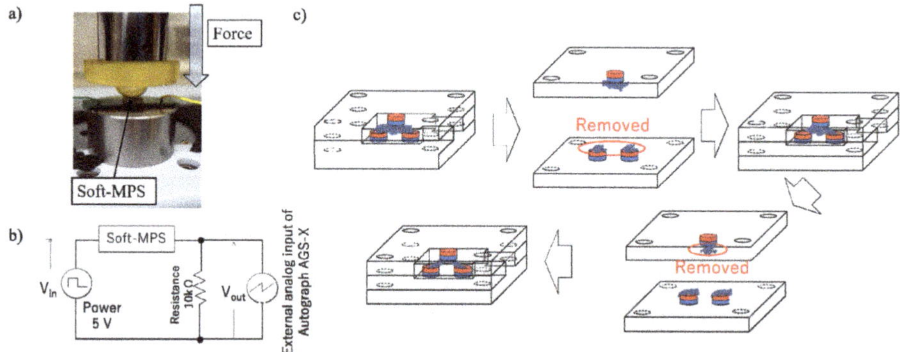

Figure 4. The experimental setup: (**a**) loading of the soft-MPS; (**b**) circuit to detect the voltage change. The voltage was measured at the resistor (10 kΩ); (**c**) removing excess iron powder.

For the soft rubber matrix, polydimethylsiloxane (PDMS) (Ecoflex 00-50, Smooth-On, Inc., Macungie, PA, USA) and Sylgard 184 (Dow Corning Corporation, Midland, MI, USA) were used. The soft-MPS using Ecoflex 00-50 has lower hardness than the one using Sylgard 184. Hence, the soft-MPS using Sylgard 184 is considered capable of measuring relatively larger loads. In this experiment, the characteristics of soft-MPSs with different hardnesses were evaluated. Using the circuit shown in Figure 4a, the change in voltage of the soft-MPS upon external pressure was measured.

The sensor manufactured using Ecoflex 00-50 is hereinafter referred to as soft-MPS (soft) and the sensor manufactured using Sylgard 184 as soft-MPS (hard).

3.1. Evaluation of Hysteresis

Changes in force and voltage were measured using Autograph at 1 mm/min giving a reciprocating displacement of 1 mm to the soft-MPS five times, and its hysteresis characteristics were measured.

The experimental results are shown in Figure 5. Figure 5a,c show the relation between the force and the voltage in soft-MPS (soft) and soft-MPS (hard), respectively. Figure 5b,d show the relationship between the force and the voltage with respect to the number of the displacement of the soft-MPS (soft) and soft-MPS (hard), respectively.

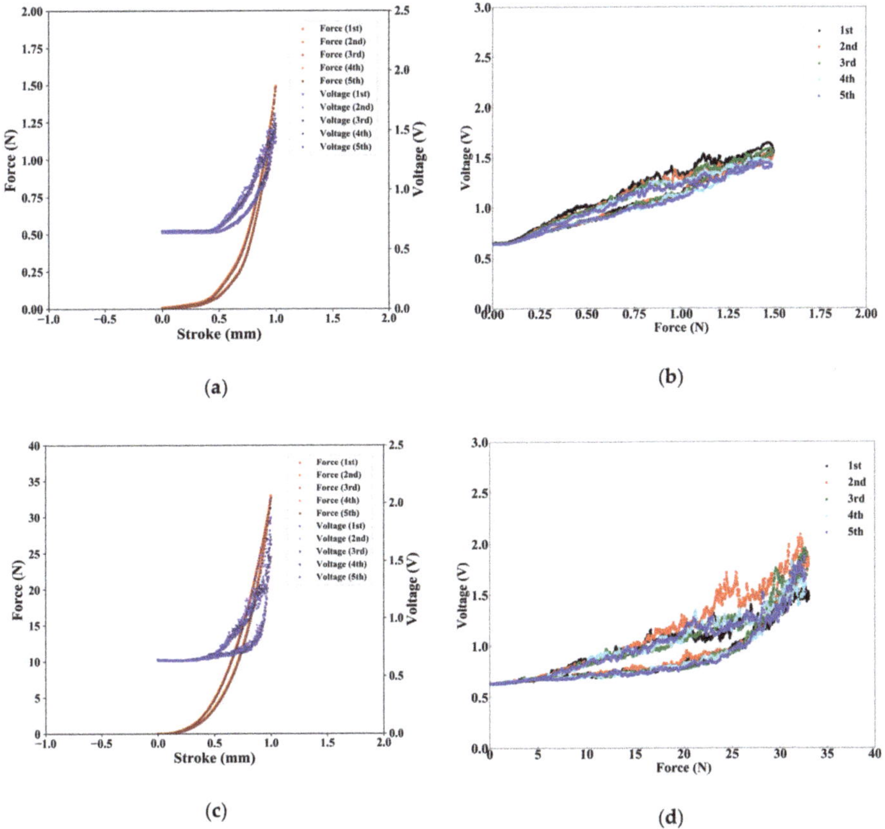

Figure 5. Results of evaluation of hysteresis. (**a**) Changes of force (from a reference value) and voltage in soft-MPS (soft) vs. stroke (1 mm/s). (**b**) Change of voltage in soft-MPS (soft) vs. force. (**c**) Changes of force and voltage in soft-MPS (hard) vs. stroke (1 mm/s). (**d**) Change of voltage in soft-MPS (hard) vs. force.

From the fact that the voltage dropped as the resistance increased, it can be seen that the resistance of both sensors decreased as force increased. The larger range of force measured by soft-MPS (hard) than soft-MPS (soft) for the same displacement was due to its larger storage modulus. Furthermore, focusing on the area where the force was applied in soft-MPS (soft), the voltage started to decrease from a small force. On the contrary, the voltage of soft-MPS (hard) started to drop from 1 N.

3.2. Evaluation of Response for Controlled Force-Input

Using Autograph, we measured the response of the soft-MPSs when applying a constant force for a fixed period of time. For soft-MPS (soft), forces of 0.2, 0.4, 0.6, 0.8, and 1.0 N were applied and then unloaded in stages. For soft-MPS (hard), forces of 1.0, 5.0, 10, 15, and 20 N were applied and then unloaded in stages. Figure 6a,c show the changes in force and voltage when force was applied to soft-MPS (soft) and soft-MPS (hard), respectively, at a displacement rate of 1 mm/min. Figure 6b,d show the corresponding trends for the soft-MPS (soft) and soft-MPS (hard), respectively, when displaced at 10 mm/min.

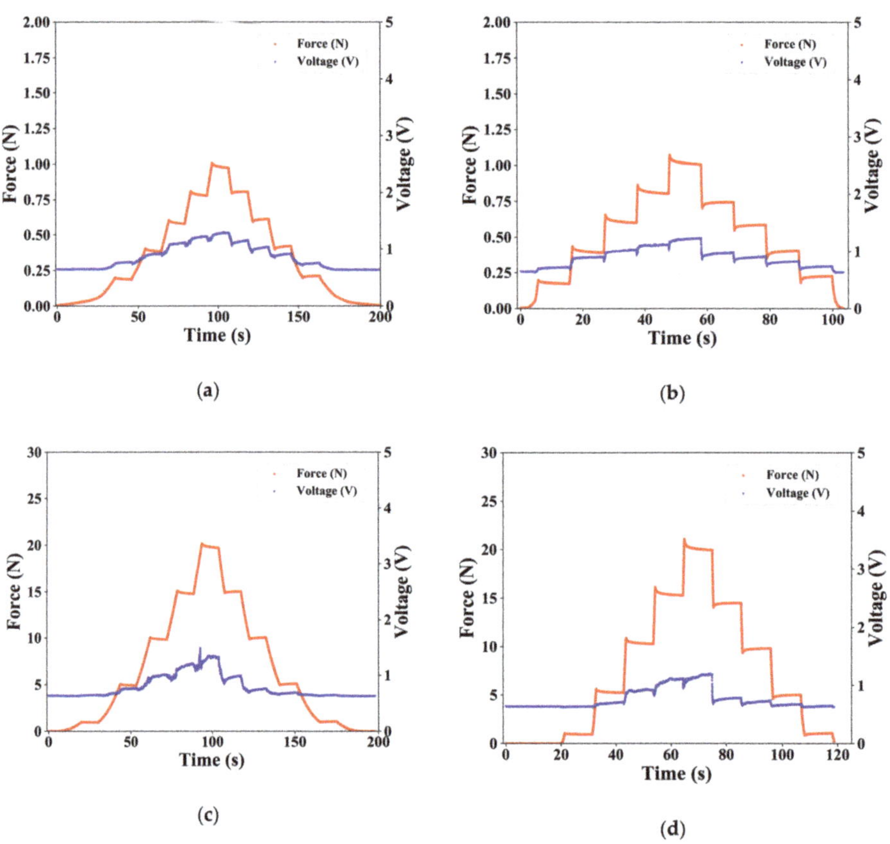

Figure 6. Results of the response for force-maintained input: (**a**,**b**) soft-MPS (soft) and (**c**,**d**) soft-MPS (hard) at testing rates of (**a**,**c**) 1 mm/s and (**b**,**d**) 10 mm/s.

For soft-MPS (soft), it can be seen that the output changed in a step-like manner in response to the applied force. For soft-MPS (hard), the response when applying a force of 5 N was small. However, when applying a force of 10 N or more, the output voltage changed in a step-like manner. In addition, when the force was unloaded from either sensor, a large change in voltage was observed compared to when the force was applied.

3.3. Evaluation of Response for a Pulsed Force

Using Autograph, changes in force and voltage were measured when a constant displacement was repeatedly applied to the soft-MPSs 50 times at 100 mm/min. Figure 7a,b show the voltage response of soft-MPS (soft) at a given displacement of 0.5 and 0.75 mm, respectively. Figure 7c,d show the voltage response of soft-MPS (hard) at a given displacement of 0.5 and 0.75 mm, respectively. From each graph, it can be seen that the voltage changed following the application of pressure in either sensor. In addition, it can be seen that the smaller the displacement, the smaller the scatter as the number of presses increased. Furthermore, from the fact that the peak of the sensor's response voltage coincided with the peak of the applied force in time, it can be concluded that the sensor showed quick response to the applied force.

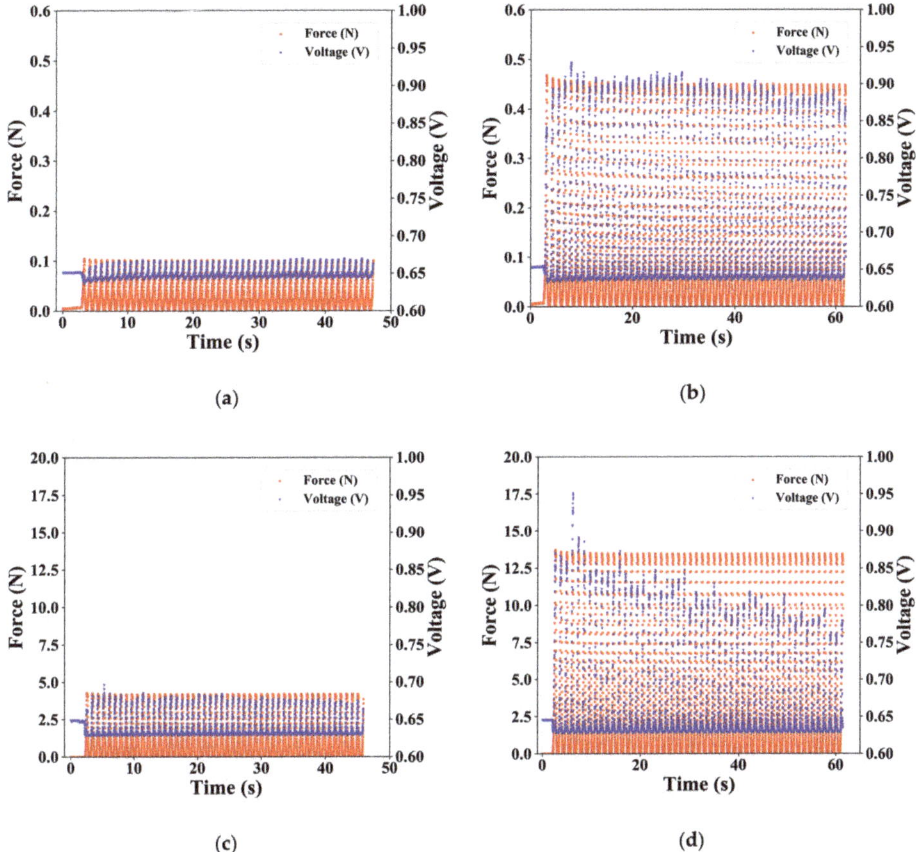

Figure 7. Results of the response for pulsed input: (**a**,**b**) soft-MPS (soft) and (**c**,**d**) soft-MPS (hard) at a pushing depth of (**a**,**c**) 0.5 mm and (**b**,**d**) 0.75 mm.

4. Discussion

4.1. Sensitivity of Soft-MPS

From Figure 5b,d, the sensitivities of the sensors can be determined as 560 mV/N for soft-MPS (soft) and 35 mV/N for soft-MPS (hard). These values largely correspond to the responses in Figure 6. In terms of sensitivity, it shows the same performance as other micro-electro-mechanical systems (MEMS) sensors [1–3].

Figure 8 presents an enlarged view of the low force region of the response voltage against applied force (Figure 5b,d). As shown in Figure 8a, the voltage of soft-MPS (soft) had an overall tendency to increase with force, and it was possible to detect small changes in force. The third cycle had the most straightforward change in resistance to force. The output decreased as the number of cycles increased. In contrast, as shown in Figure 8b, the voltage of soft-MPS (hard) showed a slight tendency to decrease with increasing force. This difference is caused by the difference in hardness of the material. Soft-MPS (soft), which is composed of Ecoflex 00-50, is relatively softer than soft-MPS (hard), which is composed of Sylgard 184. It is believed that soft-MPS (soft) has higher detection sensitivity since it is easier to deform than soft-MPS (hard). One of the reasons why the output value of the sensors varied depending on the number of trials is that the sensor does not completely return to the original position when the force is unloaded. In addition, even though the powder is fixed inside a soft material, there is a possibility that the voltage may drop off slightly when a force is applied.

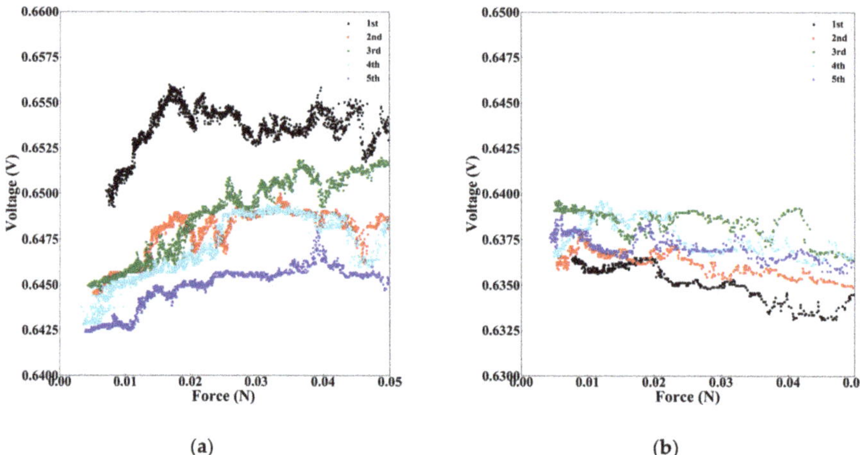

Figure 8. The response of soft-MPS in low load area. (**a**) Changes of voltage on soft-MPS (soft) vs. force. (**b**) Change of voltage on soft-MPS (hard) vs. force.

4.2. Response for a Pulsed Force

The response when a pulsed input was applied (Figure 7) appeared at first to have no delay. A closer look, however, shows that there was a time lag between the applied force and the peak of the voltage response (Figure 9). It can be seen from the graph that the time delay for the response is about 40–100 ms. The same tendency was observed in both soft-MPSs. The powder in the sensing part of the soft-MPS moves with the deformation of the flexible material (PDMS), resulting in a time delay until the internal stress changes after the force is transmitted. This is an unavoidable situation when a soft material is used.

From the response time, we estimate the response rate as 10–15 Hz. This value is not particularly good compared to other sensors [1–3] or human skin sensors. However, it is considered that the long time to transmit the force could be related to the large size of the sensor unit. We will therefore investigate whether the transmission speed can be increased by making the sensor thinner and smaller.

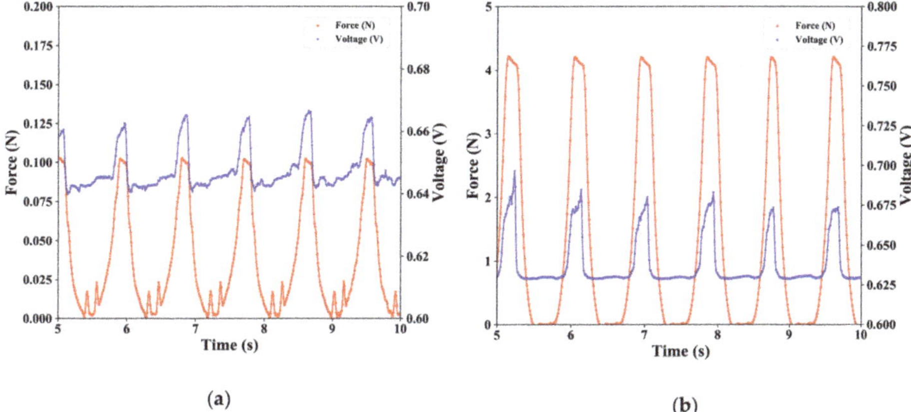

Figure 9. Response for pulsed input with a pushing depth of 0.5 mm: (**a**) soft-MPS (soft); (**b**) soft-MPS (hard).

4.3. Model of Soft-MPS

Modeling the resistance change in the soft-MPS upon the application of force is complex. In its simplest form, the mechanism of sensing depends on the fact that the compressive force changes depending on the buckling deflection based on the theory of general buckling of rectangular plates [32]. Nevertheless, in the real world, there are other mechanical forces generated from the deformation of the materials, as well as thermal forces [33]. Hence, the function of the soft-MPS is affected by a variety of parameters, and the change in resistance is caused by numerous factors in addition to strain and compressive force. Moreover, it would be necessary to estimate the mode of aggregation of the metal powder in the magnetic field as well as to analyze how the particles move in PDMS. Complete modeling of the sensor will be carried out in the future.

4.4. Arrayed Soft-MPS

The proposed sensor construction method is also effective for sensor arrays. A connection state can be achieved by utilizing the attraction between north and south poles. Moreover, a disconnection state can be achieved by utilizing the repulsion between two similar poles. It is also possible to arrange the sensor parts in an array by utilizing these properties (Figure 10). In the future, we will examine whether it is possible to detect force and position/displacement by such soft-MPS arrays.

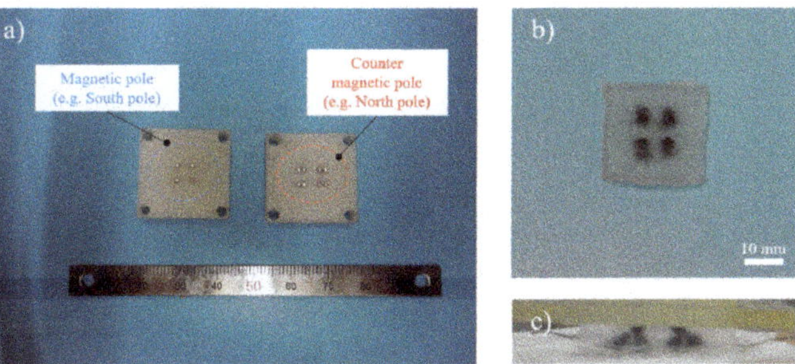

Figure 10. Arrayed soft-MPS: (**a**) mold for soft-MPS array; (**b**) picture of soft-MPS array (top view); (**c**) picture of soft-MPS array (side view).

5. Conclusions

In this study, we evaluated the manufacturing process of soft-MPSs, which are resistive tactile sensors, and their response characteristics. This sensor combines the high sensitivity of ferromagnetic powder with stability and durability by alignment along the magnetic field. Type-A soft-MPSs (i.e., with bottom electrodes) were fabricated and their characteristics were evaluated. It was confirmed that the soft-MPS (soft) (i.e., with low hardness) could detect a small force. The output voltage was evaluated for a pulsed input force, and a short time delay was observed between the input and output. In the future, modeling of soft-MPSs will be conducted to study the method of correcting for hysteresis, which is an inherent problem when using flexible materials, and distributed soft-MPSs will be developed by arraying and applied to the skin of robots. We believe that this design can be applied not only to soft materials but also to hard materials, in which tactile functions can be imparted to structural members such as resins.

Author Contributions: Conceptualization, S.N. and K.M.; Data curation, S.N.; Formal analysis, S.N.; Funding acquisition, S.N.; Investigation, S.N. and K.M.; Methodology, S.N. and K.M.; Project administration, S.N.; Resources, S.N.; Software, S.N.; Supervision, S.N. and S.S.; Validation, S.N.; Visualization, S.N.; Writing—original draft, S.N.; Writing—review and editing, S.N. and K.M.

Funding: Our experiment benefited from an instrument that was purchased using the grant JSPS KAKENHI 17K14631.

Conflicts of Interest: The authors declare no conflict of interest.

References

1. Kappassov, Z.; Corrales, J.A.; Perdereau, V. Tactile sensing in dexterous robot hands—Review. *Robot. Auton. Syst.* **2015**, *74*, 195–220. [CrossRef]
2. Yousef, H.; Boukallel, M.; Althoefer, K. Tactile sensing for dexterous manipulation. *Sens. Actuators A* **2011**, *167*, 171–187. [CrossRef]
3. Dahiya, R.S.; Metta, G.; Valle, M.; Sandini, G. Tactile sensing-from humans to humanoids. *IEEE Trans. Robot.* **2010**, *26*, 1–20. [CrossRef]
4. Wang, Z.; Zi, B.; Wang, D.; Qian, J.; You, W.; Yu, L. External force self-sensing based on cable-tension disturbance observer for surgical robot end-effector. *IEEE Sens. J.* **2019**, *19*, 5274–5284. [CrossRef]
5. Liu, X.; Zhao, F.; Ge, S.S.; Wu, Y.; Mei, X. End-effector force estimation for flexible-joint robots with global friction approximation using neural networks. *IEEE Trans. Ind. Inform.* **2019**, *15*, 1730–1741. [CrossRef]
6. Wahrburg, A.; Bös, J.; Listmann, K.D.; Dai, F.; Matthias, B.; Ding, H. Motor-current-based estimation of cartesian contact forces and torques for robotic manipulators and its application to force control. *IEEE Trans. Autom. Sci. Eng.* **2018**, *15*, 879–886. [CrossRef]
7. Viry, L.; Levi, A.; Totaro, M.; Mondini, A.; Mattoli, V.; Mazzolai, B.; Beccai, L. Flexible three-axial force sensor for soft and highly sensitive artificial touch. *Adv. Mater.* **2014**, *26*, 2659–2664. [CrossRef]
8. Cheng, M.Y.; Huang, X.H.; Ma, C.W.; Yang, Y.J. A flexible capacitive tactile sensing array with floating electrodes. *J. Micromech. Microeng.* **2009**, *19*, 115001. [CrossRef]
9. Maiolino, P.; Maggiali, M.; Cannata, G.; Metta, G.; Natale, L. A flexible and robust large scale capacitive tactile system for robots. *IEEE Sens. J.* **2013**, *13*, 3910–3917. [CrossRef]
10. Guo, S.; Kato, Y.; Ito, H.; Mukai, T. Development of rubber-based flexible sensor sheet for care-related apparatus. *SEI Tech. Rev.* **2012**, *75*, 125–131.
11. Ji, Z.; Zhu, H.; Liu, H.; Liu, N.; Chen, T.; Yang, Z.; Sun, L. The design and characterization of a flexible tactile sensing array for robot skin. *Sensors* **2016**, *16*, 2001. [CrossRef] [PubMed]
12. Nakamoto, H.; Goka, M.; Takenawa, S.; Kida, Y. Development of tactile sensor using magnetic elements. In Proceedings of the IEEE Workshop Robotic Intelligence Informationally structured Space, Paris, France, 11–15 April 2011; pp. 37–42.
13. Oh, S.; Jung, Y.; Kim, S.; Kim, S.-J.; Hu, X.; Lim, H.; Kim, C.-G. Remote tactile sensing system integrated with magnetic synapse. *Sci. Rep.* **2017**, *7*, 1–8. [CrossRef] [PubMed]
14. Kawasetsu, T.; Horii, T.; Ishihara, H.; Asada, M. Mexican-hat-like response in a flexible tactile sensor using a magnetorheological elastomer. *Sensors* **2018**, *18*, 587. [CrossRef] [PubMed]

15. Wang, H.; de Boer, G.; Kow, J.; Alazmani, A.; Ghajari, M.; Hewson, R.; Culmer, P. Design methodology for magnetic field-based soft tri-axis tactile sensors. *Sensors* **2016**, *16*, 1356. [CrossRef] [PubMed]
16. Ito, Y.; Kim, Y.; Obinata, G. Contact region estimation based on a vision-based tactile sensor using a deformable touchpad. *Sensors* **2014**, *14*, 5805–5822. [CrossRef] [PubMed]
17. Ferrier, N.J.; Brockett, R.W. Reconstructing the shape of a deformable membrane from image data. *Int. J. Robot. Res.* **2000**, *19*, 795–816. [CrossRef]
18. Begej, S. Planar and finger-shaped optical tactile sensors for robotic applications. *IEEE J. Robot. Autom.* **1988**, *4*, 472–484. [CrossRef]
19. Maekawa, H.; Tanie, K.; Komoriya, K.; Kaneko, M.; Horiguchi, C.; Sugawara, T. Development of a finger-shaped tactile sensor and its evaluation by active touch. In Proceedings of the IEEE International Conference Robotics Automation, Nice, France, 12–14 May 1992; pp. 1327–1334.
20. Chuang, C.H.; Bin Dong, W.; Bin Lo, W. Flexible piezoelectric tactile sensor with structural electrodes array for shape recognition system. In Proceedings of the 3rd International Conference Sensing Technology, Tainan, Taiwan, 30 November–3 December 2008; pp. 504–507.
21. Yu, P.; Liu, W.; Gu, C.; Cheng, X.; Fu, X. Flexible piezoelectric tactile sensor array for dynamic three-axis force measurement. *Sensors* **2016**, *16*, 6. [CrossRef]
22. Park, Y.L.; Majidi, C.; Kramer, R.; Brard, P.; Wood, R.J. Hyperelastic pressure sensing with a liquid-embedded elastomer. *J. Micromech. Microeng.* **2010**, *20*, 125029–125035. [CrossRef]
23. Engel, J.; Chen, J.; Liu, C. Development of polyimide flexible tactile sensor skin. *J. Micromech. Microeng.* **2003**, *13*, 359–366. [CrossRef]
24. Shimojo, M.; Namiki, A.; Ishikawa, M.; Makino, R.; Mabuchi, K. A tactile sensor sheet using pressure conductive rubber with electrical-wires stitched method. *IEEE Sens. J.* **2004**, *4*, 589–596. [CrossRef]
25. Shimojo, M.; Araki, T.; Ming, A.; Ishikawa, M. A high-speed mesh of tactile sensors fitting arbitrary surfaces. *IEEE Sens. J.* **2010**, *10*, 822–830. [CrossRef]
26. Guo, C.; Kondo, Y.; Takai, C.; Fuji, M. Piezoresistivities of vapor-grown carbon fiber/silicone foams for tactile sensor applications. *Polym. Int.* **2017**, *66*, 418–427. [CrossRef]
27. Shimada, K. Enhancement of MCF rubber utilizing electric and magnetic fields, and clarification of electrolytic polymerization. *Sensors* **2017**, *17*, 767. [CrossRef] [PubMed]
28. Shimada, K.; Mochizuki, O.; Kubota, Y. The effect of particles on electrolytically polymerized thin natural mcf rubber for soft sensors installed in artificial skin. *Sensors* **2017**, *17*, 896. [CrossRef] [PubMed]
29. Shimada, K.; Saga, N. Mechanical enhancement of sensitivity in natural rubber using electrolytic polymerization aided by a magnetic field and MCF for application in haptic sensors. *Sensors* **2016**, *16*, 1521. [CrossRef]
30. Zheng, Y.; Shimada, K. Research on a haptic sensor made using MCF conductive rubber. *J. Phys. Condens. Matter* **2008**, *20*, 204148. [CrossRef]
31. Nagahama, S.; Kimura, Y.; Kim, C.H.; Sugano, S. The development of magnetic powdery sensor. In Proceedings of the Proceedings IEEE Sensors 2014, Valencia, Spain, 2–5 November 2014; pp. 783–786.
32. Reddy, J.N. *Theory and Analysis of Elastic Plates and Shells*; CRC Press: Boca Raton, FL, USA, 2006.
33. Johnston, I.D.; McCluskey, D.K.; Tan, C.K.L.; Tracey, M.C. Mechanical characterization of bulk Sylgard 184 for microfluidics and microengineering. *J. Micromech. Microeng.* **2014**, *24*, 035017. [CrossRef]

Article

Hysteresis Compensation in Force/Torque Sensors Using Time Series Information

Ryuichiro Koike [1], Sho Sakaino [2] and Toshiaki Tsuji [1,*]

[1] Graduate School of Science and Engineering, Saitama University, Sakura-ku, Saitama City, Saitama 338-8570, Japan; koiryu1221@gmail.com

[2] Department of Intelligent Interaction Technologies, University of Tsukuba, Tsukuba, Ibaraki 305-8577, Japan; sakaino@iit.tsukuba.ac.jp

* Correspondence: tsuji@ees.saitama-u.ac.jp

Received: 29 August 2019; Accepted: 27 September 2019; Published: 30 September 2019

Abstract: The purpose of this study is to compensate for the hysteresis in a six-axis force sensor using signal processing, thereby achieving high-precision force sensing. Although mathematical models of hysteresis exist, many of these are one-axis models and the modeling is difficult if they are expanded to multiple axes. Therefore, this study attempts to resolve this problem through machine learning. Since hysteresis is dependent on the previous history, this study investigates the effect of using time series information in machine learning. Experimental results indicate that the performance is improved by including time series information in the linear regression process generally utilized to calibrate six-axis force sensors.

Keywords: force sensing; force sensor; tactile sensing; linear regression; hysteresis compensation

---

1. Introduction

Force control has long been a fundamental technology for robotics and is still being actively studied. Known techniques include using an external force observer [1] or series elastic actuators [2], while the most common approach utilizes force sensors. Force sensors enable the acquisition of force information in a simple manner. However, they typically face problems of hysteresis, where measurements are affected by previously applied loads. Force sensors acquire the deformation volume of a strain body using sensing elements (e.g., strain gauges [3–5], capacitance sensors [6,7], resistive sensors [8] or optical sensors [9–11]) and estimate the force applied to the deformation volume. Most of these are designed such that the deformation volume of the strain body and the force have a linear relationship, and this relationship is often estimated through linear regression. However, nonlinearity results from the physical properties of the strain body and the attachment of sensing elements. Several studies have been conducted to determine the relationship between the values obtained by the sensors on a strain body and the applied force, to discuss this nonlinearity. Yingkun et al. proposed a hybrid calibration approach combining linear regression and a support vector machine (SVM) (a form of machine learning), which improved the performance compared with calibration solely using linear regression [12]. Li et al. utilized an SVM application called support vector regression (SVR) to perform regression for calibration [13]. In addition, some previous studies have performed calibration solely using a neural network (NN) without linear regression [14–16]. Oh et al. increased the number of NN layers, with the intent of further improving the performance [17]. In general, as the number of NN layers is increased a larger amount of data is required for learning. The calibration of force sensors often involves the use of calibration devices [18]. Thus, it is not possible to prepare a large amount of data as these are operated manually.

As stated above, many studies have considered the calibration of force sensors, and, although techniques using machine learning have been proposed, these do not solve the problem of hysteresis. An example of a study addressing hysteresis is that of Tomo et al. [19]. The authors developed a three-axis skin sensor using Hall elements. However, the evaluation of its performance revealed no compensation of hysteresis. Urban et al. compensated for hysteresis using a two-axis skin sensor [20] and Horii et al. utilized a three-axis tactile sensor [21]. These were calibrated using machine learning but have not been expanded to six axes. Therefore, the purpose of this study is to perform calibration of a six-axis force sensor using machine learning and compensate for the hysteresis phenomenon. Creep also occurs when an object deforms through the continuous application of a constant load and is particularly pronounced for materials such as resins. The main difference between creep and hysteresis is that the former is dependent on both time and the previous state, while the latter depends only on the previous state. As a method to compensate for the creep phenomenon was proposed in Reference [22], and as hysteresis is a more critical issue, this study focuses on hysteresis compensation. Accounting for the fact that hysteresis is only dependent on the previous history, the technique proposed in this study compensates for errors by keeping past data on hand. Although conventional machine learning methods also show a similar effect, the method in this paper has an advantage that the hysteresis is reduced with a simple learning method with small training data.

This paper is organized as follows. In Section 2, we describe force sensors and the basic principles behind machine learning. Next, we explain the proposed technique in Section 3, and describe our experiment in Section 4. Finally, we present our conclusions in Section 5.

2. Basic Principles

2.1. Signals of a Strain Gauge-Type Force Sensor

Six-axis force sensors, measuring the three-axis directional force and three-axis moment, are the most commonly used force sensors for robots. This study utilizes the strain gauge type, while the proposed method could be utilized for other types as long as the flexure element is subject to the hysteresis or creep phenomena. Strain measurements need to be acquired at a minimum of six locations. Assuming that the strain S is measured in $i \geq 6$ locations in a strain body and the load applied to the strain body is L, S and L are expressed as follows:

$$\boldsymbol{L} = [F_x, F_y, F_z, M_x, M_y, M_z]^\top \quad (1)$$

$$\boldsymbol{S} = [S_1, S_2, \cdots, S_i]^\top \quad (2)$$

where F[N] and M[Nm] represent the force and moment produced at the sensor center O, respectively, and the subscripts x, y, and z represent the coordinates in the Cartesian coordinate system. In other words, practically speaking the acquisition of the strain S using a strain gauge and the subsequent conversion of this value to L via the function $f(\cdot)$ play the role of a force sensor. Here, $f(\cdot)$ is a function expressing the relationship between the strain S and applied load L and this function should be estimated from the results of preliminary experiments.

2.2. Linear Regression

Supposing that S and L are linearly related, the relation can be expressed as follows according to an $i \times 6$ calibration matrix C:

$$\boldsymbol{S} = \boldsymbol{C}\boldsymbol{L} \quad (3)$$

Furthermore, when the inverse or pseudo-inverse matrix of the calibration matrix C is utilized, we have

$$\boldsymbol{L} = \boldsymbol{C}^{-1}\boldsymbol{S} \quad (i = 6) \quad (4)$$

$$\boldsymbol{L} = (\boldsymbol{C}^T\boldsymbol{C})^{-1}\boldsymbol{C}^T\boldsymbol{S} \quad (i > 6) \quad (5)$$

As Equation (5) shows, the applied load $\boldsymbol{L}$ can be estimated from the strain output $\boldsymbol{S}$ acquired with the strain gauge.

Next, we describe how to determine the calibration matrix. Here, the set consisting of the applied load $\boldsymbol{L}$ and strain $\boldsymbol{S}$ at a certain time is already known and the number of samples is n. In other words, $\boldsymbol{L} = [\boldsymbol{L}^1, \boldsymbol{L}^2, \cdots, \boldsymbol{L}^n]$ and $\boldsymbol{S} = [\boldsymbol{S}^1, \boldsymbol{S}^2, \cdots, \boldsymbol{S}^n]$ have already been provided. Here, $\boldsymbol{L}^k = [F_x^k, F_y^k, F_z^k, M_x^k, M_y^k, M_z^k]^\top$ and $\boldsymbol{S}^k = [S_1^k, S_2^k, \cdots, S_i^k]^\top$. Hereinafter, for the sake of simplicity we shall only consider F_x and discuss the case with $i = 6$. Assuming the inverse matrix $\boldsymbol{C}^{-1}$ of the calibration matrix $\boldsymbol{C}$ to be

$$\boldsymbol{C}^{-1} = \begin{bmatrix} a_{11} & a_{12} & \dots & a_{16} \\ a_{21} & a_{22} & \dots & a_{26} \\ \vdots & \vdots & \ddots & \vdots \\ a_{61} & a_{62} & \dots & a_{66} \end{bmatrix} \tag{6}$$

the predicted value F_x^{k*} of the kth sample F_x^k is expressed as

$$F_x^{k*} = a_{11}^k * S_1^k + a_{12}^k * S_2^k + \cdots + a_{16}^k * S_6^k \tag{7}$$

Then, to ensure that the residual sum of squares of the actual applied load and predicted value

$$\sum_{k=0}^{n} d_k^2 = \sum_{k=0}^{n} \left(F_x^k - F_x^{k*}\right)^2 \tag{8}$$

is minimized, the constants $[a_{11}, a_{12}, \dots, a_{16}]$ are derived using the least-squares method, making it possible to determine the inverse matrix $\boldsymbol{C}^{-1}$ of the calibration matrix $\boldsymbol{C}$. Other rows are also calculated in the same token.

3. Proposed Method

3.1. Machine Learning Incorporating Time Series Information

3.1.1. Acquiring Time Series Information

The hysteresis and creep phenomena occur more prominently in resin force sensors than metal force sensors. Even if the sensor is composed of metal, hysteresis is the largest source of error and its compensation is an important issue for a stronger performance. To this end, the consideration of time series information appears to be effective. Figure 1 illustrates the strain behavior when a 10 kg weight is placed on top of a force sensor and later replaced. The result shows how the errors occur owing to the hysteresis and creep phenomena. In our previous study, we attempted to utilize regression with past data with a constant interval time [23,24] and finally concluded that it is difficult to determine the time interval, because creep and hysteresis exhibit different features. Reference [22] has shown that over 80% of the creep error can be cancelled by a single-order compensation. This study therefore focuses on hysteresis compensation. The dark thin line in Figure 1 shows the compensated result. As confirmed in Figure 1, the hysteresis effect remains after this compensation and is the most critical issue for force/torque measurement. To compensate for hysteresis, we focus on times that the force significantly changes and utilize the time series information from such instances. Figure 2 presents a conceptual diagram. In the figure, F_{th} represents the threshold value of the force and F_{current} and F_{past} represent the current force and first force to exceed the threshold value retrospective of the current value, respectively. As the time series information actually employed consists of the strain $\boldsymbol{S}$, this variable should be specifically explained. We determine the "present" and "time at which the force changes to the threshold value or higher for the first time retrospectively from the present," use these as $t_{current}$ and t_{past} and denote the difference as $t_{interval}$. In other words, the following equation holds:

$$t_{past} = t_{current} - t_{interval} \tag{9}$$

Here, when the strain at t_{past} is S_{past} and that at $t_{current}$ is $S_{current}$, the current force is expressed by

$$F_{current} = f(S_{past}, S_{current}) \tag{10}$$

Here, $f(\cdot)$ is a function expressing the relationship between the current and past strains and the current force. We express the function $f(\cdot)$ through machine learning. In this manner, we can variably handle past data from any time and not just a specific earlier time.

However, two problems emerge with this technique. The first is how to proceed when the force never retrospectively changes to the threshold value or higher. In this situation, we decided to employ the current data as the past data. In other words,

$$F_{current} = f(S_{current}, S_{current}) \tag{11}$$

The hysteresis and creep phenomena occur to a considerable extent when the force changes to a high degree but hardly occur when the change in the force is small. Thus, it is expected that if the change in the force is small a conventional technique using only the current values should be sufficient.

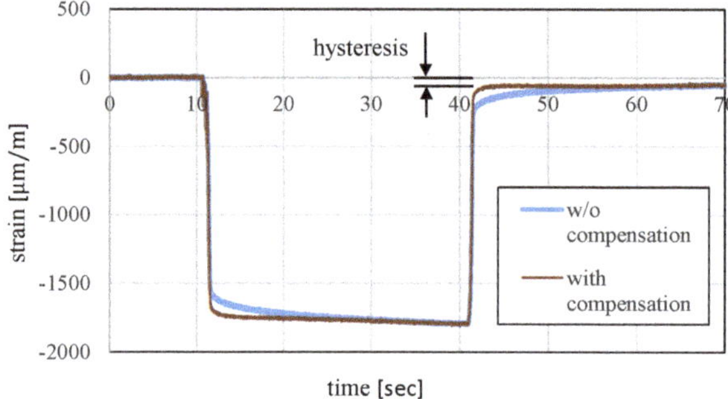

Figure 1. One-axis hysteresis and creep phenomena.

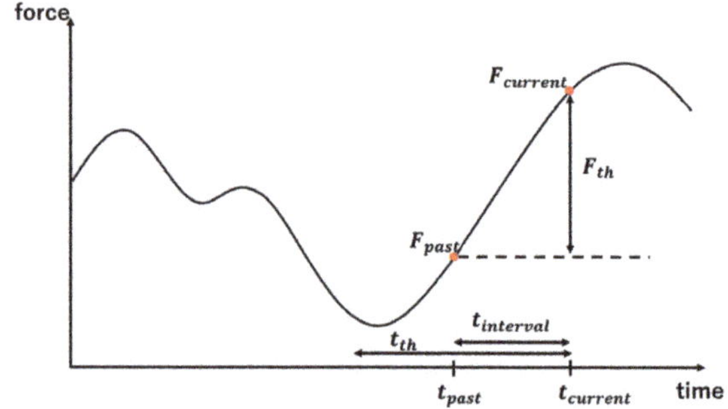

Figure 2. Conceptual diagram for utilizing time series information.

According to this process, if the change in the force is small then only the current values are utilized for force estimation. The second problem is that when retrospectively searching for changes in the force at the threshold value or higher, sometimes the search goes all the way back to the first datum in a dataset. For example, if the changes in the force following a single large change at the threshold value or higher are small. When retrospectively searching for large changes in the force under such circumstances, previous data are utilized as past data to the extent that they barely affect the current value. This is undesirable in practice. Therefore, we determined an upper limit for retrospective searches with this technique. In other words, if searching retrospectively from the present to the threshold time value t_{th} identifies no changes in force exceeding the threshold value, then the above process is performed. To summarize the above, the values used as past data S_{past} for the current data S_{current} are expressed as follows:

$$S_{\text{past}} = \begin{cases} S_{\text{past}} & (t_{\text{interval}} \leq t_{\text{th}}) \\ S_{\text{current}} & (t_{\text{interval}} > t_{\text{th}}) \end{cases} \tag{12}$$

Performing these processes at each time point results in the acquisition of time series information. Such an algorithm is implemented in our method and it is confirmed that the required calculation is sufficiently small for real-time processing in the control systems of robots.

3.1.2. Utilization of Time Series Information for Machine Learning

The machine learning approach considering time series information is illustrated in Figure 3. Linear regression, an NN and a hybrid model described in the next subsection are the candidate machine learning methods. The strain $\boldsymbol{S}$ and applied load $\boldsymbol{L}$ are provided as the input and output, respectively, for training data during training. During testing, when the strain $\boldsymbol{S}$ is entered for the input the machine learning process predicts the applied load $\boldsymbol{L}$. When using time series information, the input strain $\boldsymbol{S}$ is $\boldsymbol{S} = [\boldsymbol{S}_{\text{current}}, \boldsymbol{S}_{\text{past}}]$. Here, $\boldsymbol{S}_{\text{current}}$ and $\boldsymbol{S}_{\text{past}}$ are the current and past strains, respectively.

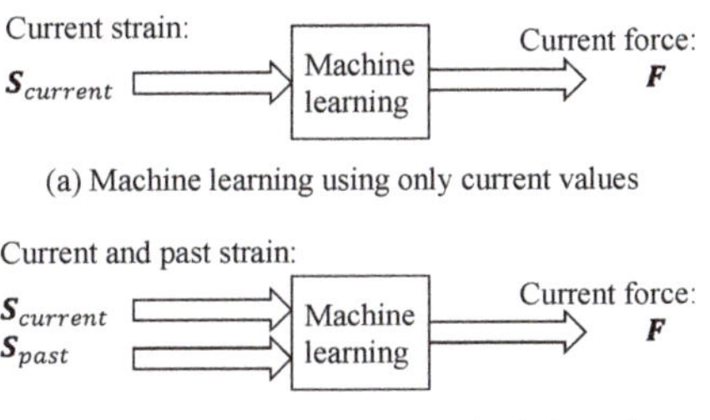

Figure 3. Machine learning inputs and outputs in this study.

3.2. NN and Hybrid Model

This study mainly utilizes linear regression, while two conventional machine learning methods are also introduced as candidates to apply our proposal. The first is an NN [15]. A commonly employed feedforward structure is introduced. In addition, the hybrid model in Reference [12] is extended to an NN. Figure 4 illustrates the hybrid model of linear regression and an NN. During training, the applied load $\boldsymbol{L}$ and strain $\boldsymbol{S}$ at a certain time are provided as training data to determine the actual relationship. By applying linear regression to these training data, it is possible to approximate the rough relationship between $\boldsymbol{L}$ and $\boldsymbol{S}$. The system learns the differences therein and the relationship with the strain $\boldsymbol{S}$ using the NN. In other words, if the linear function is $f_1(\cdot)$, then the output $\boldsymbol{L} - f_1(\boldsymbol{S})$ is learned when inputting the strain $\boldsymbol{S}$ into the NN.

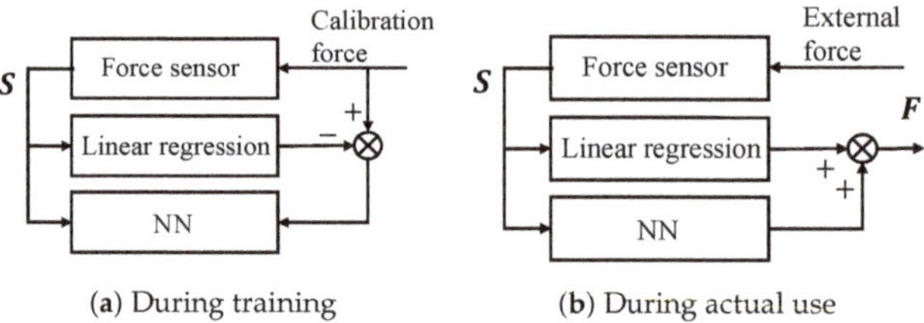

(**a**) During training (**b**) During actual use

Figure 4. Hybrid model configuration.

4. Experiment

4.1. Experimental Equipment

4.1.1. Strain Gauge and Strain Measuring Device

Figure 5 illustrates the utilized force sensor. The sensor is composed of a Keyence Corporation AGILISTA-3100 3D printer and AR-M2 modeling material [22]. AR-M2 is a resin Material, with almost the same physical properties as ABS resin. The rated force of the flexure element is 12 N and the performance was evaluated with external force smaller than the rated force. By using resin material instead of metal, it is possible to produce a lightweight and inexpensive force sensor. Because the main issue of force sensors with resin material is the error caused by hysteresis, this sensor is a good candidate to verify the proposed method.

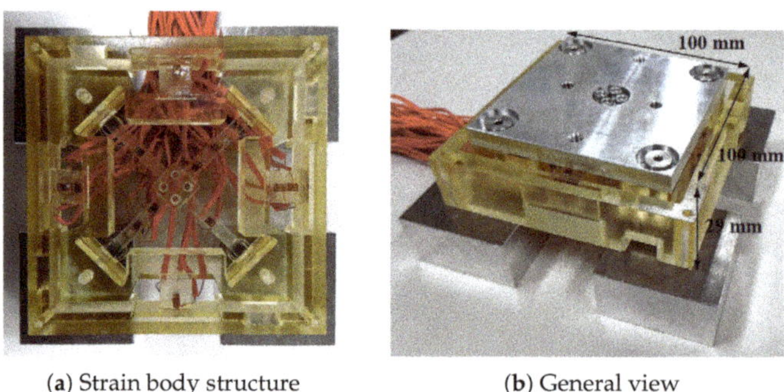

(**a**) Strain body structure (**b**) General view

Figure 5. Resin high-dynamic-range force sensor.

4.1.2. Robot Arm-Based Force Application Device

Figure 6 illustrates the force application device used for data collection. This device comprises a robot arm, force sensor, weight and strain gauge, with a strain measuring device installed within the force sensor. As the robot arm, we utilized a "MOTOMAN-MH3F" by the Yaskawa Electric Corporation. The weight for calibration was a rectangular aluminum block, with dimensions of 0.13 m $\times$ 0.13 m $\times$ 0.03 m. In addition, the holes for fixing are open at positions in the center and offset from the center by 0.00, 0.02 and 0.015 m. If the fixing is at the center hole, then no moment is applied and it is only possible to measure a three-axis force. If the fixing is at an offset position hole, then it is possible to create a moment, enabling six-axis measurement. The weight for calibration has a mass of 1.128 kg. The size of the load can be adjusted by changing the number of weights. Using these values, we computed the applied load $\boldsymbol{L}$ in an absolute coordinate system from the positioning of the robot. Sampling time of the F/T sensor was 50 ms.

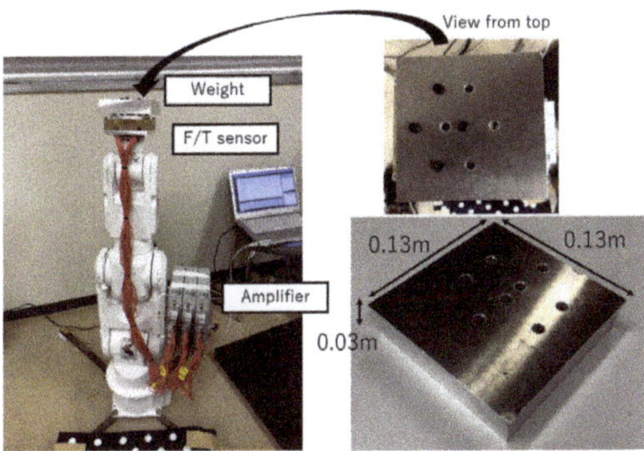

Figure 6. General view of the experimental device.

4.2. Learning

This subsection describes the regression of the applied load acquired in the previous section and its strain values using linear regression, an NN and the hybrid model in detail. We look back at past data from the current data using each technique and then utilize the data from the time at which the force first exceeded the threshold F_{th}. We measured the average hysteresis value after rated output load was applied and the result was 2.64 N. Therefore, the threshold value F_{th} was set at 3 N, around the average hysteresis. Figure 7 shows the root mean squared error (RMSE) with different F_{th}. Note that the results with $F_{th} = 0$ are equivalent to conventional LR methods without time series information. The error was the smallest when F_{th} was around 2 to 3 N and these results support the validity of above-mentioned threshold setting policy. Figure 8, the results of the preliminary test with different t_{th}, show that the RMSE becomes smaller with longer time threshold t_{th}, while longer t_{th} ends up with longer calculation time. We set t_{th} as 15 s considering the trade-off.

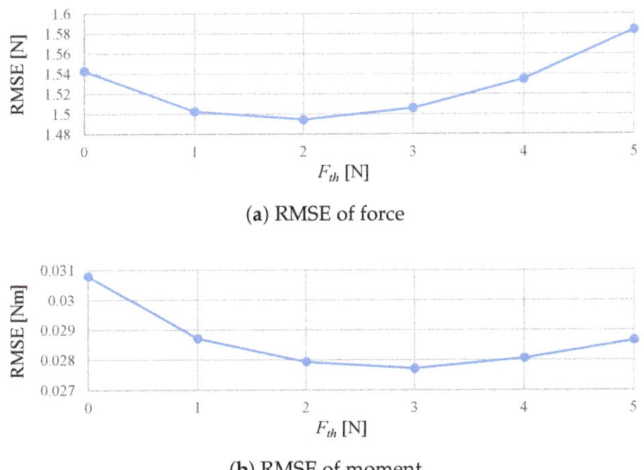

(**a**) RMSE of force

(**b**) RMSE of moment

Figure 7. Pre-examination to determine F_{th}.

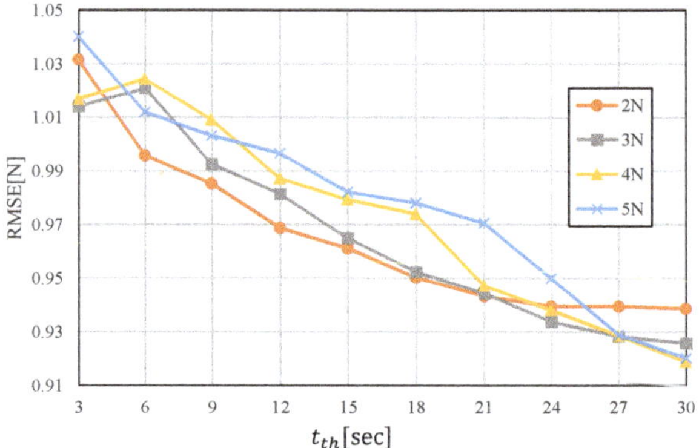

Figure 8. Pre-examination to determine t_{th}.

4.2.1. NN Structure and the Hybrid Model

A feed-forward structure with three hidden layers was introduced in this experiment as a typical NN. The input layer has a number of nodes matching the number of dimensions (six for only current values, 12 if including time series information) of the strain $\boldsymbol{S}$ and the output layer has six nodes corresponding to the dimensions of applied load $\boldsymbol{L}$. In addition, the hidden layers each have 100 nodes. The batch size was 100 and the number of epochs was 1000. Furthermore, the loss function was set as the sum of the squared errors of the force/torque responses of the six axes. The general structure of the NN in the hybrid model is the same as that described in the previous subsection.

4.2.2. Training Patterns

The NN performance varies depending on the amount of training data. The performance generally improves as more data are learned. We verified how the performances of linear regression, the NN and the hybrid model varied depending on the amount of training data.

Pattern A (amount of training data: large)
: Six different paths are given in the experiments, so that most of the operating range is covered. The trajectories are all based on sinusoidal waves with random amplitude in each direction. The frequency of the sinusoidal wave was set to 0.05 Hz to avoid error owing to inertia force error. Five of these were used for training and the other was used for testing. Each path was examined with three trajectories with different velocities. Depending on the speed of each trajectory, the recorded time differs as 95 s, 135 s, and 175 s for fast, medium and slow speeds, respectively. We acquired five paths for each of these, recorded them at a rate of 20 samples/s and performed training with a total of 202,500 samples.

Pattern B (amount of training data: small)
: The amount of data used for the training and testing of pattern A was reduced to 1/10th of the original size, simply by decimating the same trajectory. Through this operation, we created a state in which the amount of data was simply reduced and verified the precision of each technique. Depending on the speed of each trajectory, the time recorded differs as 95 s, 135 s and 175 s for fast, medium and slow speeds. We acquired five paths for each of these, recorded them at a rate of 2 samples/s and performed training with a total of 20,250 samples.

4.3. Results

4.3.1. Pattern A (Amount of Training Data: Large)

Figure 9 shows the three-axis and predicted values of the root mean square error (RMSE) of the force among the applied loads of each technique. In the figure, LR is linear regression, hybrid NN is the hybrid model, blue indicates no time series information and red indicates that time series information is included. Figure 10 shows the three-axis and predicted values of the RMSE of the moment among the applied loads of each technique. Figure 11 shows time responses of an example of a 135 s term trajectory. Here, Conv. denotes the conventional LR method while Prop. denotes the proposed method including time series information. Calc. shows the ideal external force value calculated based on kinetics. Errors of the conventional method and the proposed method are calculated as the difference between the measured value and the calculated ideal value. The errors are shown in the right graph, the enlarged drawing of the left. The responses of the proposed method show that hysteresis is the largest source of the error and the error is reduced by incorporating past information. The responses of the proposed and conventional methods from 1026 s to 1033 s match because the force variation was smaller than F_{th} then.

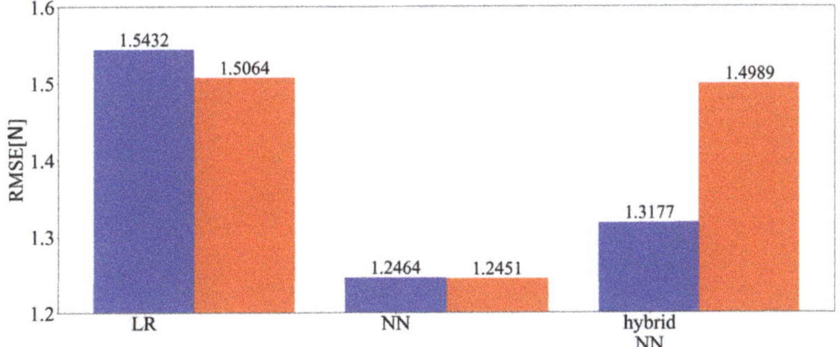

Figure 9. Three-axis root mean square error (RMSE) of the force (learned with 202,500 samples).

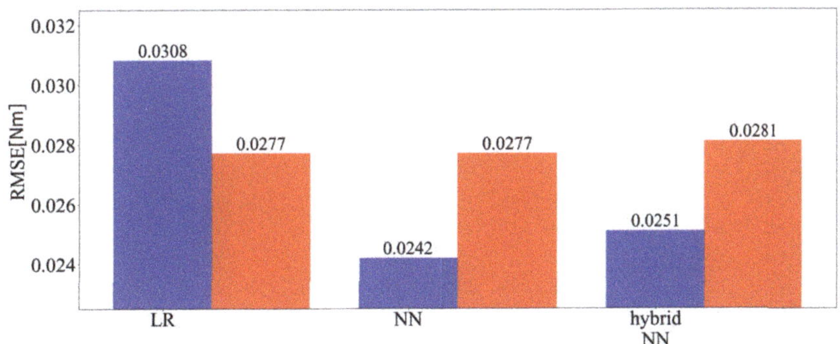

Figure 10. Three-axis RMSE of the moment (learned with 202,500 samples).

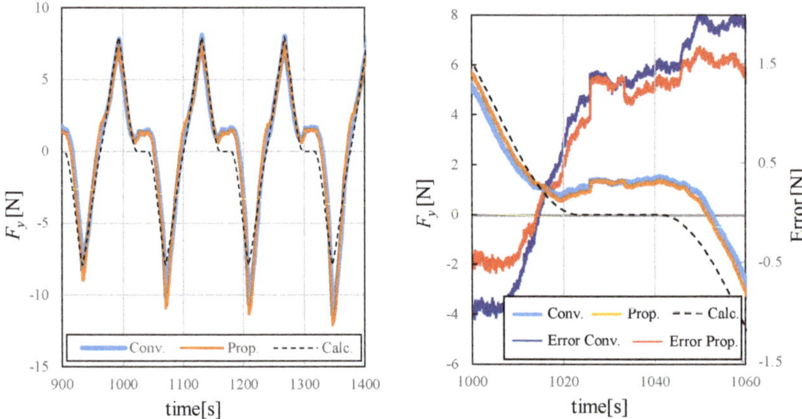

Figure 11. Example of force responses during testing.

4.3.2. Pattern B (Amount of Training Data: Small)

Figure 12 shows the three-axis and predicted values of the RMSE of the force among the applied loads of each technique. Figure 13 presents the three-axis predicted value and root-mean-square error (RMSE) of the moment for the applied loads for each technique.

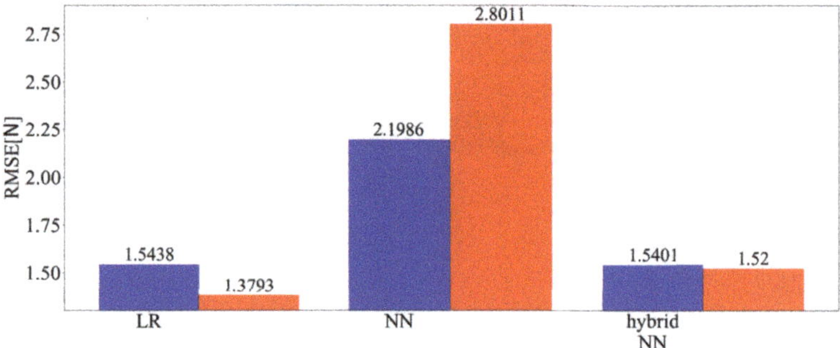

Figure 12. Three-axis RMSE of the force (learned with 20,250 samples).

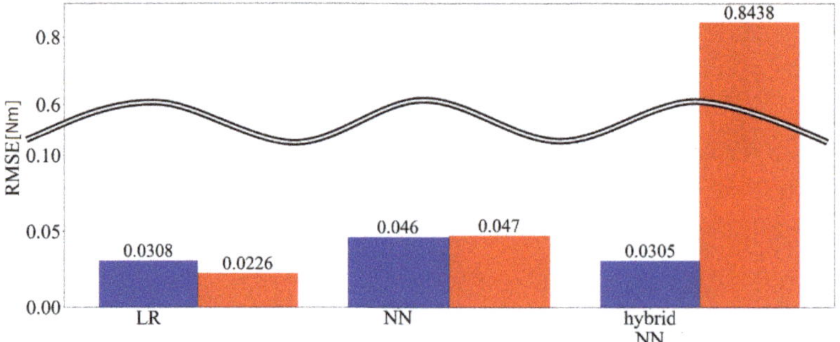

Figure 13. Three-axis RMSE of the moment (learned with 20,250 samples).

4.3.3. Discussion

When the amount of data for training was high, the inclusion of the time series information reduced the errors in the linear regression, slightly reduced the errors in the NN and increased the errors in the hybrid model. Resin force sensors are likely to be affected by the hysteresis. Therefore, the linear regression output is biased when the force either increases or decreases in the data. By including previous information, if the previous values are smaller than the present values then the proposed method corrects the force to be larger and vice versa. For the NN, as learning was performed when the force became both larger and smaller using only the current values, there was no significant change, even when past information was entered. In the hybrid model, although NN learning sections exist these did not fit well with the linear regression and the present strain output is stronger in nonlinearity than linearity. Thus, the errors tended to decrease when using only the NN.

When the amount of data was small, the inclusion of the time series information reduced the errors in the linear regression, increased the errors for the NN and increased the errors in the hybrid model. For linear regression, it is inferred that the reason is similar to that when the amount of data is high. In addition, it is also inferred that the estimation precision was lower because the amount of data was insufficient. The number of dimensions of the input doubled when the time series information was included and therefore the insufficient amount of data had an influence. Similar errors with the linear regression appeared as when the amount of data was high and the errors with the linear regression and the applied load could not be learned well with the NN as the amount of data was small.

Comparing the cases in which the amounts of data were high and low, the NN performance was the best for a high amount of data and linear regression and the hybrid model performed well when the amount of data was low. The reason for this appears to be that the NN yielded a generally inferior performance when the amount of data was low, as a large amount of data is required for training.

It was found through the examination that when adding time series information to the linear regression, the precision of estimating the applied load was improved. In addition, it was found that the linear regression precisely estimated the applied load when the amount of data was small. From this, it can be said that linear regression is effective when the amount of data cannot be fully guaranteed. In addition, it is possible to improve the performance of a force sensor with large hysteresis by estimating using linear regression instead of an NN. On the other hand, the NN performance is good when a large amount data can be prepared and sufficient data is available learning. Depending upon the application, estimation using an NN is effective if one wants to determine the performance. However, linear regression accounting for time series may be considered to be effective if we simply want to reduce the effects of hysteresis in a commercial force sensor, which requires shorter calibration efforts.

5. Conclusions

In this study, we considered time series information for the strain values converted into force information using a six-axis force sensor, to propose a technique for improving the performance. We then performed learning using three approaches: linear regression, an NN, and a hybrid model. The inclusion of time series information in the linear regression method reduced the errors. The addition of time series information in the NN and the hybrid model increased the errors. When we compared the overall performances for a large amount of data, the applied load was estimated with the most precise calibration using the NN. However, when the amount of data was small the applied load was estimated most precisely by linear regression including time series information. These results support the claim for the advantage of the linear regression method including time series information.

Author Contributions: Conceptualization, R.K., S.S. and T.T.; methodology, R.K. and T.T.; software, R.K.; validation, R.K., S.S. and T.T.; formal analysis, R.K.; writing—original draft preparation, R.K.; writing—review and editing, S.S and T.T.; supervision, T.T.; Funding acquisition, T.T.; project administration, T.T.

Funding: This paper is based on results obtained from a project commissioned by the New Energy and Industrial Technology Development Organization (NEDO).

Conflicts of Interest: The authors declare no conflict of interest.

References

1. Murakami, T.; Yu, F.; Ohnishi, K. Torque Sensorless Control in Multidegree-of-Freedom Manipulator. *IEEE Trans. Ind. Electron.* **1993**, *40*, 259–265. [CrossRef]
2. Park, Y.; Paine, N.; Oh, S. Development of force observer in series elastic actuator for dynamic control. *IEEE Trans. Ind. Electron.* **2018**, *65*, 2398–2407. [CrossRef]
3. Liang, Q.; Wu, W.; Coppola, G.; Zhang, D.; Sun, W.; Ge, Y.; Wang, Y. Calibration and decoupling of multi-axis robotic force/moment sensors. *Robot. Comput. Integr. Manuf.* **2018**, *49*, 301–308. [CrossRef]
4. Yoshikawa, T.; Miyazaki, T. A Six-Axis Force Sensor with Three-Dimensional Cross-Shape Structure. In Proceedings of the IEEE 1989 International Conference on Robotics and Automation, Scottsdale, AZ, USA, 14–19 May 1989; pp. 249–255.
5. Nishiwaki, K.; Murakami, Y.; Kagami, S.; Kuniyoshi, Y.; Inaba, M.; Inoue, H. A Six-Axis Force Sensor with Parallel Support Mechanism to Measure the Ground Reaction Force of Humanoid Robot. In Proceedings of the 2002 IEEE International Conference on Robotics and Automation (Cat. No.02CH37292), Washington, DC, USA, 11–15 May 2002; Volume 3, pp. 2277–2282.
6. Beyeler, F.; Muntwyler, S.; Nelson, B.J. A six-axis mems force–torque sensor with micro-newton and nano-newtonmeter resolution. *J. Microelectromech. Syst.* **2009**, *18*, 433–441. [CrossRef]
7. Dobrzynska, J.A.; Gijs, M. Polymer-based flexible capacitive sensor for three-axial force measurements. *J. Micromech. Microeng.* **2012**, *23*, 015009. [CrossRef]
8. Pan, L.; Chortos, A.; Yu, G.; Wang, Y.; Isaacson, S.; Allen, R.; Shi, Y.; Dauskardt, R.; Bao, Z. An ultra-sensitive resistive pressure sensor based on hollow-sphere microstructure induced elasticity in conducting polymer film. *Nat. Commun.* **2014**, *5*, 3002. [CrossRef] [PubMed]
9. Kim, J.-C.; Kim, K.-S.; Kim, S. Note: A compact three-axis optical force/torque sensor using photo-interrupters. *Rev. Sci. Instrum.* **2013**, *84*, 126109. [CrossRef] [PubMed]
10. Su, H.; Fischer, G.S. A 3-Axis Optical Force/Torque Sensor for Prostate Needle Placement in Magnetic Resonance Imaging Environments. In Proceedings of the IEEE International Conference on Technologies for Practical Robot Applications, Woburn, MA, USA, 9–10 November 2009; pp. 5–9.
11. Polygerinos, P.; Puangmali, P.; Schaeffter, T.; Razavi, R.; Seneviratne, L.D.; Althoefer, K. Novel Miniature Mri-Compatible Fiber-Optic Force Sensor for Cardiac Catheterization Procedures. In Proceedings of the IEEE International Conference on Robotics and Automation, Anchorage, AK, USA, 3–7 May 2010; pp. 2598–2603.
12. Ma, Y.; Xie, S.; Zhang, X.; Luo, Y. Hybrid calibration method for six-component force/torque transducers of wind tunnel balance based on support vector machines. *Chin. J. Aeronaut.* **2013**, *26*, 554–562. [CrossRef]
13. Li, Y.-J.; Wang, G.-C.; Yang, X.; Zhang, H.; Han, B.-B.; Zhang, Y.-L. Research on static decoupling algorithm for piezoelectric six axis force/torque sensor based on lssvr fusion algorithm. *Mech. Syst. Signal Process.* **2018**, *110*, 509–520. [CrossRef]
14. Lu, T.-F.; Lin, G.C.; He, J.R. Neural-network-based 3d force/torque sensor calibration for robot applications. *Eng. Appl. Artif. Intell.* **1997**, *10*, 87–97. [CrossRef]
15. Khan, S.A.; Shahani, D.; Agarwala, A. Sensor calibration and compensation using artificial neural network. *ISA Trans.* **2003**, *42*, 337–352. [CrossRef]
16. De Maria, G.; Natale, C.; Pirozzi, S. Force/tactile sensor for robotic applications. *Sens. Actuators A Phys.* **2012**, *175*, 60–72. [CrossRef]
17. Oh, H.S.; Kang, G.; Kim, U.; Seo, J.K.; You, W.S.; Choi, H.R. Force/Torque Sensor Calibration Method by Using Deep-Learning. In Proceedings of the IEEE 14th International Conference on Ubiquitous Robots and Ambient Intelligence (URAI), Jeju, Korea, 28 June–1 July 2017; pp. 777–782.
18. Sun, Y.; Liu, Y.; Liu, H. Analysis Calibration System Error of Six-Dimension Force/Torque Sensor for Space Robot. In Proceedings of the IEEE International Conference Mechatronics and Automation (ICMA), Tianjin, China, 3–6 August 2014; pp. 347–352.
19. Tomo, T.P.; Somlor, S.; Schmitz, A.; Jamone, L.; Huang, W.; Kristanto, H.; Sugano, S. Design and characterization of a three-axis hall effect-based soft skin sensor. *Sensors* **2016**, *16*, 491. [CrossRef] [PubMed]
20. Urban, S.; Ludersdorfer, M.; Van Der Smagt, P. Sensor calibration and hysteresis compensation with heteroscedastic gaussian processes. *IEEE Sens. J.* **2015**, *15*, 6498–6506. [CrossRef]

21. Horii, T.; Nagai, Y.; Natale, L.; Giovannini, F.; Metta, G.; Asada, M. Compensation for Tactile Hysteresis Using Gaussian Process with sEnsory Markov Property. In Proceedings of the 2014 IEEE-RAS International Conference on Humanoid Robots, Madrid, Spain, 18–20 November 2014; pp.993–998.
22. Okumura, D.; Sakaino, S.; Tsuji, T. Miniaturization of Multistage High Dynamic Range Six-Axis Force Sensor. In Proceedings of the IEEE 2019 International Conference on Robotics and Automation (ICRA), Montreal, QC, Canada, 20–24 May 2019; pp. 4297–4302.
23. Okumura, D.; Sakaino, S.; Tsuji, T. Development of a Multistage Six-Axis Force Sensor with a High Dynamic Range. In Proceedings of the IEEE 26th International Symposium on Industrial Electronics (ISIE), Edinburgh, UK, 19–21 June 2017; pp. 1386–1391.
24. Koike, R.; Sakaino, S.; Tsuji, T. Hysteresis Compensation in Force/Torque Sensor Based on Machine Learning. In Proceedings of the 44th Annual Conference of the IEEE Industrial Electronics Society (IECON2018), Washington, DC, USA, 21–23 October 2018; pp. 2769–2774.

Article

Magnetic-Based Soft Tactile Sensors with Deformable Continuous Force Transfer Medium for Resolving Contact Locations in Robotic Grasping and Manipulation

Alireza Mohammadi [1,2,*], Yangmengfei Xu [1], Ying Tan [1], Peter Choong [2,3] and Denny Oetomo [1,2]

1. Department of Mechanical Engineering, The University of Melbourne, Parkville, VIC 3040, Australia; yangmengfeix@student.unimelb.edu.au (Y.X.); yingt@unimelb.edu.au (Y.T.); doetomo@unimelb.edu.au (D.O.)
2. Australian Research Council Centre of Excellence for Electromaterials Science, Wollongong, NSW 2500, Australia; pchoong@unimelb.edu.au
3. Department of Surgery, University of Melbourne, St Vincent's Hospital, Fitzroy, VIC 3065, Australia

* Correspondence: alireza.mohammadi@unimelb.edu.au

Received: 2 October 2019; Accepted: 8 November 2019; Published: 12 November 2019

Abstract: The resolution of contact location is important in many applications in robotics and automation. This is generally done by using an array of contact or tactile receptors, which increases cost and complexity as the required resolution or area is increased. Tactile sensors have also been developed using a continuous deformable medium between the contact and the receptors, which allows few receptors to interpolate the information among them, avoiding the weakness highlighted in the former approach. The latter is generally used to measure contact force intensity or magnitude but rarely used to identify the contact locations. This paper presents a systematic design and characterisation procedure for magnetic-based soft tactile sensors (utilizing the latter approach with the deformable contact medium) with the goal of locating the contact force location. This systematic procedure provides conditions under which design parameters can be selected, supported by a selected machine learning algorithm, to achieve the desired performance of the tactile sensor in identifying the contact location. An illustrative example, which combines a particular sensor configuration (magnetic hall effect sensor as the receptor, a selected continuous medium and a selected sensing resolution) and a specific data-driven algorithm, is used to illustrate the proposed design procedure. The results of the illustrative example design demonstrates the efficacy of the proposed design procedure and the proposed sensing strategy in identifying a contact location. The resulting sensor is also tested on a robotic hand (Allegro Hand, SimLab Co) to demonstrate its application in real-world scenarios.

Keywords: contact location in tactile sensor; soft tactile sensors; deformable continuous force transfer medium; array of discrete tactile sensors; soft robotics; robotic hand and gripper

1. Introduction

The resolution of contact locations is an important feature required in many robotics applications involving grasping and manipulation [1,2]. In the context of robotic grasping and manipulation, the knowledge of the location of contact with the object (i.e., where contact force is applied by the object to the robotic hand/gripper) is very important in validating the grip configuration, evaluating the satisfaction of wrench closure condition, allowing the robot to adjust its grip on an object to reject disturbances and to ensure that the object remains within the manipulable part of the hand/fingers during manipulation [2–9]. Due to the complexity of the manipulation tasks, model-based methods utilising force/torque sensors to detect the location of the applied contact force have not been shown to be robust [10,11]. It is, therefore, necessary for tactile sensors to be developed to identify locally the contact locations of the object with the robot/gripper.

There is a wide range of different tactile sensors depending on their modality to obtain contact information [12]. In terms of the medium of detection mechanism for contact localisation, tactile sensing methods may be classified into two main approaches: (1) using an array of discrete individual tactile receptors, where the contact location is identified by which individual receptor is triggered, and (2) using a continuous medium which deforms due to the contact and alters the transmitted signal to the sensing mechanism.

In the array of discrete sensor approach, each individual sensor form an independent receptor and its signal is individually processed. In this type of sensors, each receptor unit only takes charge of contact detection in one specific region and ideally, one can neither influence the results of the other sensor units nor be affected by others. The array of discrete sensor can be realised using one of many sensing modality, for example: the strain gauge sensor skin [13,14], array of capacitive sensors [15], tactile array based on barometric measurements [16] and magnetic-based soft distributed skin [17]. These sensors have been successfully used in many robotic hands for manipulation tasks. The main advantage of this method is its ability to detect multiple contact points simultaneously. The main weakness with the array-based tactile sensors is that the number of individual sensors increases for a higher spatial resolution, which introduces a physical limit to the resolution of the contact location, increased cross-talk between sensing elements when packed closer to each other and relatively complex circuitry [12]. The increase in the number of sensors for higher resolution also means increased cost. In addition, most of the array-based tactile sensors are manufactured in a flexible flat shape which cannot be easily formed into the desired shapes.

In the second approach, the sensor mechanism consists of a continuous force transfer medium, a signal source and the corresponding signal receptors whose responses vary as a function of the deformation of the medium under the contact force. The receptor is usually embedded on the rigid member of the robotic hand or finger, in physical contact with the deformable medium but does not move with the deformable medium. Upon contact with the robotic gripper or finger, the contact force causes deformation on the medium relative to the receptors and causes a change in the signal that is picked up by the receptor. For example, OptoForce [18] is a commercial tactile sensor based on this concept. An elastic semi-sphere dome is used as force transmission medium, with a light source and receptor under the dome. When force is applied to the deformable medium, the surface deforms and causes variation in the reflected light beams which are detected with infrared sensors as receptors.

In contrast to many discrete tactile sensors needed, the sensors with continuous medium require fewer receptors and processing units at the cost of requiring more complex algorithms to achieve the contact force localisation. Hence, the resolution is not restricted by the size or the quantity of the receptor units. Under such a situation, the number of receptor units is a function of the number of independent readings required to obtain a unique solution to solve for the unknowns, which in our case is the location of the contact point expressed in a given coordinate frame. Such a setting also provides higher flexibility in design of the desired shape, curvature and volume of the sensor for different robotic applications, often achievable through the moulding of the deformation medium to the intended shape. As expected, it is challenging for such a sensor to discern more than one contact force (at one contact location) per individual piece of continuous medium. This is regarded as an acceptable limitation in this paper. For example, for a robotic grasping problem, most objects are large enough to result only in one contact point per finger segment of the robotic hand/gripper.

In the proposed approach of using continuous deformable medium, there are two widely used signal modalities: optical and magnetic. Applied force causes the light transmission medium to deform resulting in a variation in the reflected light beams [19] which allows the applied contact force to be quantified. OptoForce, which has been set to measure the magnitude of the contact force (in 3D) but not the contact location, is an optical tactile sensor. TacTip [20] is another example of an optical sensor that operates using an embedded camera to track the deformations of the pins on the inside surface of its deformable dome-shaped membrane. This sensor can provide both the contact force location and the magnitude of the contact force. A variety of TacTip sensors are developed for robotic applications [21], however, these sensors require rigid camera systems that would be difficult to integrate in each phalange of multi-fingered robotic hands, especially in soft robotic hands [22,23]. Vision-based sensing also conventionally requires relatively higher computational costs.

Magnetic-based tactile sensor, which consists of permanent magnets, Hall effect sensors, and a soft medium is another example of a tactile sensor with continuous medium. Deformations on the soft medium due to the applied force will result in displacement of the permanent magnet which can be measured by magnetic field sensors as receptors [24,25]. These sensors are compact, low cost, highly sensitive and easy to integrate in robotic systems as their soft medium can be designed in different shape and size [12,19,26]. Most of current magnetic-based soft tactile sensors are aiming to provide the force measurement at a priori known contact location and the sensor is calibrated for that point of contact. For instance, in [27] a pyramid shape of the force transfer medium is used to ensure that the sensor contacts the objects with its tip. As a result, these sensor designs cannot provide the contact force location on the sensor.

This paper focuses on the systematic design and characterisation procedure when the location of the contact force is needed for a large class of magnetic-based tactile sensors with a continuous force-transfer deformation medium. We consider a general class of configurations of sensors including the arrangement of the multiple permanent magnets (as the source of the signal to be deformed upon contact) and Hall effect sensors (as receptors), general shape, size and stiffness of the soft medium. Given the desired configuration of the sensor, the location of the contact force can be calculated. Such a procedure requires (1) a reliable mapping between the contact force location and the resulting displacement of the deformable medium (within which a set of magnetic source in the form of permanent magnets are embedded) and (2) the mapping between the resultant magnetic field (and its changes due to the contact force) and measurements from Hall effect sensors serving as the receptors. Both mappings are highly nonlinear and sensitive to the configuration of the sensors. While the equation of the resultant magnetic field is readily available, a model-based method is neither practical nor robust in determining the accurate deformation of the continuous medium. The alternative approach is using data-driven methods [27,28] which do not require the model of deformable medium and can provide direct relation between the magnetic field variation from the contact force and Hall effect sensor measurements.

The objective of this paper is to propose a systematic design procedure for the design of the soft tactile sensor in locating the contact force location. This design and characterisation procedure provides conditions under which one can combine different sensor configurations, supported by a selected (basic) machine learning algorithm to achieve the desired performance of the tactile sensor. The sensing performance is given in the accuracy of contact force location detection for a given shape and size of the deformable force transfer medium and a given spatial resolution of the sensing location. Moreover, the conditions provide a systematic approach in finding the minimum number of required permanent magnets and Hall effect sensors and optimum stiffness of the soft medium. An illustrative example, which combines a particular sensor configuration and a specific data-driven algorithm, is used to illustrate the proposed design procedure.

However, the relation between the external contact force and the resulting displacement of the deformable medium (and hence, that of the permanent magnet(s)) is challenging to obtain. Finite element analysis (FEA) has been employed to describe this relation. However, these models are not suited for real-time computation. The alternative approach is using data-driven models. These models do not require the model of deformable medium and can provide a direct relation between the magnetic field variation due to the contact force and Hall effect sensor measurements. Nevertheless, the sensor design in these papers are presented for a particular shape and size of the sensor and particular arrangement of the permanent magnets and Hall effect sensors.

2. The Proposed Sensor Design Procedure

2.1. Sensor Configuration

Figure 1 shows the general configuration of the proposed sensor. It consists of multiple permanent magnets as sources of a magnetic field, multiple Hall effect sensors as the receptors to measure the magnetic field, and a deformable medium for force transfer. The applied force F will result in the deformation of the medium and a change in the pose (position and orientation) of the permanent magnets embedded in it.

We assume that the following desired performance characteristics and physical properties are given in advance as design requirements:

- the shape and size of the sensor unit (the base plate and the deformable medium)
- the contact surface area on the medium (A [mm^2])
- the desired spatial resolution of the identified contact location, defined as the number of grid cells (or the dimension of the grid cell) as shown in Figure 1 (k)
- the minimum detectable applied force on sensor ($F_{min}[N]$)
- the permanent magnet properties including size and magnetic moment ($\mathbf{m}$ [Nm/T])
- and the specifications of the Hall effect sensor, such as the sensitivity ($S_H[T]$), saturation ($B_{sat}[T]$), minimum measurable magnetic field ($B_{min}[T]$).

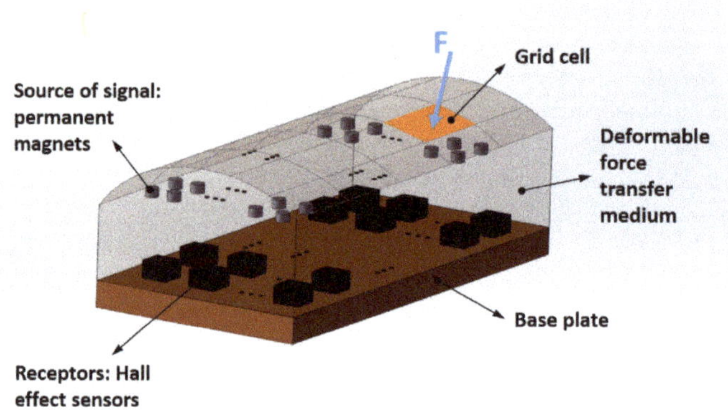

Figure 1. The general case configuration of the proposed tactile sensor with n permanent magnets as the sources of the signal and m Hall effect sensors as the signal receptors, embedded in the deformable force transfer medium. The intended spatial resolutions of the contact point localisation procedure are defined by the size of the grid cell.

Based on the above information, it is desired to find the conditions under which the minimum number of permanent magnets required, the optimum stiffness of the deformable medium and the minimum number of Hall effect sensors required, can be obtained. The knowledge obtained is then used to resolve the contact location of the applied force on the sensor through a data-driven contact localisation algorithms.

It should be noted that in the proposed procedure, it is assumed that only one contact point is applied to one sensor unit. We justify the assumption by the argument that when used as the tactile sensor for a robotic hand to grasp objects, the variations in the surface topology and shape of the common objects are generally large enough to only result in one contact location per sensor unit, even if the contact area could be large to be considered as a point contact.

It is initially defined that n permanent magnets and m Hall effect sensors are employed. The magnetic field density of the permanent magnets can be expressed as [29]:

$$\mathbf{B}_{PM} = \frac{\mu_0}{4\pi}\Big(\frac{3(\mathbf{m}.\hat{\mathbf{r}})\hat{\mathbf{r}} - \mathbf{m}}{|\mathbf{r}|}\Big), \tag{1}$$

where $\mu_0 = 4\pi \times 10^{-7}$ is the permeability of free space, $\mathbf{m}$ is the magnetic moment of the permanent magnet, $\mathbf{r}$ is the distance from center of permanent magnet and $\hat{\mathbf{r}}$ is the unit vector in the direction of $\mathbf{r}$. The superposition of the magnetic field of the permanent magnets measured by a Hall effect sensor (H_j) is:

$$\mathbf{B}_{@H_j} = \sum_{i=1}^{n} \mathbf{B}_{PM}(\mathbf{r}_{ij}), \tag{2}$$

where $j \in [1, m]$, $\mathbf{r}_{ij}$ is the distance of permanent magnet i from the Hall effect sensor j and it is a function of the force-transfer medium shape and size. An applied force to the sensor unit would deform the medium, in which the permanent magnets are embedded, which in turn causes a change in the magnetic field measured at the Hall effect sensors.

Condition 1. *If no force is applied on the sensor, then there exists one $j \in [1, m]$ such that:*

$$B_{min} < |\mathbf{B}_{@H_j}| < B_{sat}.$$

This condition states that the superposition of magnetic field of all permanent magnets is greater than minimum measurable magnetic field of at least one Hall effect sensor and less than the saturation value of that sensor.

Condition 2. *If F_{min} is applied in each of k grid cells of the sensor, then for all cells of the sensor surface, $1, ..., k$, there exists one $j \in [1, m]$ such that:*

$$\Delta\mathbf{B}_{@H_j} > S_H.$$

This condition guarantees that when the minimum force is applied to each grid cell, the variation of the magnetic field is more than the sensor sensitivity. This condition is needed to ensure that at least one Hall effect sensor will be affected by the applied contact force.

Remark 1. *It should be noted that using a large number of permanent magnets does not necessarily result in satisfaction of condition 1 as superposition of the magnetic field of permanent magnets can cancel out the effect of each other.*

Remark 2. *Conditions 1 and 2 are necessary but not sufficient for the desired performance of the sensor in locating the contact location since the sensor performance is also related to the machine learning algorithm used to provide the mapping between the Hall effect sensor readings and the applied force, which are highly non-linear.*

2.2. Contact Localisation Algorithm

After the design of sensor configuration with the desired number of permanent magnets and Hall effect sensors based on required conditions, we need a contact localisation algorithm for the specific sensor that results from the design exercise. If a data-driven algorithm is used, then it is required to find the relationship between the external force contact location and Hall effect sensor measurements. In other words, the input to this data-driven algorithm is the Hall effect sensor measurements and the output is the contact point location (which grid cell) on the surface of the sensor. In order to collect the required data for such an algorithm, we divide the surface of the force transfer medium to a number of grid cells depending on the desired resolution of the sensor and label them from 1 to k.

Then, contact forces with different magnitudes are applied to the sensor and collect the corresponding measured magnetic fields from Hall effect sensors.

After collecting the required data, one of many classification algorithms can be employed for training the data-driven algorithm. Depending on the selected algorithm, the parameters of the algorithm can be tuned to a desired performance. This will be explained with an example in Section 3.2.

3. Illustrative Example and Experimental Validation

In this section, an example configuration of the proposed tactile sensor is designed, fabricated and used to illustrate the evaluation of the Conditions 1 and 2 and implementation of the contact location resolution algorithms.

3.1. Sensor Design and Fabrication

Figure 2 shows the schematic arrangement of three permanent magnets and two Hall effect sensors in the soft tactile sensor. The deformable force-transfer medium was considered as a semi-cylinder constructed over a 40 mm × 20 mm base plate as shown in Figure 2. The permanent magnets were neodymium block magnets with a size of 2 × 2 × 1 mm height and magnetic moment of 1.4 Nm/T. These magnets were placed at the middle axis of the sensor with 10 mm distance from each other and 10mm from the sensor base plate. The Hall effect sensors are MLX90393 tri-axis magnetic sensors from Melexis Inc. (Leper, Belgium) providing magnetic field measurements in three axes thorough I2C fast mode protocol. The Hall effect sensors were placed in 20 mm distance from each other and 10 mm from two ends of the cylinder.

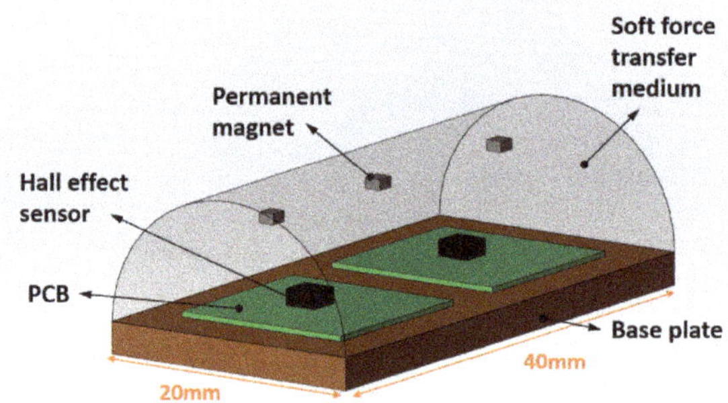

Figure 2. A specific design of the soft tactile sensor with three permanent magnets and two Hall effect sensors embedded inside a semi-cylinder force transfer medium made of Polyurethane (VytaFlex 20).

The sensor is fabricated using a two-stage moulding process as shown in Figure 3. At first stage, the PCB of Hall effect sensors was adhered to the base plate and then assembled with a 3D printed mould which included an extruded negative of the permanent magnets shape to create a placeholder for the magnets. Polyurethane rubber (VytaFlex20 from Smooth-On Inc. (Macungie, PA, USA)) with Shore hardness of 20 A and tensile strength of 200 psi) was then poured into the mould and de-moulded after curing. At the second stage, the magnets are placed in their placeholders and the second 3D printed mould is assembled and the second layer of the polyurethane rubber is injected in the mould.

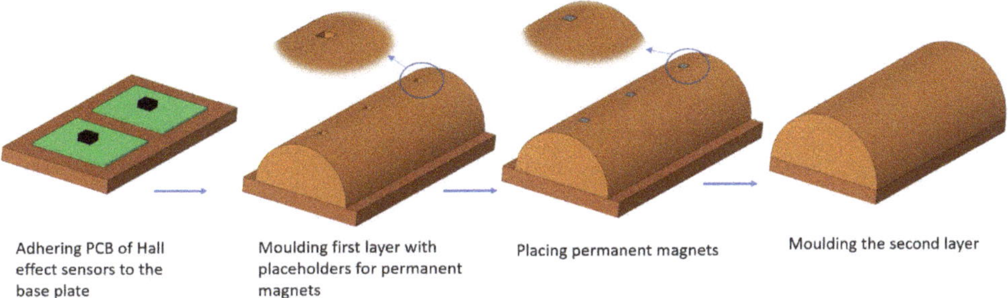

Figure 3. Sensor fabrication procedure.

Given this specific size and shape of the sensor and the stiffness of the deformable medium, condition 1 was evaluated for this particular configuration.

To do so, the magnetic field readings in Hall effect sensors 1 and 2 were obtained when no force was exerted on the senor. The values were found to be

$$\begin{aligned} 10\,\mu\mathrm{T} &< |\mathbf{B}_{@H1}| = 3300\,\mu\mathrm{T} < 50\,\mathrm{mT}, \\ 10\,\mu\mathrm{T} &< |\mathbf{B}_{@H2}| = 3650\,\mu\mathrm{T} < 50\,\mathrm{mT}, \end{aligned}$$

therefore, it satisfies Condition 1. To verify Condition 2, the maximum desired resolution of the sensor surface was assumed to be a grid with 25 cells as shown in Figure 4. A contact force was exerted on each of these cells and the minimum force that results in variation of the magnetic field which can be detected with one of two Hall effect sensors was measured. As a result, a matrix with 25 values representing the grid cell is produced as:

$$F_{min} = \begin{bmatrix} 1 & 0.5 & 0.5 & 0.5 & 0.5 \\ 0.5 & 1 & 0.5 & 0.5 & 0.5 \\ 0.5 & 0.5 & 0.5 & 0.5 & 0.5 \\ 0.5 & 1 & 0.5 & 0.5 & 0.5 \\ 1.5 & 1 & 0.5 & 1 & 1.5 \end{bmatrix}$$

Therefore, the minimum contact force required to satisfy Condition 2 is the maximum value in the F_{min} matrix which is $max(F_{min}) = 1.5N$. If it is desired to discern the contact location of forces with lower magnitude, the following parameters should be altered in the sensor design:

- If $max(F_{min})$ is high, we need to change the stiffness of the soft force transfer medium in order to have a larger displacement of permanent magnets;
- If a few number of cells have high F_{min} value, then we need to add more permanent magnets close to those cells;
- If a block of adjacent cells have high F_{min} value, then we need to add extra Hall effect sensor(s) close to those cells.

Since both conditions are satisfied, it can be concluded that by using an appropriate classification algorithm, the desired performance of the sensor can be achieved. To this end, a sufficient amount of data needs to be collected for the training and validation of the classification algorithm selected.

3.2. KNN Algorithm for Contact Localisation

In this paper, K-nearest neighbours (KNN) algorithm was selected, which is one of the most common classification algorithms. It should be noted that the proposed framework is not restricted to a specific classification algorithm and depending on the amount of collected data, speed of algorithm, etc., an appropriate algorithm can be employed. In the KNN classifier, for each set of training data one label is assigned. Every time a new datum point needs to be classified, the algorithm will select K closest samples in terms of feature similarity in the training data set. Finally, these new data points will be classified through a majority vote among its nearest K neighbours. As this method is based on the raw data, it allows the existence of noise in the training data, which can account for small dynamics complexities in the sensor such as the hysteresis of the deformable medium. The other advantage of the KNN algorithm which makes it suitable for our application is that it is a non-parametric technique and does not make any assumptions on the underlying data distribution such as linear relation. Therefore, KNN is one of the first choices for a classification study when there is little or no prior knowledge about the distribution of the measured data.

3.3. Data Collection Method

To collect the required data, the experiments were conducted using a single-axis force testing machine (Mecmesin MultiTest 2.5i Testing Centre 2500 with 100N load cell) to apply an external force on the surface of the sensor as shown in Figure 4. The sensor was fixed using a vice and a 3D printed probe was used to apply the force over different grid cells of the sensor. Each datum point is an array of the magnetic field values of the two Hall effect sensors, $(B_{x1}, B_{y1}, B_{z1}, B_{x2}, B_{y2}, B_{z2})$. The data were transferred to a PC using an Arduino Uno microcontroller.

In order to evaluate the performance of the contact localisation algorithm in different real-world scenarios, we will collect three sets of data to investigate the effect of the following three cases:

- Sensor resolution: to investigate the accuracy of the KNN contact point localisation algorithm in different resolutions of the sensor, the surface of the sensor is divided to 6, 15 and 25 grid cells. This results in 6, 15 and 25 classes in the KNN algorithm, where each grid cell corresponds to a single class. The force is applied in the normal direction on each grid cell with magnitudes varying from 1.5 N to 10 N and $(B_{x1}, B_{y1}, B_{z1}, B_{x2}, B_{y2}, B_{z2})$ is obtained. This process is repeated five times for each cell. For resolutions 6, 15 and 25, the total number of sample points collected are 2400, 6000 and 10,000, respectively. The total collected data is partitioned randomly in 20–80%, 50–50% and 80–20% ratio for training and testing. In random sampling, every observation in the main data set has an equal probability of being selected for the partition data set. The performance of the algorithm is investigated in these three training-testing ratios. We also investigate the effect of different K values (K = 1, 5, 10) in the accuracy of the contact point localisation.
- Contact forces in different directions: the previous set of data are collected where the contact forces are normal to the surface of the sensor. Although, in real-world scenarios, forces can be applied in any directions. Therefore, in this case, forces are applied to each grid cell in random directions. These forces, however, are confined inside a cone with 45 degrees angle with the cell surface. For the data collection for this case, the sensor resolution 15 and 3D printed probes with a tip diameter of 2 mm and 6 mm are used. The KNN algorithm was set to 50% training data and K = 5.
- Contact surface area: to explore the effect of contact surface area on the performance of the sensor, four probes with different tip sizes (2 mm, 6 mm, 10 mm and 14 mm) as shown in Figure 4 are used to collect the data. This represents the variations in the contact surface area that may result in a practical implementation. For example, when the tactile sensor is mounted on a robotic hand and grasp an object such as a mug, where the contact between the sensor and the mug is not just a point, but a surface area. The data collection procedure was similar to the previous data set collection, in different directions with sensor resolution 15, 50% training data and K = 5.

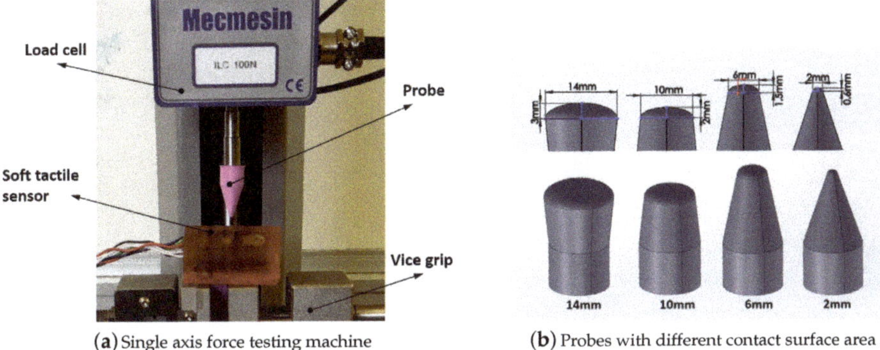

(a) Single axis force testing machine

(b) Probes with different contact surface area

Figure 4. Experimental setup for data collection using a single-axis force testing machine to apply desired force on the sensor with different probes.

4. Results and Discussion

In this section, the performance of the sensor is investigated for using a combination of the proposed design configuration in Section 3.1 and the KNN classification algorithm in Section 3.2. The performance of the sensor is reported in terms of the accuracy of contact localisation. When considering sensors with different resolutions (different grid cell), the average Euclidean error is also considered. Euclidean error (EE) [28] is a metric to weigh the cost associated with each misclassified class by considering its 2D Euclidean distance from its true response class.

4.1. Sensor Resolution

The results of contact localisation accuracy in different resolutions of the sensor are shown in Figure 5 for different training data percentage and different K values of the KNN algorithm. The results show that the training data percentage and K value of the KNN algorithm do not significantly affect the accuracy of contact localisation. Therefore, we can use 20% training data with K = 1, which will result in a significant reduction in the training and response time. The minimum accuracy of all different combinations is more than 92%. As expected, the localisation accuracy of different resolutions decreases with increasing the resolution although this reduction is in an acceptable range (maximum 7% reduction). The localisation accuracy in resolutions 6 and 15 are similar.

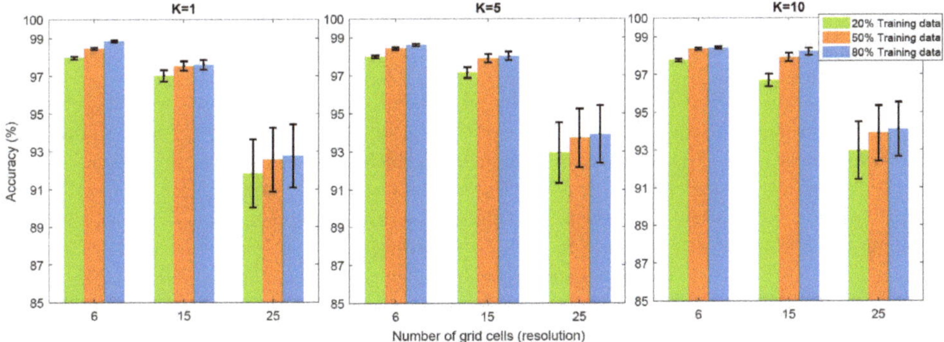

Figure 5. Accuracy of contact force location detection in terms of surface resolution, size of training data and K values of K-nearest neighbours (KNN) algorithm.

In order to determine which cells of the sensor contribute more to the misclassification percentage, localisation accuracy of each cell of the sensor in different resolutions are shown in Figure 6 for the case of K = 5 and 50% training data. The results show that the cells far from permanent magnets and Hall effect sensors (e.g., cell 21 and 25 in the resolution = 25) had lower accuracy and the middle axis of the sensor on which permanent magnets are placed had the highest accuracy. Therefore, with adding extra permanent magnets close to the cells with lower accuracy we can get higher overall detection accuracy.

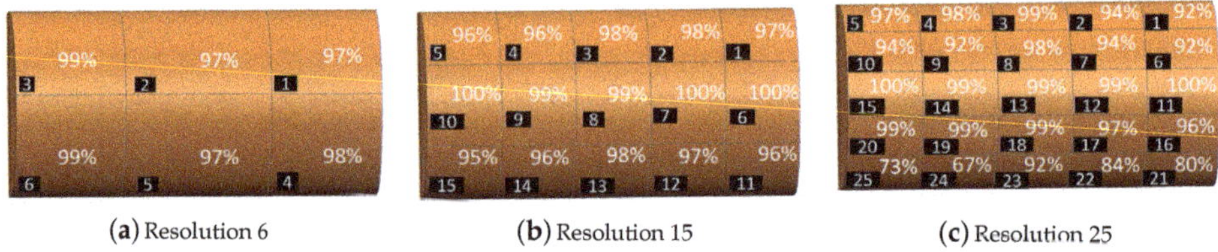

(**a**) Resolution 6 (**b**) Resolution 15 (**c**) Resolution 25

Figure 6. Accuracy of contact force location detection for each cell of sensor grid at different resolutions with K = 5 and 50% training data.

Moreover, KNN classifier confusion matrix is shown in Figure 7 for the 25 grid cells resolution with K = 5 and 50% training data. The diagonal of the matrix shows the percentage of the correct classification and off-diagonal elements show misclassifications (false positive in the upper-right and false negative in the lower left).

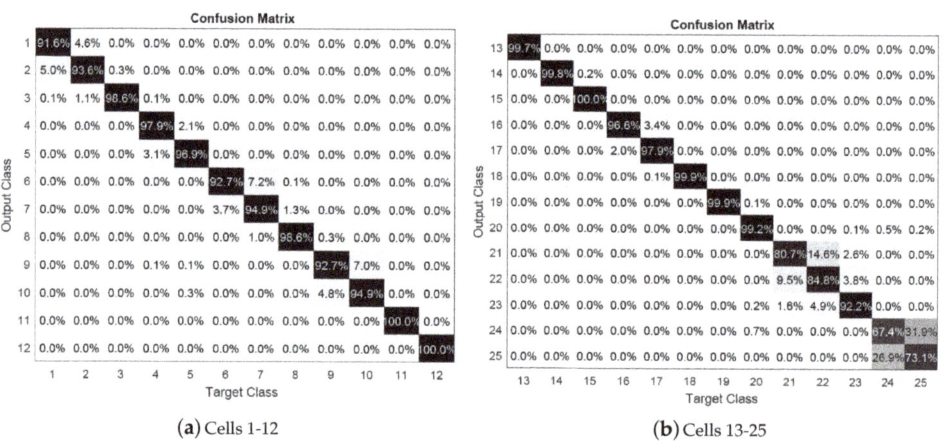

(**a**) Cells 1-12 (**b**) Cells 13-25

Figure 7. Accuracy of contact force location detection for resolution of 25 cells.

In order to find the Euclidean distance between different classes of the sensor, the 3D grid cells of the sensor surface are projected to the 2D base of the sensor and the distance between the centre of each cell from the centre of the other cell is considered as Euclidean distance. Average Euclidean error (AEE) is defined as the accumulation of all the EEs divided by the number of classes (grid cells of the sensor). Figure 8 depicts the AEE results for different training data percentage, different resolution of the sensor and different K values of the KNN algorithm. Since standard deviations are less than 0.01 mm for all results, they have not been shown in the bar plots. AEE results show that increasing the training data percentage will result in a slight decrease in the AEE. Similarly, increasing the K value in KNN algorithm will decrease the AEE marginally. The interesting point in the results of AEE is related to the resolution of the sensor where the 15 grid cell resolution has the minimum AEE in comparison to resolution 6 and 25. This is due to the trade-off between grid cell size and the accuracy of the contact localisation. In other words, the misclassified grid cell in 25 grid cell resolution will contribute smaller penalty values to AEE compared to those in 6 and 15 grid cell accuracy, but it has lower contact localisation accuracy as well. Therefore, this trade-off will result in minimum AEE in resolution 15. This is also in agreement with the results of contact localisation accuracy in the previous section whereby resolution 15 has the optimum performance in having a good trade-off between resolution and contact localisation accuracy.

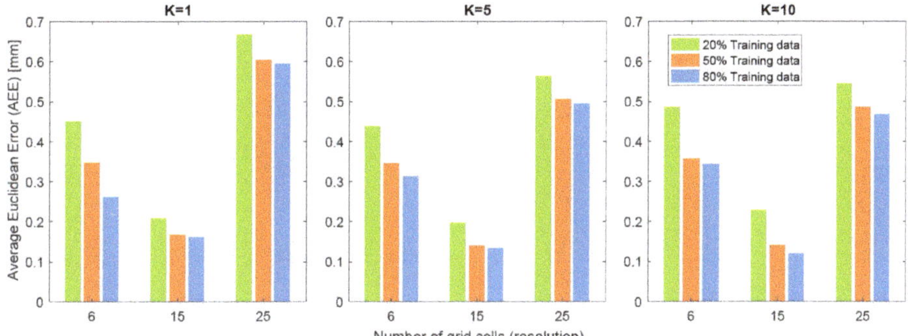

Figure 8. Average Euclidean error (AEE) in terms of surface resolution, size of training data and K values of KNN algorithm.

4.2. Contact Forces in Different Directions

In this case, data sets were collected with forces applied in random contact directions. The data is partitioned for training of the classifier and to validate the resulting classifier. Two probe sizes used were 2 mm and 6 mm. When the data set was collected using 2 mm probe for both training and validation, an accuracy of 92% was achieved. Figure 9 shows the contact localisation accuracy with 2 mm probe for individual grid cells of the sensor. When 6 mm probe was used for both training and validation, an accuracy of 96% was achieved. When both 2 mm and 6 mm probes were used (randomly distributed) in the same data set for both training and validation, an average accuracy of 93% was achieved. Therefore, given a reasonable amount of data, the data-driven classifier approach was found to provide acceptable performance in the resolution of the contact location when dealing with arbitrary direction of the applied force. Using other advanced learning algorithms or regression methods can potentially reduce the amount of data needed to be collected.

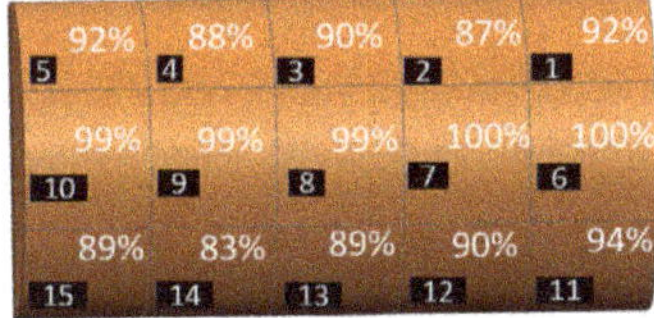

Figure 9. Contact localisation accuracy for individual grid cells of the sensor with contact forces in normal and non-normal directions using 2 mm and 6 mm probes.

4.3. Contact Surface Area

The experiments are conducted with contact forces in the normal direction with four different sizes of the probes as shown in Figure 4. The surface of the larger probes (diameters 10 mm and 14 mm) does not fully touch the surface of the sensor when applied in a non-normal direction. As a result, the effect of contact surface area on the accuracy of the contact localisation is only investigated in normal direction. The results are shown in Table 1. In the cases where the same probe is used for training and testing data, the accuracy of contact localisation is more than 98% (the accuracy percentage of the model on the diameter of the Table 1). This indicates that irrespective of the probe sizes, the accuracy of the contact localisation is highly dependent on the consistency of the data used for training and testing the model.

Additionally, Table 1 shows relatively high accuracy for the cases when different probe sizes were used for training and for testing. The performance consistent excludes that for the 2 mm probe data. In this case, deformations of the soft force transfer medium are small due to the small probe diameter (in comparison to other probes) which results in small variations in the permanent magnet's displacement and hence magnetic field measurements. Therefore, the behaviour of the 2 mm probe cannot be grouped with other probes. In other words, there is a lower limit to the size of the probe used in either training or testing data.

Finally, if we train and test the classifier with all the observations from all probe sizes, the overall accuracy is 97%. This shows that the accuracy of the contact localisation when trained and tested with the data collected from all probes (the integrated dataset) is still similar to those with the individual probe (subset of the integrated dataset).

Table 1. Contact point localisation accuracy using probes with different sizes.

Training Dataset	Testing Dataset			
	2 mm Probe	**6 mm Probe**	**10 mm Probe**	**14 mm Probe**
2 mm probe	98%	63%	64%	55%
6 mm probe	45%	98%	83%	73%
10 mm probe	49%	80%	99%	92%
14 mm probe	44%	71%	95%	99%
All probes	97%			

5. Demonstration on Allegro Robotic Hand

To demonstrate the application of the proposed tactile sensor design in real-world scenarios, we mounted the sensor on the Allegro robotic hand (SimLab Co., Seoul, Korea). One of the fingertips of the hand is replaced by our soft tactile sensor as shown in Figure 10. The sensor has the same configuration of the permanent magnets and Hall effect sensors as described in Section 3.1, the only difference is a slight change in the shape of the soft force transfer medium in order to make it similar to the other fingertips of the hand. This change also facilitates the grasping ability of the finger. The procedure of data collection with forces in normal and non-normal directions and with different probe sizes was performed for this sensor while the resolution is 15.

We used the Allegro hand with integrated soft tactile sensor to grasp three objects (a Rubik's cube, a 3D printed sphere and a wireless mouse) and visually checked the correctness of the contact location detection. Each object has been grasped 50 times and the accuracy of the contact location prediction for the Rubik's cube, 3D printed sphere and wireless mouse were 94%, 96% and 94%, respectively. A video showing real-time contact point detection is available as Supplementary Materials to this paper.

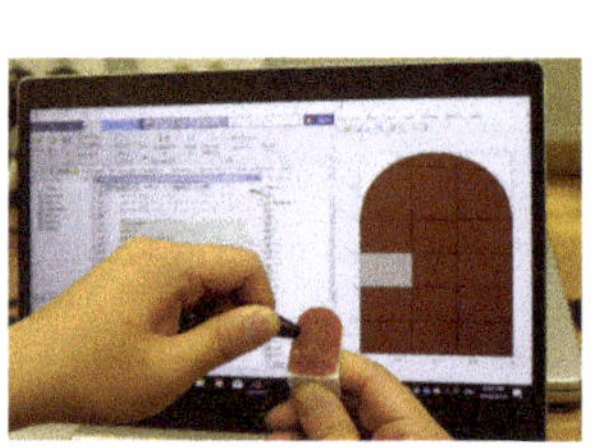

(**a**) Soft tactile sensor

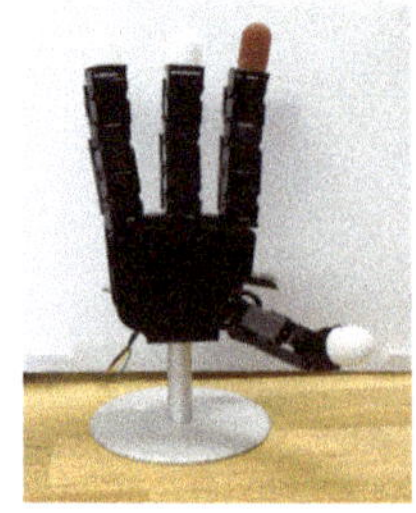

(**b**) Sensor mounted on Allegro hand

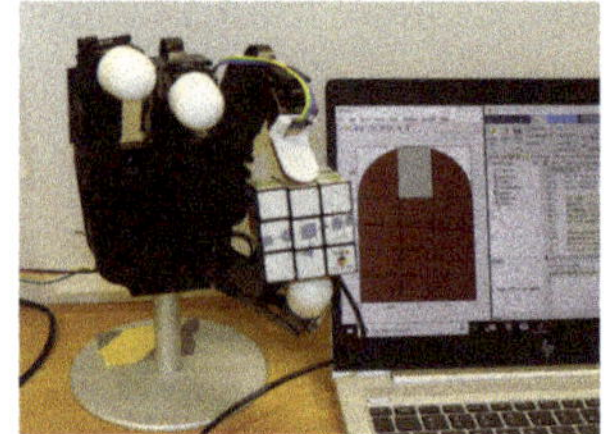

(**c**) Grasping objects

Figure 10. Demonstration of the proposed tactile sensor application in detection of the contact location of an object with a robotic hand.

6. Conclusions

In this paper, a systematic design and characterisation procedure for magnetic-based soft tactile sensors with continuous force transfer medium is presented with the purpose of resolving the location of the contact force. The required conditions for the desired configuration of the sensor are presented. An illustrative example of the sensor configuration with three permanents magnets and three Hall effect sensors and classification method of KNN algorithm demonstrated the minimum accuracy of 92% in contact force localisation, given different resolutions of the sensor, different contact surface areas and with forces applied in random contact directions.

Supplementary Materials: The following are available online at http://www.mdpi.com/1424-8220/19/22/4925/s1.

Author Contributions: Conceptualization, A.M., Y.X. and D.O.; methodology and experiments, Y.X.; writing–original draft preparation, A.M.; writing–review and editing, D.O. and Y.T.; supervision, P.C. and D.O. and Y.T.; funding acquisition, P.C. and D.O.

Funding: This work is supported significantly by the Valma Angliss Trust.

Conflicts of Interest: The authors declare no conflict of interest.

References

1. Kerr, J.; Roth, B. Analysis of multifingered hands. *Int. J. Robot. Res.* **1986**, *4*, 3–17. [CrossRef]
2. Fungtammasan, P.; Watanabe, T. Grasp input optimization taking contact position and object information uncertainties into consideration. *IEEE Trans. Robot.* **2012**, *28*, 1170–1177. [CrossRef]
3. Buss, M.; Hashimoto, H.; Moore, J.B. Dextrous hand grasping force optimization. *IEEE Trans. Robot. Autom.* **1996**, *12*, 406–418. [CrossRef]
4. Nahon, M.; Angeles, J. Real-time force optimization in parallel kinematic chains under inequality constraints. *IEEE Int. Conf. Robot. Autom.* **1991**, *8*, 439–450. [CrossRef]
5. Shaw-Cortez, W.; Oetomo, D.; Manzie, C.; Choong, P. Tactile-based blind grasping: A discrete-time object manipulation controller for robotic hands. *IEEE Robot. Autom. Lett.* **2018**, *3*, 1064–1071. [CrossRef]
6. Shaw-Cortez, W.; Oetomo, D.; Manzie, C.; Choong, P. Robust Object Manipulation for Tactile-based Blind Grasping. *Control Eng. Pract.* **2019**, *92*, 104136. [CrossRef]
7. Shaw-Cortez, W.; Oetomo, D.; Manzie, C.; Choong, P. Control Barrier Functions for Mechanical Systems: Theory and Applications to Robotic Grasping. *IEEE Trans. Cont. Syst. Tech.* **2019**, in press.
8. Eden, J.; Lau, D.; Tan, Y.; Oetomo, D. Unilateral manipulability quality indices: generalised manipulability measures for unilaterally actuated robots. *ASME J. of Mech. Design* **2019**, *14*, 092305-13. [CrossRef]
9. Kawamura, A.; Tahara, K.; Kurazume, R.; Hasegawa, T. Dynamic grasping for an arbitrary polyhedral object by a multi-fingered hand-arm system. In Proceedings of the 2009 IEEE/RSJ International Conference on Intelligent Robots and Systems (IROS), St. Louis, MO, USA, 10–15 October 2009; pp. 2264–2270.
10. Son, J.S.; Cutkosky, M.R.; Howe, R.D. Comparison of contact sensor localization abilities during manipulation. *Robot. Auton. Syst.* **1996**, *17*, 217–233. [CrossRef]
11. Molchanov, A.; Kroemer, O.; Su, Z.; Sukhatme, G.S. Contact localization on grasped objects using tactile sensing. In Proceedings of the 2016 IEEE/RSJ International Conference on Intelligent Robots and Systems (IROS), Daejeon, Korea, 9–14 October 2016; pp. 216–222.
12. Kappassov, Z.; Corrales, J.A.; Perdereau, V. Tactile sensing in dexterous robot hands. *Robot. Auton. Syst.* **2015**, *74*, 195–220. [CrossRef]
13. Katragadda, R.B.; Xu, Y. A novel intelligent textile technology based on silicon flexible skins. *Sens. Actuators A Phys.* **2008**, *143*, 169–174. [CrossRef]
14. Noda, K.; Hashimoto, Y.; Tanaka, Y.; Shimoyama, I. MEMS on robot applications. In Proceedings of the International Solid-State Sensors, Actuators and Microsystems Conference, Denver, CO, USA, 21–25 June 2009; pp. 2176–2181.
15. Kerpa, O.; Weiss, K.; Worn, H. Development of a flexible tactile sensor system for a humanoid robot. In Proceedings of the IEEE/RSJ International Conference on Intelligent Robots and Systems (IROS), Las Vegas, NV, USA, 27–31 October 2003; Volume 1, pp. 1–6.

16. Tenzer, Y.; Jentoft, L.P.; Howe, R.D. The feel of MEMS barometers: Inexpensive and easily customized tactile array sensors. *IEEE Robot. Autom. Mag.* **2014**, *21*, 89–95. [CrossRef]
17. Tomo, T.P.; Regoli, M.; Schmitz, A.; Natale, L.; Kristanto, H.; Somlor, S.; Jamone, L.; Metta, G.; Sugano, S. A new silicone structure for uSkin—A soft, distributed, digital 3-axis skin sensor and its integration on the humanoid robot iCub. *IEEE Robot. Autom. Lett.* **2018**, *3*, 2584–2591. [CrossRef]
18. Koike, M.; Saga, S.; Okatani, T.; Deguchi, K. Sensing method of total-internal-reflection-based tactile sensor. In Proceedings of the 2011 IEEE World Haptics Conference, Istanbul, Turkey, 21–24 June 2011; pp. 615–619.
19. Wang, H.; Totaro, M.; Beccai, L. Toward perceptive soft robots: Progress and challenges. *Adv. Sci.* **2018**, *5*, 1800541. [CrossRef] [PubMed]
20. Winstone, B.; Griffiths, G.; Melhuish, C.; Pipe, T.; Rossiter, J. TACTIP—Tactile fingertip device, challenges in reduction of size to ready for robot hand integration. In Proceedings of the 2012 IEEE International Conference on Robotics and Biomimetics (ROBIO), Guangzhou, China, 11–14 December 2012; pp. 160–166.
21. Ward-Cherrier, B.; Pestell, N.; Cramphorn, L.; Winstone, B.; Giannaccini, M.E.; Rossiter, J.; Lepora, N.F. The tactip family: Soft optical tactile sensors with 3d-printed biomimetic morphologies. *Soft Robot.* **2018**, *5*, 216–227. [CrossRef] [PubMed]
22. Mohammadi, A.; Lavranos, J.; Choong, P.; Oetomo, D. X-Limb: A soft prosthetic hand with user-friendly interface. In *International Conference on NeuroRehabilitation*; Springer: Berlin/Heidelberg, Germany, 2018; pp. 82–86.
23. Godfrey, S.B.; Zhao, K.D.; Theuer, A.; Catalano, M.G.; Bianchi, M.; Breighner, R.; Bhaskaran, D.; Lennon, R.; Grioli, G.; Santello, M.; et al. The SoftHand Pro: Functional evaluation of a novel, flexible, and robust myoelectric prosthesis. *PLoS ONE* **2018**, *13*, e0205653. [CrossRef] [PubMed]
24. Tomo, T.; Somlor, S.; Schmitz, A.; Jamone, L.; Huang, W.; Kristanto, H.; Sugano, S. Design and characterization of a three-axis hall effect-based soft skin sensor. *Sensors* **2016**, *16*, 491. [CrossRef] [PubMed]
25. Wang, H.; De Boer, G.; Kow, J.; Alazmani, A.; Ghajari, M.; Hewson, R.; Culmer, P. Design methodology for magnetic field-based soft tri-axis tactile sensors. *Sensors* **2016**, *16*, 1356. [CrossRef] [PubMed]
26. Funabashi, S.; Yan, G.; Geier, A.; Schmitz, A.; Ogata, T.; Sugano, S. Morphology-Specific Convolutional Neural Networks for Tactile Object Recognition with a Multi-Fingered Hand. In Proceedings of the 2019 International Conference on Robotics and Automation (ICRA), Montreal, QC, Canada, 20–24 May 2019; pp. 57–63.
27. Dwivedi, A.; Ramakrishnan, A.; Reddy, A.; Patel, K.; Ozel, S.; Onal, C.D. Design, modeling, and validation of a soft magnetic 3-D force sensor. *IEEE Sens. J.* **2018**, *18*, 3852–3863. [CrossRef]
28. Russo, S.; Assaf, R.; Carbonaro, N.; Tognetti, A. Touch Position Detection in Electrical Tomography Tactile Sensors Through Quadratic Classifier. *IEEE Sens. J.* **2018**, *19*, 474–483. [CrossRef]
29. Jackson, J.D. *Classical Electrodynamics*; Wiley: New York, NY, USA, 2017.

Article

Towards Tangible Vision for the Visually Impaired through 2D Multiarray Braille Display

Seondae Kim [1], Yeongil Ryu [2], Jinsoo Cho [2,*] and Eun-Seok Ryu [1,*]

1 Department of Computer Education, Sungkyunkwan University, Seoul 03063, Korea; ele7004@skku.edu
2 Department of Computer Engineering, Gachon University, Seongnam-si, Gyeonggi-do 13120, Korea; yiryu@gc.gachon.ac.kr
* Correspondence: jscho@gachon.ac.kr (J.C.); esryu@skku.edu (E.-S.R.); Tel.: +82-31-750-5515 (J.C.); +82-2-760-0677 (E.-S.R.)

Received: 25 September 2019; Accepted: 27 November 2019; Published: 3 December 2019

Abstract: This paper presents two methodologies for delivering multimedia content to visually impaired people with the use of a haptic device and braille display. Based on our previous research, the research using Kinect v2 and haptic device with 2D+ (RGB frame with depth) data has the limitations of slower operational speed while reconstructing object details. Thus, this study focuses on the development of 2D multiarray braille display using an electronic book translator application because of its accuracy and high speed. This approach provides mobility and uses 2D multiarray braille display to represent media content contour more efficiently. In conclusion, this study achieves the representation of considerably massive text content compared to previous 1D braille displays. Besides, it also represents illustrations and figures to braille displays through quantization and binarization.

Keywords: haptic telepresence; braille device; braille application; visually impaired people

1. Introduction

According to recent statistical research, the number of people who are visually impaired is steadily increasing. It categorized 36,000,000 people as legally blind, approximately 217,000,000 people had moderate to severe visual impairment, and 188,000,000 had mild vision impairment. Generally, people who have visual impairments have a decreased ability to see to a degree that causes problems not fixable by usual means, such as glasses. Moreover, as the population and the age increase, the number of people who lose their sight increases proportionally. The United States Census Bureau stated that the population of the United States is increasing every year [1] and as it grows older, the number of diseases affecting vision is also increasing [2]. Table 1 indicates the correlation between age and prevalence of vision disorders [3].

To solve the practical difficulties of visually impaired people, several researchers have studied smart canes, smart cameras with various sensors, braille displays, etc. In general, smart canes consist of electronic navigation aids attached to canes for visually impaired people for detecting obstacles and aiding in practical road navigation. In addition, many studies that focused on canes have applied electro-tactile devices, vibration sensors, ultrasonic sensors, and auditory feedback for assistive devices [4]. The methods using cameras contribute to vision sensing, text reading, guidance, navigation, and obstacle detection. In particular, Microsoft Kinect is significantly useful for obtaining a 3D depth image or video for detecting 3D obstacles in the real world. Kinect has infrared (IR) sensors and an RGB camera. To telepresent 3D shape information, haptic devices are useful for depicting 3D objects. For a considerable amount of time, braille displays or reading materials have been studied for interpreting literal content (e.g., news, books, documents) and convert it to braille.

Table 1. Global numbers affected and the prevalence of vision impairment (approximate values) according to age and sex in 2015 (data are expressed in % (80% uncertainty interval) or number (80% uncertainty interval)).

Sex and Age (Years)	World Population (Millions)	Blind		Moderate and Severe Vision Impairment		Mild Vision Impairment	
Men		**Prevalence (%)**	**Number (Millions)**	**Prevalence (%)**	**Number (Millions)**	**Prevalence (%)**	**Number (Millions)**
Over 70	169	4.55	7.72	20.33	34.53	14.05	23.85
50–69	613	0.93	5.69	6.78	41.57	6.46	39.65
0–49	2920	0.08	2.46	0.74	21.66	0.81	23.61
Women							
Over 70	222	4.97	11.06	21.87	48.71	14.57	32.45
50–69	634	1.03	6.52	7.48	47.46	6.99	44.35
0–49	2780	0.09	2.56	0.82	22.68	0.89	24.64

This research attempts to express a tactual vision through a haptic device to depict 3D objects. Further, this paper shows braille applications for a next-generation braille display that has 2D multiarray braille cells to describe media content (e.g., text content, photographs, paintings) with audio feedback for multimedia information delivery. The contributions of this study toward sharing multimedia content with visually impaired people are as follows:

- Firstly, this study presents a haptic telepresence system that the visually impaired people can use to feel the shape of a remote object in real time. The server captures a 3D object, encodes a depth video using high efficiency video coding (HEVC)(H.265/MPEG-H) and sends it to the client. The 2D+depth video is originally reconstructed in 30 frames per second (fps) but the visually impaired people do not need such a high frame rate. Thus, the system downsamples the frame rate according to human feedback; the hpatic device has the button to let the system know the user finished exploring the frame. It can express a 3D object to visually impaired people. However, the authors realized the limitations of this methodology; consequently, the focus of this study was turned to the research on a next-generation braille display. The limitations of haptic telepresence will be discussed in Section 6.

- This study also presents the 2D multiarray braille display. For sharing multimedia content, it presents a braille electronic book (eBook) reader application that can share a large amount of text, figures, and audio content. In the research steps, there are a few types of tethered 2D multiarray braille display devices; they are developed to simulate the proposed software solution before developing a hardware (HW) 2D braille device. The solution in HW device is that does not need the tethered system.

In this study, only the eBook reader application is considered. The authors hope that this research on the eBook-based braille display will help in designing other applications based on the multiarray braille display. This method also has its strengths and weaknesses.

This paper consists of multiple sections. Section 2 introduces related work with varying methods to help visually impaired people. There are related studies that present other viable support mechanisms for visually impaired people, such as tangible devices, alternative visions, and auditory feedback. Subsequently, both the haptic telepresence system and the eBook reader application based on the braille display interface are specifically discussed in Sections 3 and 4. In Section 5, the implementation and experimental results are presented. This study discusses the limitations and current challenges in Section 6. Finally, Section 7, concludes the contribution of this paper.

2. Related Work

This section introduces various assistive devices for the visually impaired. There are several helpful devices using ultrasonic sensors, vision, and the sense of touch. Figure 1 illustrates the criteria considered by the visually impaired people while purchasing assistive devices. Each device has different objectives practically. For example, most of the visually challenged people can use a smart cane to recognize certain obstacles. It is a useful and an indispensable item when they are walking. Conversely, haptic devices or braille displays are not indispensable products. The necessity is different for each person corresponding to their jobs. Comparatively, the braille display is more useful than a haptic device. It provides media information through the internet and it is also easy to use.

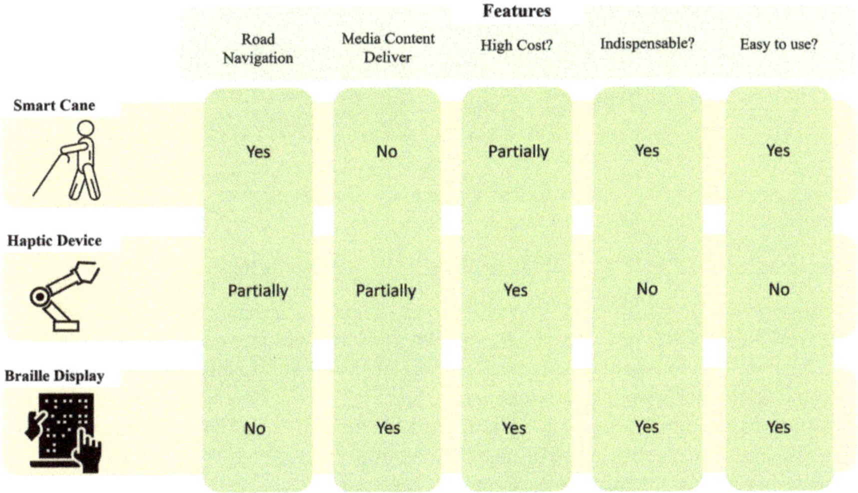

	Features				
	Road Navigation	Media Content Deliver	High Cost?	Indispensable?	Easy to use?
Smart Cane	Yes	No	Partially	Yes	Yes
Haptic Device	Partially	Partially	Yes	No	No
Braille Display	No	Yes	Yes	Yes	Yes

Figure 1. Comparison according to the features of assistive devices for the visually impaired.

2.1. Smart Canes Using Various Sensors

Until now, one of the most popular studies concerning people with visual impairments was regarding the smart cane. Smart canes have been studied in various ways. Laser sensors, ultrasonic sensors, and cameras are usually used to upgrade canes and provide more accurate guide services. One of the earliest studies on smart canes used a laser sensor. Bolgiano et al. [5] are considered the pioneers on research regarding smart canes. They developed a smart cane with relatively high energy conversion efficiency, because the smart canes operated with a lower battery capacity previously. The technology was coupled with the time of use to build a better pulse system that achieved a significantly small duty cycle of approximately one to one million.

The ultrasonic sensor was beneficially utilized in smart canes, which has resulted in several studies using ultrasonic sensors in smart canes. Borenstein et al. [6] developed a smart cane with ultrasonic sensors, as shown in Figure 2a, which could detect at an angle of 180° horizontally in the forward direction; thus, it was possible to distinguish between walls on the left and right of the person wielding the cane from objects 120° vertically in the forward direction. The obstacle recognition method was developed using ultrasonic waves; consequently, rapid processing was possible and feedback on obstacles was provided in real time to match the speed at which the human walked. Yi et al. [7] developed a lighter and more convenient smart cane. Although this cane demonstrates certain blind spots, such as locations just beside a user, it is easy to carry and solves the limitation in determining obstacles above the height of the user. A study conducted by Wahab et al. [8] used ultrasonic sensors, a buzzer, a vibrator, and a fuzzy controller to customize the smart cane. This cane can be easily folded and provides feedback to users when obstacles are present through the vibrator and buzzer. It also has a water sensor to recognize more than 0.5 cm of water, which can warn users.

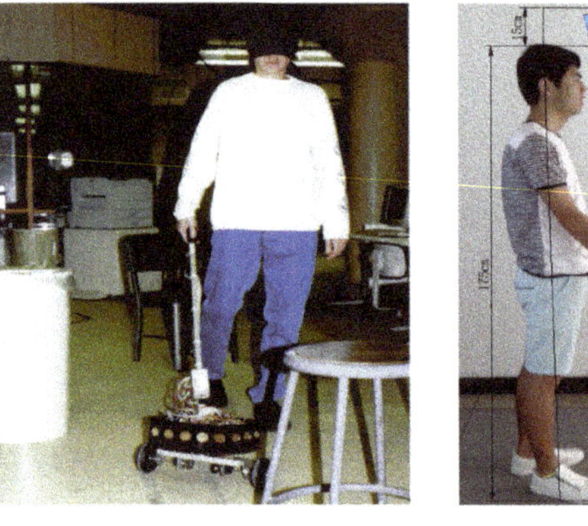

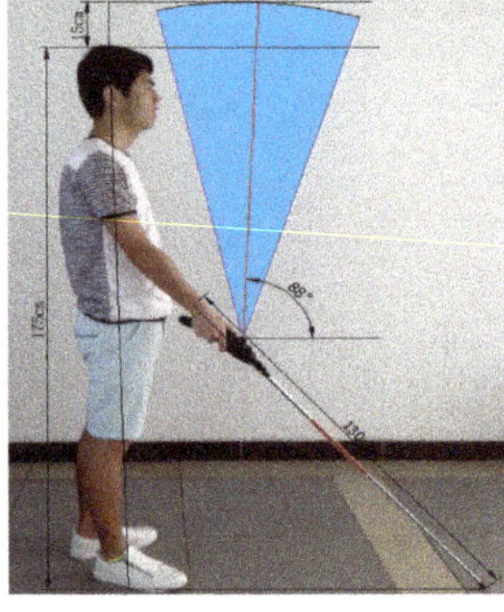

(**a**) (**b**)

Figure 2. (**a**) Blindfolded graduate student. This student walks through an obstacle course using the guide cane (**b**) Detection range of the sensor. It covers a range above and in front of the user [7].

2.2. Communication of 2D or 3D Objects Using Haptic Device or Depth Camera

There are methods that use haptic telepresence devices, which can express tactual perception using 2D+depth data. Park et al. [9–13], who collaborated with our team, developed the robot system to remotely observe the visual attractions and evade obstacles at art galleries or museums. Figure 3 illustrates a telepresence robot and a haptic interaction system. Previously, Park at al. studied remote robots that provide feedback to users and haptic devices that can be used for exploration. They focused on the following: (i) implementation of depth video projection system; (ii) development of a video processing method; and (iii) enhancing 2D+ depth map quality.

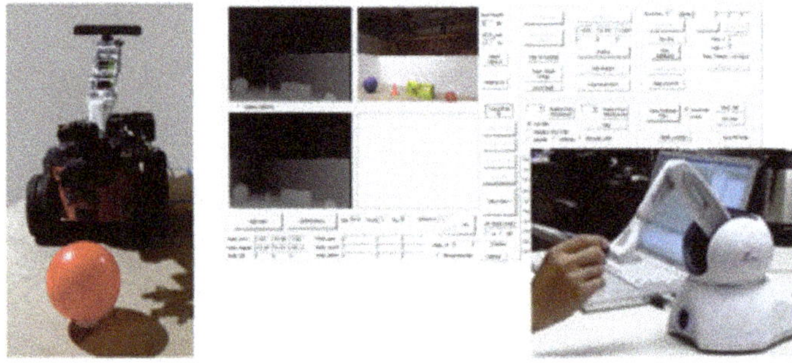

Figure 3. Telepresence robot with multimodal interface for visually impaired people [9].

There are several studies using depth cameras. Hicks et al. [14] studied a link between depth cameras and a head-mounted display (HMD) which enabled people with low vision, i.e., not complete visual loss, to detect obstacles. They studied the images captured using a depth camera, identified the obstacles, and communicated the presence of obstacles using light-emitting diodes (LEDs). A low-resolution display illustrated the distance to nearby objects on a scale of brightness. Hong et al. [15] identified obstacles recognized by the cameras installed in car to enable visually impaired people to drive. Tactile actuators are installed on the handle and gloves to indicate the location of obstacles identified through video recognition. Kinateder et al. [16] avoided unnecessary equipment by replacing only the eyeglasses with augmented reality (AR) glasses for the visually impaired, as shown in Figure 4 for the visually imparied. The HoloLens (Microsoft) is a head-mounted AR system that can display 3D virtual surfaces within the physical environment. The major limitation of the HoloLens system is that it updates distance information at only up to 1 Hz; consequently, it may be difficult to perceive fast-moving objects. The authors expect that the lag and limited range of the mapping, which are clear limitations of the device, will improve with the next generations of HMDs.

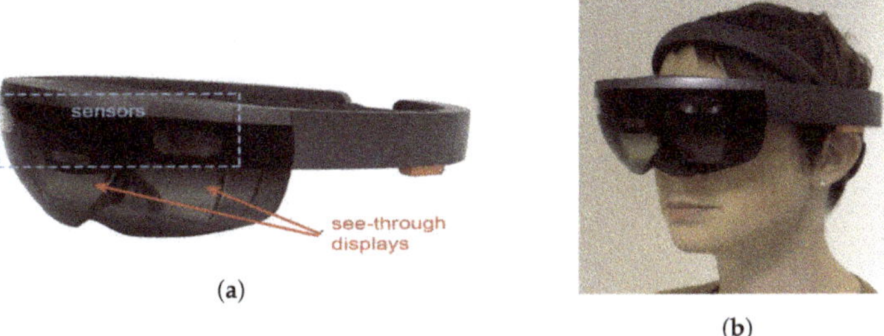

(a)

(b)

Figure 4. HoloLens hardware (**a**) HoloLens has binocular displays (red arrows), sensors (blue dashed box), and an onboard operating system (**b**) Users wear HoloLens by tightening an adjustable band around the head and position the screen in front of their eyes [16].

2.3. *Information Transfer Using Braille Device and Assistive Application*

This topic has been studied more than other areas so far; moreover, it is a significantly effective method for visually impaired people to obtain useful information from books. Oliveira et al. [17] presented a text-entry method based on the braille alphabet. Further, this method avoids multi-touch gestures in favor of a simple single-finger interaction, featuring fewer and larger targets. Velázquez et al. [18] studied a computer-based system that automatically translates any eBook into braille. This application is called TactoBook, which displays the contents of eBooks converted into braille and can be transferred to prototype devices through a universal serial bus port. While there are certain inconveniences while transferring data from eBooks to a braille display through a separate storage device, it has simple interfaces for users and is easy to carry. In addition, Goncu and Marriott [19] produced an eBook image creator model based on iBooks (an iPad application), which utilizes graphic content for people with low vision.

Bornschein et al. [20,21] studied tactile graphics to create figures by using braille to introduce information effectively for visually impaired people. Further, several previous projects address the quantitative aspects, such as the support of sighted graphics producers, to accelerate the direct transformation. By capturing the depth map from the IR sensor of Kinect v2, the silhouette extracted from objects can be utilized to create graphical shapes. Figure 5a illustrates the 2D braille workstation of Bornschein et al. Thus, the visually impaired people could communicate with non-visually impaired people through graphical expressions. Byrd et al. [22] presented a conceptual braille display similar to the display used in this study, which is illustrated in Figure 5b. It has a 4×28 LED braille display, Perkins-style braille keyboard, scroll wheel, and navigation buttons. Moreover, it supports Bluetooth communication with a PC for remote information exchange. However, this study is focused on only braille hardware and does not consider the software, such as the operating system (OS) or applications that are run in the braille display.

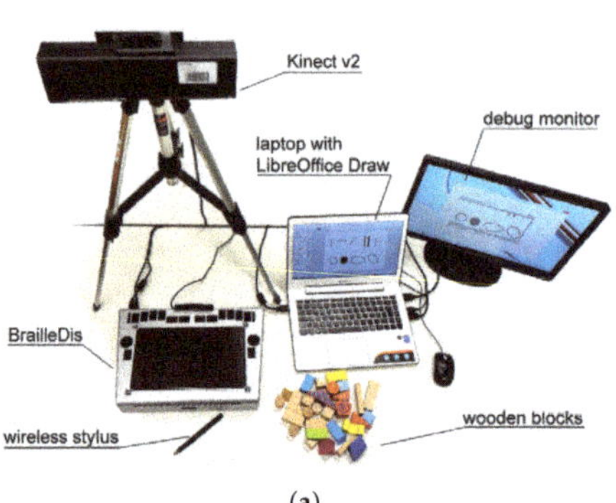

(a)

(b)

Figure 5. Related braille display works. (**a**): The drawing workstation with laptop computer, monitor, BrailleDis 6240 tactile pin-matrix device, ToF 3D camera system with wooden blocks for capturing, and wireless digitizer stylus [20]. (**b**): Final prototype internal components. The LED display is on the top, with the micro-controllers and keyboard below [22].

Additionally, several other researchers conducted studies to help visually impaired people. Bae [23] conducted a study to better serve visually impaired people who want to read utilizing the digital accessible information system (DAISY), which is a eBook standard for handicapped people. Further, Kim et al. [24,25] implemented a DAISY viewer into the smartphone. The DAISY v2.02 and v3.0 users are supported using smartphones so that people with reading disabilities can easily read DAISY eBooks anywhere. Harty et al. [26] studied a DAISY mobile application to allow access to DAISY-standard eBooks using Android OS. It supports DAISY v2.02 partially. The present study examined the DAISY v2.02 reader application based on this research. Moreover, it is difficult to obtain detailed information regarding the Android OS-based players that fully support DAISY v3.0 eBooks. Mahule [27] studied a DAISY v3.0 application based on Android smartphones; however, it has not been completely developed. Thus, this study redesigned his research for implementation. It is described in detail in Section 4. Further, significant studies were conducted by other companies; however, no useful information was released because of their intellectual property rights. The next-generation braille display presented in this paper is also a commercial product. To set better research goals, this paper introduces the current commercially popular braille displays. We clarify that the presented information is based on the published information in their manuals and websites [28–32]. Table 2 lists a comparison of their products with features.

Table 2. Comparison of popular mobile braille displays.

Product	Special Feature	Media Support	Braille Cells	Cost (USD)	OS	Release (Year)
Blitab [28]	Displaying braille image	Image and audio	14×23	Unknown	Android	Unknown
BrailleSense Polaris [29]	Office and school-friendly	Audio-only	32	5795.00	Android	2017
BrailleSense U2 [29]	Office and school-friendly	Audio-only	32	5595.00	Windows CE 6.0	2012
BrailleNote Touch 32 braille Notetaker [30]	Smart touchscreen keyboard	Audio-only	40	5495.00	Android	2016
Brailliant B 80 braille display (new gen.) [31]	Compatibility with other devices	Text-only	80	7985.00	Mac/iOS/ Windows	2011

Finally, this study collaborated with the InE lab at Gachon University to develop the next-generation display. They have developed a braille OS and its additional applications. Their system was utilized in this study for developing a braille eBook reader for multimedia information delivery.

Park et al. [33] developed a method for automatically translating the scanned images from print books into electronic braille books with reduction in translation time and cost required for producing braille books. In addition, when compared to the traditional methods of obtaining information by only using braille text, the aforementioned studies have attempted to accurately transfer 2D graphics, and photographic data. These studies have researched the conversion of 2D graphics to braille data display using multiarray braille simulated in smartphones. Similarly, this study has researched an eBook reader application for a 2D multiarray braille display.

3. 3D Haptic Telepresence System

3.1. Architecture of 3D Haptic Telepresence System

Previous studies [34–37] have continuously attempted to enable visually impaired people to access visual information. As discussed in Section 1, the number of people who have poor vision or have lost their eyesight is increasing worldwide every year. Further, it is difficult for them to obtain visual multimedia from information technology devices, such as PCs, smartphones, and televisions. Park, a co-researcher [9–13], developed methods for touching 3D shapes through depth capturing and its reconstruction. However, these methods are considerably slow for real-time haptic telepresence. Thus, this study focused on real-time 3D haptic telepresence with moderate quality. Section 3.1 introduces a haptic interaction system by capturing a depth map, which upgraded the processing speed. This section introduces the 3D haptic telepresence in real time using Kinect v2 and a haptic device.

The studies conducted by Park [9–13] were focused on capturing 2D+ depth images and rendering using a haptic device. However, this study performs haptic rendering through a 2D+ depth video. The Kinect server can capture 3D spatial information of an object using an IR projector and a camera. The extracted information is processed and converted to a 3D point cloud form, which is then compressed through the latest compression technique, HEVC. The compressed image is delivered to a client through wireless communication, and the client receives the compressed depth map data. The received 2D image is expressed to a graphical user interface (GUI) in real time; further, the 3D point cloud information is expressed using a haptic device through a force rendering. Figure 6 shows architecture of 3D Haptic Telepresence System.

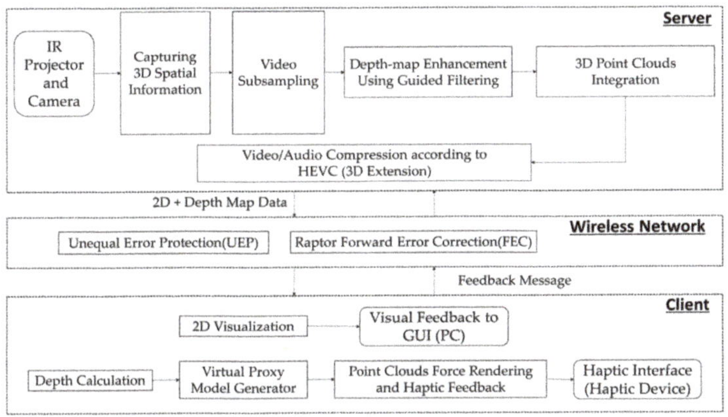

Figure 6. Telepresence system architecture for visually impaired people.

3.2. 3D Spatial Information Capture with Depth Map Using IR Projector and Camera

The 3D spatial information capture module was performed using the Microsoft Kinect software development kit (SDK) [38,39]. The Kinect SDK captures 2D+ depth images. The size of the 2D images are up to 1920 × 1080 (full high definition) pixels. The 3D depth map is up to 512 × 424 (close to standard definition) pixels. However, the haptic device that we studied has a limited resolution (320 × 240). This implies that the information projected to the 3D point cloud need be downsampled to a lower resolution. In addition, the 3D spatial information can be captured by using either stereo/multiview matching from RGB images or an IR projector and camera. Although the stereo/multiview matching can render more details than IR sensors, its computational complexity is also significantly higher than IR sensors. Thus, this capture module of the proposed system is implemented using IR sensors to provide real-time 2D+ depth video streaming service.

The depth map from IR sensors is generally noisy and not sufficiently accurate. It suffers from noise, such as jitter and unequalized edges, because of the distorted light wave fluctuation. Therefore, guided filtering, which is a linear translation-variant filtering process, is applied to the depth map improvement module. The guided filtering process smooths the image while preserving edges. Moreover, it is used to equalize the noises in the depth map [40,41]. Further, the guided filtering is lightweight, removes artifacts more efficiently, and provides better smoothing than bilateral filtering. Figure 7 shows sample results of the depth map enhancement using guided filtering in this work. As shown in Figure 7a, the original IR image was captured using Kinect v2; subsequently, Figure 7c illustrates the depth map image from the Kinect v2 SDK. Figure 7b shows the filtered data with guided filtering. From this guided filter image, point clouds can be constructed using the metadata of the depth map. This 3D metadata involves the coordinates (horizontal, vertical, and depth values) for the depth map projection. To make it easier to understand, in Figure 8, there are two grid maps (grayscale and color) obtained using Kinect v2 SDK. From the grid, objects that are closer (white or red) and farther (black or green) and the approximate shape of each object can be observed. This 3D metadata from the depth map can be used for creating a point cloud for haptic device projection though the OpenHaptics and CHAI3D library [42,43].

(**a**) IR image

(**b**) Depth map image

(**c**) Enhanced depth map image

Figure 7. Depth map enhancement using guided filtering.

(**a**) Mesh (grayscale)

(**b**) Mesh (color)

Figure 8. Mesh map(3D view) from Kinect v2 using Kinect Studio v2.0.

3.3. Real-Time Video Compression and Transmission to Haptic Device

To transmit multimedia combined with 3D spatial information in the form of a point cloud, 2D true-color image, and audio, compression is crucial in networks with limited bandwidth. The captured spatial data are too big to transfer to client in real time. To transmit spatial data efficiently, HEVC is used [44] to enable advanced compression. It reduces more than 50% of the bitrate when compared to existing video coding standards. Park et al. [9] studied haptic rendering with a 320 × 240 point cloud, which was transformed into a 3D grid map, translated into a disparity map for visualization, and calibrated for distance in case of scaling. In contrast, this research provides the depth map from IR sensors at approximately 30 fps. Although this research was directly scaled for transmission, more effective streaming of 3D images is possible using scalable HEVC [45,46].

Additionally, HEVC considers a 3D extension for depth map compression, which also includes multiview video compression. The 3D HEVC extension uses the view concept instead of the frame concept, because 2D frames are captured at a fixed angle, but 3D frames have multiple angles. Thus, 3D compression uses the view concept for interprediction and intraprediction, view synthesis, and dealing with the prediction unit, error correction technique, and motion compensation methodology, by aligning the motion vectors by view. The sequences on 2D and 3D HEVC are quite similar; however, there are additional techniques involved in the 3D HEVC, such as view synthesis prediction to reduce the inter-view redundancy and partition-based depth intra coding (e.g., depth modeling modes). This study processed the depth video/image view according to the feedback from the user by using an original 3D HEVC extension. When users want to view the next depth image (view), as shown in Figure 6, it appropriately encodes the next 2D+ depth view for the client [44,47–49].

This study assumes that a server system can be connected using a wireless network. For reliable delivery, the network module includes rate control, Raptor forward error correction, and unequal error protection to improve the quality of service. In particular, these techniques are very useful to protect data and provide robust data transmission over the wireless network [50,51]. In addition, they enable the optimization of the data packet size. The compressed data involve large buffer streams generated from a depth sensor. After the server completes the transmission of a clean 2D+ depth map data to the client using the transmission control protocol/internet protocol, the haptic device can depict the appearances of objects [43]. The server and client exchange their data, and data can be transferred steadily. Then, the client side proceeds to decompress video data.

3.4. Real-Time Haptic Interaction Using 2D+ Depth Video

To optimize haptic feedback, a haptic interaction module is implemented using the CHAI3D library and OpenHaptics [42,43]. CHAI3D provides several kinds of examples to utilize haptic devices. As shown in Figure 9, this haptic device has a haptic pencil on the arm. It describes the edges of object to users; moreover, it has a button for requesting the next depth frame. Furthermore, the OpenHaptics toolkit is a framework that supports low-level controls in the haptic device. This study has applied haptic projection examples and their 3D projection algorithms, such as manipulation of a haptic device, correcting boundaries, coordinate scaling. Applications also include 3D designed visualization and simulation. When the client receives the 2D+ depth video data, it is processed using a virtual proxy algorithm. The occupied and unoccupied points can be used to calculate and express the volume of captured objects with virtual proxy force. The force rendering that is used for depicting the objects is termed virtual proxy force. It is similar to a spring–damper model, as expressed by (1).

$$\vec{F}_{feedback} = k(\vec{P}_{proxy} - \vec{P}_{probe}) + b(\vec{V}_{proxy} - \vec{V}_{probe}) \tag{1}$$

Given the position vector of the proxy $\vec{P}_{proxy}$, the position vector of the probe $\vec{P}_{probe}$, and velocities of the proxy and the probe $\vec{V}_{proxy}$ and $\vec{V}_{probe}$, respectively, a virtual proxy force feedback $\vec{F}_{feedback}$ is composed of a penetration depth term (with a spring constant k) and damping term (with a damping constant b). $\vec{P}_{proxy}$ and $\vec{P}_{probe}$ are static values that represent the object shape and stiffness on the haptic device. The damping term is a dynamic vale that conveys the movement of the probe. This model is widely adopted in haptic rendering. When translating the depth map, data points consist of a virtual proxy model within neighboring 3D points. These points, called haptic interaction points, are used to calculate 3D surface estimation [9]. Furthermore, the feedback could be shown when the user wants to recognize the subsequent frames; generally, an input is provided using a button on the pencil of the haptic device. This real-time interaction allows the system to have robust data translation, i.e., generating visualization feedback or representing dynamic movements in real time.

Figure 9. PHANTOM Omni haptic devices from SenAble Technologies. The company provided drivers and related software to utilize them.

4. Electronic Book (eBook) Reader Application for 2D Braille Display

4.1. Design of 2D Multiarray Braille Eisplay and Its Architecture of eBook Reader Application

This section presents the design of the next-generation braille display device and its eBook reader application. Moreover, it explains the method used to convert the massive text to a braille display page, and the method used for braille figure conversion. So far, most of the mobile braille displays were composed of single-lined braille cells, which demonstrate the following disadvantages: (i) when the braille text shown in the braille cells is limited to a single line, it is inconvenient to read a book; (ii) images or videos are expressed with some of the text sentences; and (iii) an assistant of a visually impaired person must know the braille letters to sufficiently understand the multimedia content. The first problem occurs because braille cells are expressed in a single line; therefore, it is difficult to understand certain text (e.g., poetry, quotation) in a single reading. In addition, if there is a significant amount of text content, a certain button has to be pressed repeatedly to turn over the pages. For example, as shown in Figure 10, a large amount of text can be expressed simply for normal people whereas the number of braille cells for the same text is physically increased. Consequently, when visually impaired people read this text, there will be a difference in the reading level when compared to non-visually impaired people.

When visually impaired people choose a braille device, they generally consider mobility, Perkins-style keyboards, different numbers of cells (32 is the popular choice), a supportive system for the braille device, etc. However, the panning problem is the primary issue while obtaining media information. When a given text is too long to fit in the display screen, the user has to "pan" either to the left or the right to continue reading along the line. For example, while using a 20-cell braille display, to read a 72-character line, the reader will have to pan at least twice to read the entire text [52]. This study utilizes the braille OS designed by InE lab, which is based on a 2D multiarray braille display, because it partially solves the second problem. For the third problem, this study provides screen mirroring between visually impaired and non-visually impaired people.

In this study, the braille display is designed to solve the aforementioned problems. As shown in Figure 11, 12 × 12 braille cells are located on the braille pad. The volume keys located beside it can be used to increase or decrease the volume of audio or text-to-speech (TTS). Using a long click, the speed of the TTS can be increased or decreased. It has page keys for turning a page to the next or previous page, when reading the texts that have pages similar to an eBook. In addition, six function keys, the backspace key, enter key, space key, directional keys, shift key, control key, device power on/off key, and TTS key exist on the braille pad. These keys perform various roles, such as playing or pausing TTS/audio file and saving/fetching/deleting bookmarks.

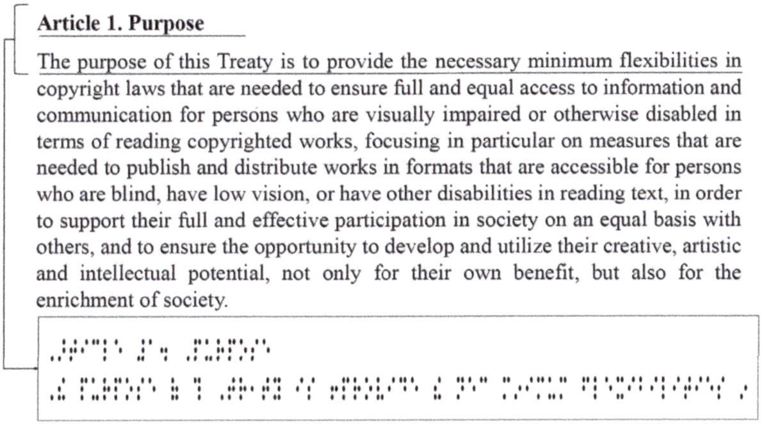

Article 1. Purpose
The purpose of this Treaty is to provide the necessary minimum flexibilities in copyright laws that are needed to ensure full and equal access to information and communication for persons who are visually impaired or otherwise disabled in terms of reading copyrighted works, focusing in particular on measures that are needed to publish and distribute works in formats that are accessible for persons who are blind, have low vision, or have other disabilities in reading text, in order to support their full and effective participation in society on an equal basis with others, and to ensure the opportunity to develop and utilize their creative, artistic and intellectual potential, not only for their own benefit, but also for the enrichment of society.

Figure 10. Quantitative comparison between text and braille content.

Volume Keys
2D Multiarray Braille Display(12 × 12)
Previous Page Key
Next Page Key
Function Keys(F1~F6)
Space Bar Key
Backspace Key
Enter Key
Shift Key
Ctrl Key
Device Power Key(On/Off)
TTS Key (On/Off)
Directional Keys (4way)

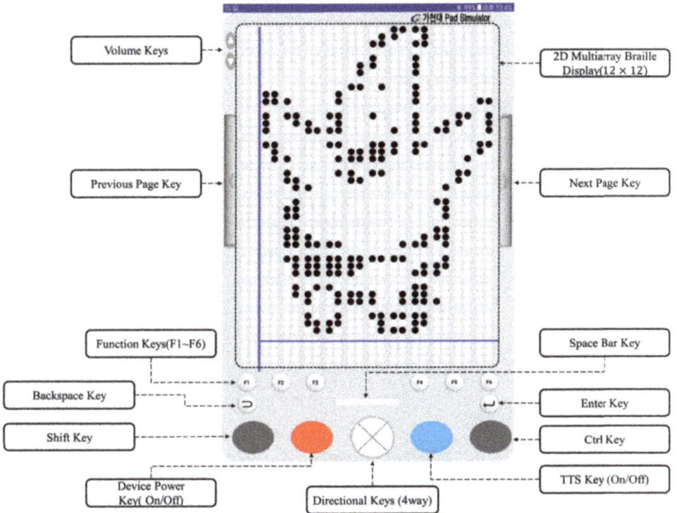

Figure 11. Design of braille pad simulator. There are keyboards and 2D multiarray braille display (12 × 12 braille cells). It is operated on a tablet PC.

4.2. Modules and Application on Braille OS

This braille device was developed by a joint research team. The braille OS, image viewer, and other applications were developed at the InE lab at Gachon University. As mentioned before, they had already developed a technology that can convert visual illustrations to tactual braille images. As illustrated in Figure 12, there are modules of the braille display simulator that can assist visually impaired people. This study conducted the research among the visually impaired people regarding the braille eBook reader application module. As shown in Figure 13, the visual content, such as text and image contents, are expressed through the tablet that has a multiarray braille display.

The computation related to the expression of visual/audio content is primarily performed using a smartphone. The media content is tangibly expressed through the 2D multiarray braille display. Further, the video content is shown using a cover image, such as a thumbnail. Moreover, the mirroring between the braille display device (in this study, it was analyzed using the braille pad simulator application on the tablet) and a regular smartphone application can solve the aforementioned third problem. A smartphone can express its original language.

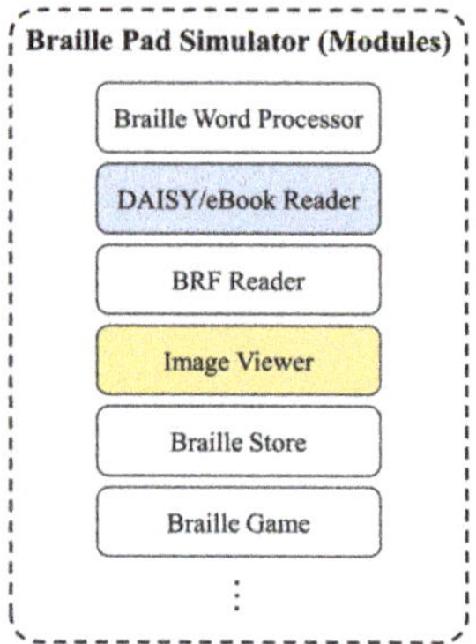

Figure 12. Representative modules on braille simulator. This study treats the digital accessible information system (DAISY)/electronic book (eBook) reader only. In addition, the image viewer is partially used to display braille figures.

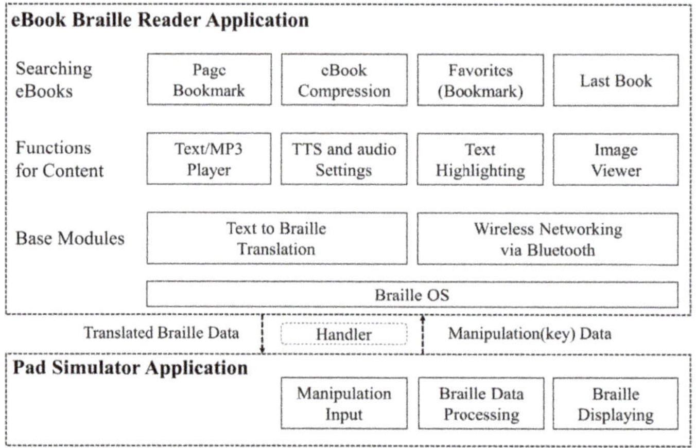

Figure 13. Proposed eBook braille reader application architecture for 2D multiarray braille display.

Figure 13 illustrates the structure of the eBook reader application studied in this research. This application is implemented based on DAISY v2.02 and v3.0 [53], which are specialized standards for disabled people. Moreover, the international electronic publication (EPUB) standards, i.e., EPUB 2.0 and 3.0 [54] are also available. Every book under this standard is implemented to be compatible with the braille OS. The books are compressed in different backgrounds; page management is also available for 2D multiarray braille display. Further, the content display, text highlight, audio player, TTS, the speaking rate, and the pitch of TTS can be controlled by users. In addition, the following functions are also provided in this braille eBook reader application: a function for searching a book stored in the storage of the smartphone, saving/opening a book that was read most recently, bookmark function, favorites function, book or bookmark deletion function, and book compression/decompression function for efficient use of storage space. All contents are translated to braille and sent to the braille tablet. The braille translation is performed using a braille translation module (engine) developed by Dot Inc., which is company that manufactures assistive devices. It can translate both English and Korean for the braille devices.

4.3. Wireless Mirroring between Braille Pad and Smartphone

The braille OS is installed on both the smartphone and pad simulator (tablet). Messages are produced on the eBook reader application on the smartphone; moreover, those messages are transmitted to the braille pad simulator. The messages for sharing media data are transmitted and received through a wireless network. The Android OS message object is used for this function, which is provided by default in the Android OS [55]. This object facilitates sending and receiving messages between devices by using a handler, and the key code used for a message is defined. The key code is transmitted from the braille pad simulator, which is used by visually impaired users. This study defined the data of the key with 16 bits.

Moreover, the braille OS is a background application. When the on/off button is pressed on the pad, the Android activity of the smartphone [56] is automatically displayed. Subsequently, the visually impaired users can remotely control the smartphone with the tablet by using the braille pad. Then, the eBook application is activated, as shown in Figure 13. All the data of the eBook is sent to the smartphone at a rate of a byte per braille cell. This is because a single braille cell has six or eight points (varies according to the format; eight cells were used in this paper), which have binary values, and is expressed in a byte (8 bits). As the display contains 12×12 2D multiarray braille cells, the number of braille cells available in total is 144; therefore, to efficiently express the data on each braille cell, a transmission unit of a byte array is used. In this architecture, the visually impaired people can read the translated multimedia content and send feedback by using the keyboard recursively, as shown in Figure 14.

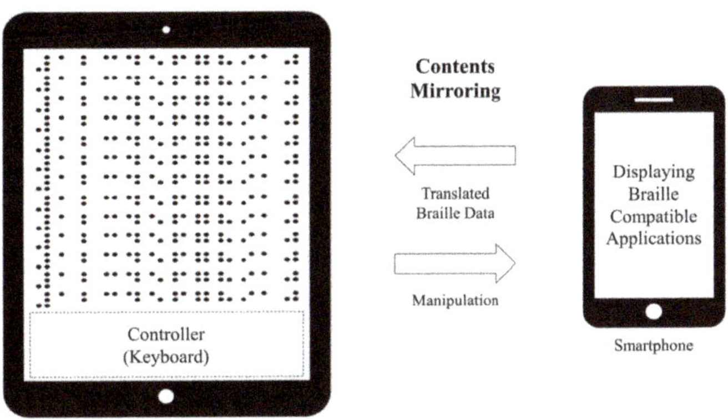

Figure 14. Screen mirroring between the smartphone and the tablet. This mirroring can share every activity on the braille operating system (OS). The smartphone displays original multimedia contents for non-visually impaired people. The tablet display translates multimedia content converted to braille for visually impaired people.

4.4. Extraction and Translation Method for 2D Multiarray Braille Display

For the implementation, this study partially used open source softwares [26,27] to extract the content of the eBook. Because they are not perfectly implemented for parsing media data in the DAISY eBooks, additional implemenation was conducted in this study; (i) a compatibility correction with the braille OS and (ii) the extraction of media content by parsing each standard. In DAISY and EPUB, the text, audio, picture, photograph, and book information can be extracted for each standard based on the markup language using tags [57].

In the relatively simple DAISY v2.02, there exists a navigation control center document and the synchronized multimedia integration language (SMIL) file corresponding to each chapter or section; moreover, the text, image, and audio contents are tagged in the SMIL file [58]. DAISY v3.0 has a less complex structure that can support more image and audio files, which was achieved by enhancing the navigation function one level higher by using the navigation control for the extensible markup language (XML) file, which is referred to as NCX, and strengthening the multimedia functions [59]. Table 3 lists the differences between DAISY v2.02 and v3.0 standards. The structure of DAISY v3.0 is similar to the EPUB 2.0 and 3.0 standards [35]. As mentioned previously, the parsing methodology varies according to the standards; moreover, the references of media file tags are also completely different. Therefore, the eBook reader also varies according to the standard so that we put together all of standards in one eBook reader application. In addition, DAISY v2.02 has no bookmark tool for the disabled. Therefore, this study implements the bookmark feature using .bmk files, which involve metadata, such as page index and chapter number of MP3 files. In addition, for audio or image content, an audio player and image viewer corresponding to the eBooks were implemented to provide visual/audio content, separately.

Table 3. Comparison of digital accessible information system v2.02 and v3.0 standard (media files involve text, image, and audio file).

Standard	DAISY v2.02	DAISY v3.0
Document	based-on HTML	based-on XML
Configuration	Media files, SMIL, and NCC	Media files, SMIL, NCX, XML, and OPF
Reference file	NCC file	OPF file
Metadata Tag	<title>, <meta>	<metadata>
Context Tag	<h>	<level1>

Conversely, DAISY v3.0 does not utilize the NCX for the implementation of the application and extracts content accurately by using the tag-parsing method, which searches at a level directly below the NCX file. The media data parsing, which uses a tagging-based method, is implemented using the follow steps. (i) A validity test of the XML file is required to process the DAISY v3.0 eBook using conditional statements. (ii) The level of the XML document of the book is extracted by searching with depth attributes. (iii) When an author creates a DAISY v3.0 book, the eBook file is composed by classifying parts, chapters, and sections using level tags. Therefore, every tag is collected and then, the chapter list and all the contents of the book are extracted. As this method outputs an XML document without going through the navigator, the computation speed is much faster while the compatibility in this study is maintained. Because the method using DAISY v3.0 open source does not perform the optimization properly, a few seconds are required to perform the braille translation. This study resolved this problem by direct tagging without using a NCX file when the current activity is moved to the next chapter activity on the eBook reader application. Therefore, the outputs of the text content are displayed in less than 1 s, which is relatively much faster.

Using epublib, an open library that supports EPUB 2.0 and 3.0, a book created using the EPUB standard can be extracted on the Android OS. Epublib is designed to facilitate metadata extraction using several tags [60]. Through the epublib library that is converted into the Android standard, parsing is performed on the braille OS by following the usual EPUB parsing method using the NCX [61]. In contrast to the open source of DAISY v3.0, the method using epublib can extract the entire content of the EPUB based on the hyper-text markup language (HTML) document without any compatibility problem. Furthermore, because it has been developed over a long time, this open source has been well optimized and debugged. Epublib extracts the contents of the eBook to a HTML document. EPUB and DAISY are based on the style of tag-based documents. Therefore, for extracting text content from books of these formats, tag tracking and tag-based content extraction is required. By using jsoup open library, the tags of corresponding content are parsed, then all the contents are extracted [62]. The extracted content are then divided into text, audio, and image contents, which are allocated to the corresponding player or viewer.

Additionally, in Algorithm 1, the text content of eBooks of different standards are distinguished sequentially. The text content that are organized by chapter are converted into braille data as follows. (i) The carriage return characters and spaces in the text are deleted, and the words are inputted in the byte array. (ii) The divided words are converted into braille. The translated words are stored with spacing in the braille buffer page, and then data are accumulated per word. When the converted data exceed the size of the braille display page size, the buffer page flushes the accumulated braille data to a braille page. Finally, if the book is completely translated to braille, the finished braille translation data are saved and outputted on the braille display. In this algorithm, row and col refer to the horizontal and vertical cell sizes of the braille display; further, words are distinguished and added to each row of the braille array. Through row and col, the maximum amount of text that can be shown on the braille display is presented. Then, the braille content of the buffer is saved, and all the variables used for allocating a page are initialized to prepare for the subsequent braille page. Even if the braille display size or source language of the original text is changed, the braille array is assigned to fit the cells in accordance with the byte array by assigning the variables through this algorithm. Through this algorithm, the text content of the book are divided into words and converted to *rows* × *cols* braille cells.

Algorithm 1: Text to braille conversion and text/braille page division for text content of an eBook

```
Input: Text content of an eBook
Output: braille eBook and highlighted textbook
Function pageDivision(textContent):
    for 0 to length of textContent do
        Initialize text buffer page and braille buffer page
        if translated text buffer page + one translated-additional word > max cells then
            Translate and save text buffer page to braille buffer page
            Save text buffer page to highlighted textbook
            Arrange braille buffer page to braille eBook
            Initialize braille buffer page and text buffer page
        else
            Add a spacing word and an additional word to text buffer page
    end
    if translated text buffer page + one translated-additional word < max cells then
        Translate and save the rest of text buffer page to braille buffer page
        Save text buffer page to highlighted textbook
        Arrange braille buffer page to braille eBook
        Save braille eBook and highlighted textbook
        Initialize braille buffer page and text buffer page
    return braille eBook and highlighted textbook
End Function
```

4.5. Braille Image Translation Based on 2D Multiarray Braille Display

This section presents a braille figure conversion system, which was developed by InE Lab and Professor Cho at Gachon University, who conducted this joint research and has presented novel methodologies for automatic text and graphics translation to braille. Their methodology was used for figure conversion [33,63,64]. Generally, the experts on braille graphics consider three factors when constructing braille graphics: image complexity classification, tactile translation of low- or high-complexity images, and image segmentation. For a complex classification, images are considered as complex, excluding simple graphic images, such as graphs and tables. When comparing the gray-level histograms of simple and complex images, the differences are apparent. Because visually impaired people usually touch the braille display from the top-left side to the bottom-right side sequentially, it is difficult for them to understand high-complexity images immediately, even though such images are presented significantly well. In addition, visually impaired people cannot perceive other image features, such as color and texture, by touching braille cells. Therefore, the edges of the central object (mostly region of interest (ROI)) were extracted and presented on the braille device, as illustrated in Figure 15.

The central object calculates the color differences in an image and labels them for extraction. It is assumed that the central object is in the center of the image, and its objectives are to: (i) simplify color values by quantization and (ii) obtain color similarities and extract color values by separating each color in the image. Generally, based on photography and triangulation, the largest area within these labeled areas is designated as central label.

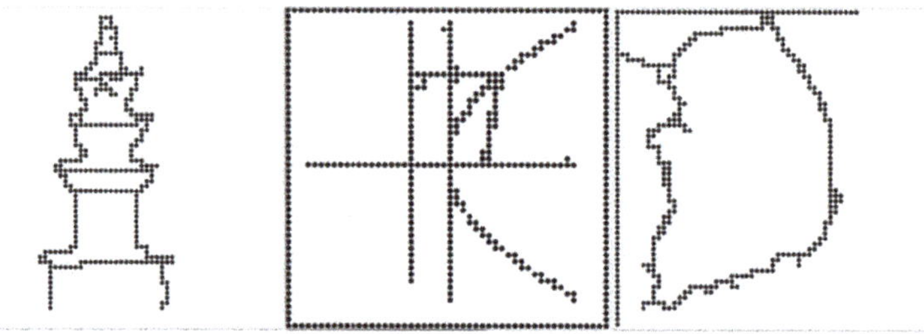

Figure 15. Implementation of braille figure within 50 × 50 multiarray braille display [33]. Converted figure on the left: tower; center: graph; right: map.

The edges of the central object area were modified to binary data and the border of the binary figure was extracted. However, according to the survey listed in Table 4, tactile graphics must be expressed simply [33,63,64]. Otherwise, the information might not be recognized accurately by the users. In the binary image, there are certain irregular sharp lines in the contours. Cell lines on the edges should be simplified to avoid recognition problems. Therefore, to remove these noises, expansion, and segmentation were conducted on the binarized image to create a simplified contour image. Then, corner points were extracted using the Shi–Tomasi algorithm [65]. Consequently, unnecessary corner points were removed to simplify the contour image. Finally, this simplification method increased the recognition rate of the central object by visually impaired people.

Table 4. Survey results on tactile graphic recognition characteristics of the visually impaired (15 participants were completely blind, 10 participants had low vision) [33].

Item	Recognition Characteristics
Image details	Expressing excessive detail can cause confusion in determining the direction and intersection of image outlines. Therefore, the outline in both low-and high-complexity images should be expressed as simply as possible to increase the information recognition capabilities.
High-complexity image with a central object	For an image containing a primary object, the background and surrounding data should be removed and only the outline of the primary object should be provided to increase recognition capabilities.
High-complexity image without a central object	For an image without a primary object, such as a landscape, translating the outline does not usually enable the visually impaired to recognize the essential information.

5. Implementation and Results

5.1. Haptic Telepresence System

In this study, an experiment was performed for the method of delivering objects tactually to visually impaired people by using a haptic device via Kinect v2 SDK (C# Language) and the PHANTOM Omni haptic device manufactured by SensAble Technologies. A laptop was used to implement the client role of the haptic device and a desktop was also used for the server and GUI expression. Visual studio and C++ were used for this implementation. However, in this haptic research, although the research team invited certain visually impaired people for participating in this haptic telepresence test, they were unwilling owing to various reasons, such as their schedule, their lack of interest, or their apprehensions regarding the study. Thus, to investigate the practical applicability of the haptic system introduced in this study, a test was conducted with 10 non-visually impaired participants, who were blindfolded. In general, because of the lack of certain developed abilities (i.e., recognition through touch), it was expected that the blindfolded participants would show lower haptic recognition rate than visually impaired people. Figure 16 shows our experiment.

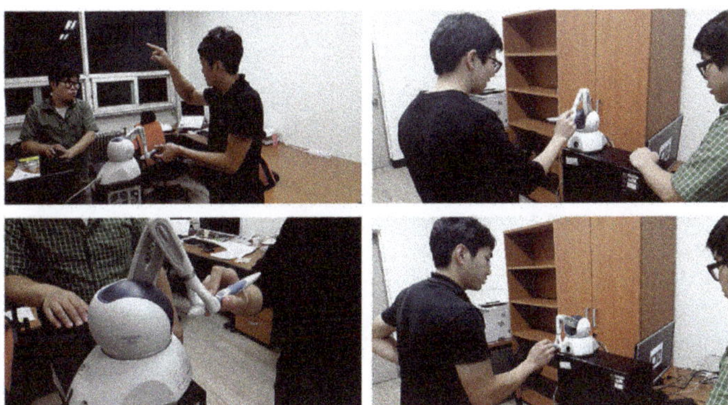

Figure 16. Subjects were trained on the utilization of haptic device. Haptic recognition rate had a very close relationship with the time of use.

Prior to the experiment, the experimental subjects were trained on how to use the haptic device and how an object was projected. The researchers provided random 3D shapes to subjects. Subsequently, the recognition rate, recognition time, 3D view reconstruction time, and the feedback of the subjects regarding this experiment were analyzed. The experiment was conducted as described: (i) a set of solid figures [(a) sphere, (b) cone, (c) cylinder, (d) pyramid, and (e) cube, were provided to 10 testers; (ii) the director provided one of the sets to testers randomly; (iii) the testers chose a randomly presented solid figure through tactual cues from the haptic device. Consequently, the reconstruction time was less than 1 s. It varied according to the complexity or size of the shapes. As shown in Figure 17, the experimental result shows that certain testers recognized the objects quickly, in approximately 20–30 s, while certain subjects required over 90 s. On average, the experiment subjects required approximately 35–70 s to recognize an object. This result is similar to the experimental results presented by Park et al. [10], because visually impaired people and non-visually impaired people have different levels of understanding regarding haptic devices, as was used in this experiment; consequently, a large recognition gap is recognized. If the participants were well-trained, the test result could be enhanced.

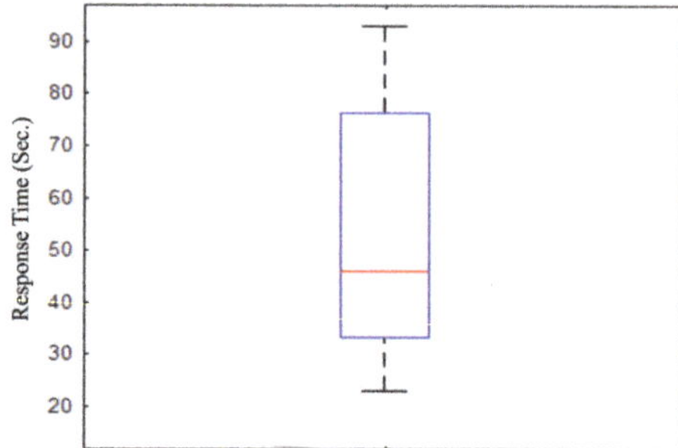

Figure 17. Response time of the proposed real-time 3D haptic telepresence system for 10 blindfolded participants (non-visually impaired people). The length of the box represents the 25th to 75th interquartile range. The interior red line represents the median. Vertical lines issuing from the box extend to the minimum and maximum values of the analysis variable [36].

5.2. Braille eBook Reader Application Based On 2D Multiarray Braille Display

This study also attempted to develop a 2D braille display that was more practical and intuitive compared to haptic devices. This research made an effort to eliminate the existing disadvantages and provide convenient use while maintaining the superior features of the most popular braille device. It focused on the interaction between visually impaired people and their assistants or guardians as well, to enable smooth mirroring. Further, the information shared with the smartphone is used by non-visually impaired people. In addition, it also presents the result of the braille figure conversion system on a braille pad simulator.

In this study, the experiment was performed by using a combination of a tablet and smartphone, which replaced the braille display device, as shown in Figure 18. The tablets used in this study were Samsung Galaxy Tab S and Galaxy Tab S2 8.0; moreover, Samsung Galaxy S7 Edge, S6, and Note 5, and LG G4 smartphones were used. Their OS was Android 6.0 Marshmallow. The development was conducted using the lowest SDK version of 19 for the Android application programming interface. For the Android SDK, Android Studio 2.0 and 3.0 were used, and the experiment was conducted on a laptop computer, which had Intel i3 CPU, 4 GB RAM, and solid-state drive of 500 MB/sec read/write speed. The sample files used for DAISY books were sample books provided by the DAISY Consortium and the EPUB sample files were provided by Pressbooks [66,67]. Figure 18 shows that a tablet and a smartphone were connected by Bluetooth communication, and text and braille were expressed while exchanging data through the braille OS. Figure 18b shows the braille content that a user is reading on the tablet, which was a text of DAISY v2.02 translated to braille. This illustrates an actual execution on the tablet and smartphone; moreover, the tablet screen was modified to show only the content of the multiarray braille display.

Figure 19 shows examples using an eBook of the EPUB standard. Figure 19b shows all the chapters listed in a sample eBook is normally outputted. In addition, Figure 19b also illustrates that the text content of an eBook is normally outputted through the text player. Figure 20 shows the implemented bookmark page. DAISY v2.02 and EPUB 2.0 have no bookmark feature; however, DAISY v3.0 and EPUB 3.0 have this feature. This study provides the bookmark feature for both DAISY v2.02 and EPUB 2.0 standards.

(**a**) English

(**b**) Korean

Figure 18. Text player with digital accessible information system (DAISY) v2.02 (English), DAISY v3.0 (Korean) eBook [37].

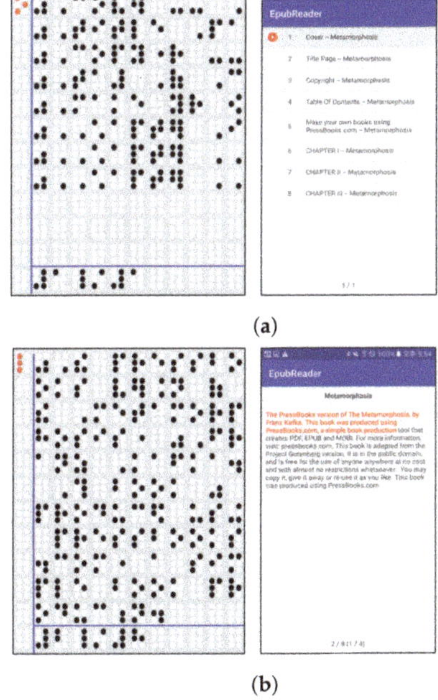

(**a**)

(**b**)

Figure 19. Text player with sample eBook of EPUB standard [37]. (**a**) The chapter list of electronic publication (EPUB) [37]; (**b**) The text player of EPUB.

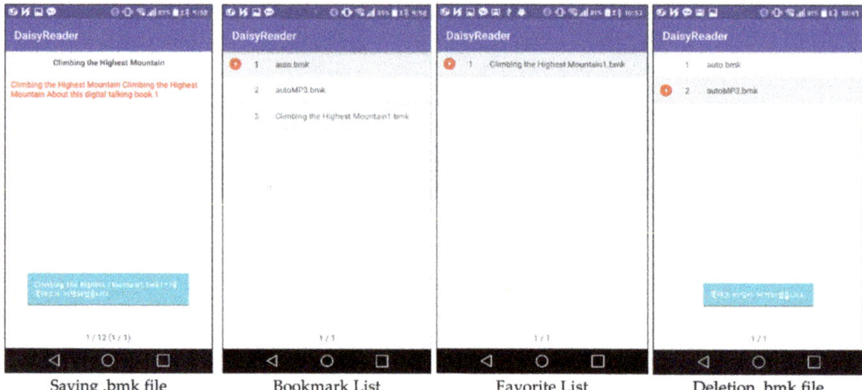

Figure 20. Bookmark and favorites placeholder in the braille eBook reader application.

Figure 21 shows the sequence of processes performed when visually impaired people are reading a book with a smart phone example. In the main page, a search for books can be performed, the last book can be opened, or a favorite bookmark can be selected. When the opened last book or a favorite book is selected, the text player is executed immediately. When "search for books" is selected, the book list is opened, and this displays the list of books in the smartphone storage. Moreover, if the searched book contains music, either the audio mode or the text mode is selected, and the selected mode is executed. In this study, the modes are executed based on chapters (usually audio files are divided according to chapters). According to the selected part of the book, the book content is played. Park and Jung et al. [33,63,64,68–70], who collaborated with our team, conducted research pertaining to the processing of images in books, and developed a method where when a user presses a certain key in the image viewer, the book image list is displayed. In this study, if a tag of an image file is attached to a chapter that a user is currently reading, it can be tagged and extracted; thus, an image file list is created. When the user enters the image viewer by using the list, the image-oriented objects are selected and expressed, as shown in Figure 22.

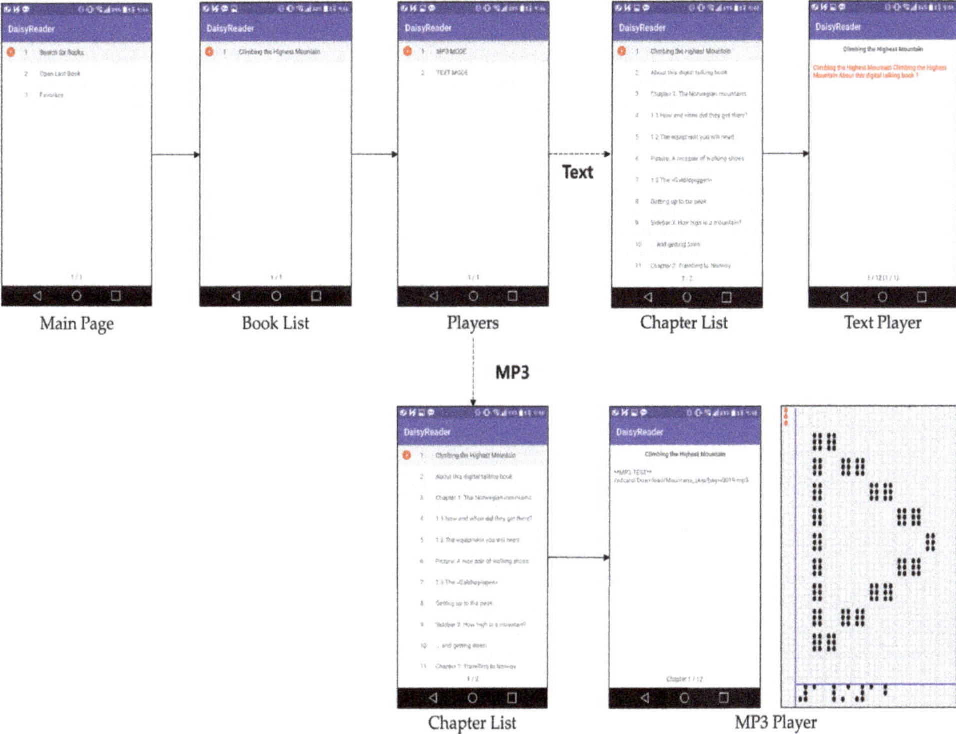

Figure 21. Sample progress in an eBook reader application [37].

Figure 22. Implementation of figure conversion on a 12 × 12 braille pad simulator. Converted figure on the left: cat character [71]; center: like logo [72]; right: chat logo [73]. This conversion uses six digits of a braille cell per braille figure.

Table 5 and Figure 23 present the comparison of related work in this area. Certain research focused on visually impaired people and/or people having reading disabilities. The DAISY standard, especially for people with reading disabilities, was used in most studies. Excluding EPUB, DAISY is an essential eBook standard for visually impaired people. In addition, we realized that most studies tended to develop software based on Android OS for mobility. Most Windows OSs, except Windows CE 6.0, are used for operating desktop computers. Further, some applications could not be implemented on a braille display; therefore, we could not compare the panning problem. Byrd's method and Blitab have no information regarding DAISY and EPUB. Specifically, Byrd's implementation uses the Windows OS, and it supports nonvisual desktop access (NVDA), developed by NV Access. NVDA allows blind and visually impaired people to access and interact with the Windows OS and many third-party applications [74]. The study by Bornschein et al. [20] compared the methods for displaying braille figures, which contributed to the education of visually impaired people; however, they concluded that the BrailleDis device was not popular as it was a tethered display; consequently, users may have felt uncomfortable while using it. Unfortunately, the proposed braille display device has not been fully developed yet. Nevertheless, to reduce the issues with the currently marketed braille displays, we resolved the limitations of the panning problem and braille figure support.

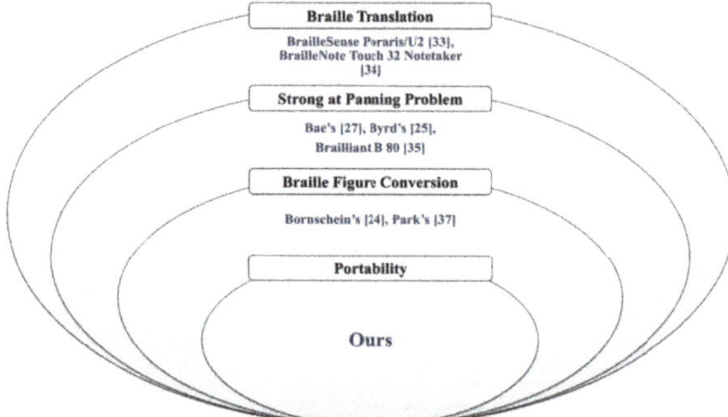

Figure 23. Cumulative Ben diagram based on Table 5. There are reference numbers on the diagrams, and they are classified by features. eBook standards and OS features are excluded.

Table 5. Comparison of assistive applications and devices in this study (DAISY: v2.02 and v3.0, EPUB: 2.0 and 3.0); the ∘ means 'support', and the × means 'not support'.

Research	eBook Support (DAISY/EPUB)	Portability	Braille Translation	Braille Figure Conversion	Panning Problem	OS
Application						
Bae's [23]	DAISY	Weak	∘	×	Strong	Windows
Kim's [24,25]	DAISY	Strong	×	×	-	Android
Bornschein's [21]	Unknown	Weak	∘	∘	Strong	Windows
Bornschein's [20]	Unknown	Weak	∘	∘	Strong	Windows
Goncu's [19]	EPUB	Strong	×	×	-	iOS
Harty's [26]	DAISY(v2.02)	Strong	×	×	-	Android
Mahule's [27]	DAISY(v3.0)	Strong	×	×	-	Android
Braille device						
Byrd's [22]	Unknown	Strong	∘	×	Strong	Windows (NVDA)
Park's [33]	Both	Weak	∘	∘	Strong	Windows
BrailleSense Polaris [29]	Both	Strong	∘	×	Weak	Android
BrailleSense U2 [29]	Both	Strong	∘	×	Weak	Windows CE 6.0
BrailleNote Touch 32 Braille Notetaker [30]	Both	Strong	∘	×	Weak	Android
Brailliant B 80 (new gen.) [31]	Both	Strong	∘	×	Strong	-
Ours	Both	Strong	∘	∘	Strong	Android

6. Limitation and Discussion

6.1. Haptic Telepresence

Unfortunately, it was observed that subjects felt uncomfortable while using the 3D haptic device. They reported that it was confusing to classify similar shapes (e.g., cone and pyramid). Although the recognition rate by subjects was not low, it took longer than the anticipated time. The large differences were due to the different skill levels of individuals as well as the different levels of understanding of instructions. This research was a pre-study with non-impaired persons because recruiting the visually impaired people for this specific project was difficult. When easy-to-understand objects were expressed by subsampling of figures or long shapes, subjects demonstrated no problems in recognizing them. However, when the given shape was of a small object or was a complex shape (i.e., human-like depth map, as shown in Figure 24, it was more difficult for the subjects to recognize the shapes using the haptic device with depth maps. Although this method transfers the point cloud video in real-time without issues, when a video is produced with an IR sensor, the resolution tends to be lower. Because of these reasons, limitations were revealed while expressing and recognizing a small or complex object. Moreover, the haptic device is expensive; further, because it is large and heavy, it is difficult to carry. This implies that new technology is required to deliver more detailed information to express an object precisely and tactually and to improve mobility. Because of these problems, this research changed its direction to 2D multiarray braille display.

Additionally, as shown in Figure 25, the original objective of this research was the "haptic telepresence television for visually impaired people." It can enable visually impaired people to meet others remotely using certain applications, such as Skype (from Microsoft), or observe the hand-held screen of a TV. However, as the research progresses, this research team was skeptical about the effectiveness of the haptic device, as it had demonstrated unclear results regarding practical haptic telepresence. This implied that a novel technology was required to capture the object and deliver more details while expressing an object tactually. Currently, it is difficult to express tiny or complex objects using a haptic device.

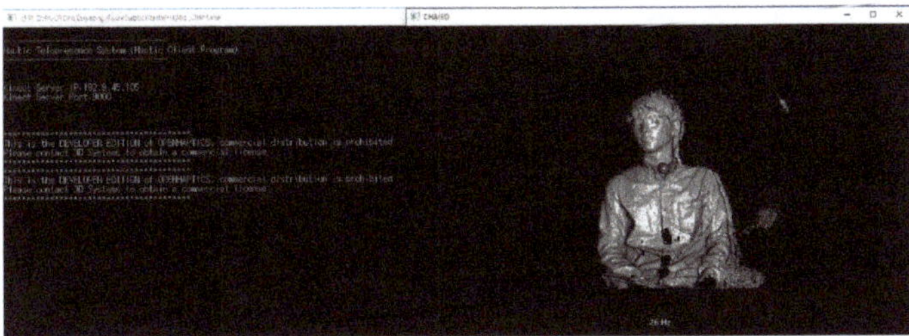

Figure 24. Human recognition through proposed haptic telepresence (client side).

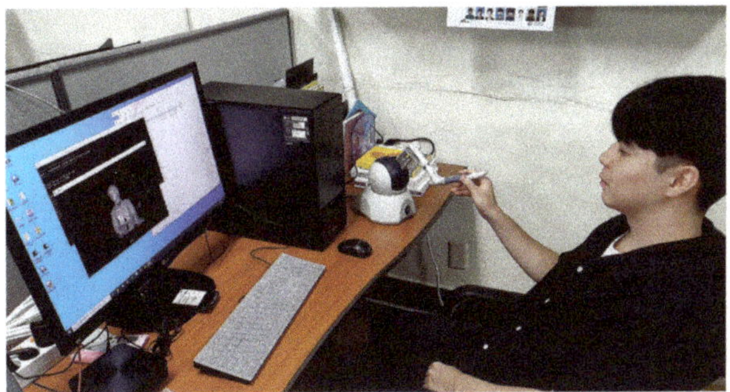

Figure 25. Subject is testing the recognition of a human using the proposed haptic telepresence system.

6.2. Braille eBook Reader Application

The braille display device demonstrates a superior performance than the haptic device while delivering media information. However, several components have to be implemented without helpful information while designing the braille display device. Unlike other software, the software required for this device has not been developed well. The research on open source softwares or software/applications related to DAISY, braille refreshable format, and EPUB standards are considerably scarce. In addition, it is still difficult to let visually impaired people touch and feels images or videos through the braille pad. In this study, a video was expressed with only the cover image of the video. Further, images expressed on the braille pad simulator were more unclear than the images appearing in teaching materials or books. Originally, the main objective of this study was to provide ways to educate the visually impaired people. A new format of tactile cells is required to express photographs and pictures that contain many objects on a small braille display. As shown in Figure 26, the braille figure looks like a raster image. It has several more artifacts than the original vector image, because the number of pixel and braille dots are significantly different. The original figures are depicted imperfectly within limited number of dots in a braille display.

Figure 26. Limitation of braille figure conversion. Green circles indicate the same position between the original and the converted braille image. The braille figure shows unclear boundaries in 12×12 braille cells.

Specifically, this study was unable to prove the developed applications in less developed prototypes. Although this study considered a method for expressing a video on the braille pad as well, this requires further research on the reaction speed of actuators, the power consumption, design, etc.. While considering the braille hardware, there are several challenges that have to be solved in terms of software, such as video sampling, filtering, and streaming. Hence, this study attempted to use a method for expressing a typical cover image of the video. This also was difficult to express properly because cover images have to be expressed through accurate filtering and conversion. Therefore, more studies are required for adequate expression of images and videos on a braille display or other tactile display. Assuming that advances will be made in hardware using superior technologies, the final objective of this study is to output videos on the multiarray braille device. The authors hope that liquid- and solid-like display devices will be studied for a novel tactile display.

7. Conclusions

This paper explains two main systems for visually impaired people: (i) converting a video and delivering tactile information remotely by using a 3D haptic device, and (ii) delivering braille data and multimedia content by using a 2D multiarray braille display. The first study enables visually impaired people to feel 2D and 3D object contour that cannot usually be felt remotely. To enhancing the object recognition performance, a new 2D braille system is designed in the second study. This study overcomes the limitations of traditional one-dimensional braille devices. In the implemented system, the text content is expressed within multiple lines, and it enables the visually impaired people to feel eBooks more efficiently. Moreover, a study provides the way to express images on the braille display by extracting central-labeled object.

Implemented 2D multiarray braille system displays massive textual content and other media content simultaneously. It increases the accessibility of visually impaired people to multimedia information and affects the 3D and 2D haptic information delivery for the education of visually impaired people. This research is work in progress and has the purpose to explore the possibilities to implement new devices for the benefit of the visually impaired.

Author Contributions: Conceptualization, data curation, writing—original draft, data curation, and investigation: S.K.; data curation and investigation: Y.R.; Project administration, supervision, and writing—review and editing: J.C. and E.-S.R.

Funding: This study was supported by the Ministry of Science and ICT, Korea, under the Information Technology Research Center support program (IITP-2019-2017-0-01630) supervised by the Institute for Information & Communications Technology Promotion (IITP).

Conflicts of Interest: The authors declare no conflict of interest.

Abbreviations

The following abbreviations are used in this manuscript:

DAISY	Digital accessible information system
HEVC	High efficiency video coding
SHVC	Scalable high efficiency video coding
LED	Light emitting diode
IR	Infrared
EPUB	Electronic publication
TTS	Text-to-speech
NCC	Navigation control center
NCX	Navigation control center for extensible markup language
XML	Extensible markup language
SMIL	Synchronized multimedia integration language
HTML	Hyper text markup language
SDK	Software development kit

References

1. Colby, S.L.; Ortman, J.M. *Projections of the Size and Composition of the US Population: 2014 to 2060: Population Estimates and Projections*; US Census Bureau: Washington, DC, USA, 2017.
2. He, W.; Larsen, L.J. *Older Americans with a Disability, 2008–2012*; US Census Bureau: Washington, DC, USA, 2014.
3. Bourne, R.R.; Flaxman, S.R.; Braithwaite, T.; Cicinelli, M.V.; Das, A.; Jonas, J.B.; Keeffe, J.; Kempen, J.H.; Leasher, J.; Limburg, H.; et al. Magnitude, temporal trends, and projections of the global prevalence of blindness and distance and near vision impairment: A systematic review and meta-analysis. *Lancet Glob. Health* **2017**, *5*, e888–e897. [CrossRef]
4. Elmannai, W.; Elleithy, K. Sensor-based assistive devices for visually-impaired people: Current status, challenges, and future directions. *Sensors* **2017**, *17*, 565.
5. Bolgiano, D.; Meeks, E. A laser cane for the blind. *IEEE J. Quantum Electron.* **1967**, *3*, 268–268. [CrossRef]

6. Borenstein, J.; Ulrich, I. The guidecane-a computerized travel aid for the active guidance of blind pedestrians. In Proceedings of the ICRA, Albuquerque, NM, USA, 21–27 April 1997; pp. 1283–1288. [CrossRef]
7. Yi, Y.; Dong, L. A design of blind-guide crutch based on multi-sensors. In Proceedings of the 2015 12th International Conference on Fuzzy Systems and Knowledge Discovery (FSKD), Zhangjiajie, China, 15–17 August 2015; pp. 2288–2292.
8. Wahab, M.H.A.; Talib, A.A.; Kadir, H.A.; Johari, A.; Noraziah, A.; Sidek, R.M.; Mutalib, A.A. Smart cane: Assistive cane for visually-impaired people. *arXiv* **2011**, arXiv:1110.5156.
9. Park, C.H.; Howard, A.M. Robotics-based telepresence using multi-modal interaction for individuals with visual impairments. *Int. J. Adapt. Control Signal Process.* **2014**, *28*, 1514–1532.
10. Park, C.H.; Ryu, E.S.; Howard, A.M. Telerobotic haptic exploration in art galleries and museums for individuals with visual impairments. *IEEE Trans. Haptics* **2015**, *8*, 327–338. [CrossRef]
11. Park, C.H.; Howard, A.M. Towards real-time haptic exploration using a mobile robot as mediator. In Proceedings of the 2010 IEEE Haptics Symposium, Waltham, MA, USA, 25–26 March 2010; pp. 289–292.
12. Park, C.H.; Howard, A.M. Real world haptic exploration for telepresence of the visually impaired. In Proceedings of the Seventh Annual ACM/IEEE International Conference on Human-Robot Interaction, Portland, OR, USA, 2–5 March 2012; pp. 65–72. [CrossRef]
13. Park, C.H.; Howard, A.M. Real-time haptic rendering and haptic telepresence robotic system for the visually impaired. In Proceedings of the World Haptics Conference (WHC), Daejeon, Korea, 14–17 April 2013; pp. 229–234.
14. Hicks, S.L.; Wilson, I.; Muhammed, L.; Worsfold, J.; Downes, S.M.; Kennard, C. A depth-based head-mounted visual display to aid navigation in partially sighted individuals. *PLoS ONE* **2013**, *8*, e67695.
15. Hong, D.; Kimmel, S.; Boehling, R.; Camoriano, N.; Cardwell, W.; Jannaman, G.; Purcell, A.; Ross, D.; Russel, E. Development of a semi-autonomous vehicle operable by the visually-impaired. In Proceedings of the IEEE International Conference on Multisensor Fusion and Integration for Intelligent Systems, Daegu, Korea, 16–18 November 2008; pp. 539–544.
16. Kinateder, M.; Gualtieri, J.; Dunn, M.J.; Jarosz, W.; Yang, X.D.; Cooper, E.A. Using an Augmented Reality Device as a Distance-based Vision Aid—Promise and Limitations. *Optom. Vis. Sci.* **2018**, *95*, 727.
17. Oliveira, J.; Guerreiro, T.; Nicolau, H.; Jorge, J.; Gonçalves, D. BrailleType: Unleashing braille over touch screen mobile phones. In Proceedings of the IFIP Conference on Human-Computer Interaction, Lisbon, Portugal, 5–9 September 2011; Springer: Berlin/Heidelberg, Germany, 2011; pp. 100–107.
18. Velázquez, R.; Preza, E.; Hernández, H. Making eBooks accessible to blind Braille readers. In Proceedings of the IEEE International Workshop on Haptic Audio visual Environments and Games, Ottawa, ON, Canada, 18–19 October 2008; pp. 25–29. [CrossRef]
19. Goncu, C.; Marriott, K. Creating ebooks with accessible graphics content. In Proceedings of the 2015 ACM Symposium on Document Engineering, Lausanne, Switzerland, 8–11 September 2015; pp. 89–92.
20. Bornschein, J.; Bornschein, D.; Weber, G. Comparing computer-based drawing methods for blind people with real-time tactile feedback. In Proceedings of the 2018 CHI Conference on Human Factors in Computing Systems, Montreal, QC, Canada, 21–26 April 2018; p. 115.
21. Bornschein, J.; Prescher, D.; Weber, G. Collaborative creation of digital tactile graphics. In Proceedings of the 17th International ACM SIGACCESS Conference on Computers & Accessibility, Lisbon, Portugal, 26–28 October 2015; pp. 117–126. [CrossRef]
22. Byrd, G. Tactile Digital Braille Display. *Computer* **2016**, *49*, 88–90.
23. Bae, K.J. A Study on the DAISY Service Interface for the Print-Disabled. *J. Korean Biblia Soc. Libr. Inf. Sci.* **2011**, *22*, 173–188.
24. Jihyeon, W.; Hyerina, L.; Tae-Eun, K.; Jongwoo, L. *An Implementation of an Android Mobile E-book Player for Disabled People*; Korea Multimedia Society: Busan, Korea 2010; pp. 361–364.
25. Kim, T.E.; Lee, J.; Lim, S.B. A Design and Implementation of DAISY3 compliant Mobile E-book Viewer. *J. Digit. Contents Soc.* **2011**, *12*, 291–298.
26. Harty, J.; LogiGearTeam; Holdt, H.C.; Coppola, A. Android-Daisy-Epub-Reader. 2013. Available online: https://code.google.com/archive/p/android-daisy-epub-reader (accessed on 6 October 2018).
27. Mahule, A. Daisy3-Reader. 2011. Available online: https://github.com/amahule/Daisy3-Reader (accessed on 6 October 2018).
28. BLITAB Technology. Blitab. 2019. Available online: http://blitab.com/ (accessed on 10 January 2019).

29. HIMS International. BrailleSense Polaris and U2. 2018. Available online: http://himsintl.com/blindness/ (accessed on 18 October 2018). [CrossRef]
30. Humanware Store. BrailleNote Touch 32 Braille Notetaker. 2017. Available online: https://store.humanware.com/asia/braillenote-touch-32.html (accessed on 10 January 2019).
31. Humanware. Brailliant B 80 Braille Display (New Generation). 2019. Available online: https://store.humanware.com/asia/brailliant-b-80-new-generation.html (accessed on 8 June 2019).
32. Humanware. *Brailliant BI Braille Display User Guide*; Humanware Inc.: Jensen Beach, FL, USA, 2011.
33. Park, T.; Jung, J.; Cho, J. A method for automatically translating print books into electronic Braille books. *Sci. China Inf. Sci.* **2016**, *59*, 072101. [CrossRef]
34. Kim, S.; Roh, H.J.; Ryu, Y.; Ryu, E.S. Daisy/EPUB-based Braille Conversion Software Development for 2D Braille Information Terminal. In Proceedings of the 2017 Korea Computer Congress of the Korean Institute of Information Scientists and Engineers, Seoul, Korea, 18–20 June 2017; pp. 1975–1977.
35. Park, E.S.; Kim, S.D.; Ryu, Y.; Roh, H.J.; Koo, J.; Ryu, E.S. Design and Implementation of Daisy 3 Viewer for 2D Braille Device. In Proceedings of the 2018 Winter Conference of the Korean Institute of Communications and Information Sciences, Seoul, Korea, 17–19 January 2018; pp. 826–827.
36. Ryu, Y.; Ryu, E.S. Haptic Telepresence System for Individuals with Visual Impairments. *Sens. Mater.* **2017**, *29*, 1061–1067.
37. Kim, S.; Park, E.S.; Ryu, E.S. Multimedia Vision for the Visually Impaired through 2D Multiarray Braille Display. *Appl. Sci.* **2019**, *9*, 878.
38. Microsoft. Kinect for Windows. 2018. Available online: https://developer.microsoft.com/en-us/windows/kinect (accessed on 3 October 2018).
39. Microsoft. Setting up Kinect for Windows. 2018. Available online: https://support.xbox.com/en-US/xbox-on-windows/accessories/kinect-for-windows-setup (accessed on 4 October 2018).
40. Kaiming, H.; Jian, S.; Xiaoou, T. Guided Image Filtering. In *Proceedings of the Computer Vision—ECCV 2010*; Daniilidis, K., Maragos, P., Paragios, N., Eds.; Springer: Berlin/Heidelberg, Germany, 2010; pp. 1–14. [CrossRef]
41. Atılım Çetin. Guided filter for OpenCV. 2014. Available online: https://github.com/atilimcetin/guided-filter (accessed on 25 May 2019). [CrossRef]
42. CHAI3D. CHAI3D: CHAI3D Documentation. 2018. Available online: http://www.chai3d.org/download/doc/html/wrapper-overview.html (accessed on 5 October 2018).
43. 3D Systems, Inc.. OpenHaptics: Geomagic® OpenHaptics® Toolkit. 2018. Available online: https://www.3dsystems.com/haptics-devices/openhaptics?utm_source=geomagic.com&utm_medium=301 (accessed on 5 October 2018). [CrossRef]
44. Sullivan, G.J.; Ohm, J.R.; Han, W.J.; Wiegand, T. Overview of the high efficiency video coding (HEVC) standard. *IEEE Trans. Circuits Syst. Video Technol.* **2012**, *22*, 1649–1668. [CrossRef]
45. Roh, H.J.; Han, S.W.; Ryu, E.S. Prediction complexity-based HEVC parallel processing for asymmetric multicores. *Multimed. Tools Appl.* **2017**, *76*, 25271–25284.
46. Ryu, Y.; Ryu, E.S. Video on Mobile CPU: UHD Video Parallel Decoding for Asymmetric Multicores. In Proceedings of the 8th ACM on Multimedia Systems Conference, Taipei, Taiwan, 20–23 June 2017; pp. 229–232. [CrossRef]
47. Sullivan, G.J.; Boyce, J.M.; Chen, Y.; Ohm, J.R.; Segall, C.A.; Vetro, A. Standardized extensions of high efficiency video coding (HEVC). *IEEE J. Sel. Top. Signal Process.* **2013**, *7*, 1001–1016. [CrossRef]
48. Zhang, Q.; Chen, M.; Huang, X.; Li, N.; Gan, Y. Low-complexity depth map compression in HEVC-based 3D video coding. *EURASIP J. Image Video Process.* **2015**, *2015*, 2.
49. Fraunhofer.; HHI. 3D HEVC Extension. 2019. Available online: https://www.hhi.fraunhofer.de/en/departments/vca/research-groups/image-video-coding/research-topics/3d-hevc-extension.html (accessed on 16 October 2019).
50. Shokrollahi, A. Raptor codes. *IEEE/ACM Trans. Netw. (TON)* **2006**, *14*, 2551–2567.
51. Ryu, E.S.; Jayant, N. Home gateway for three-screen TV using H. 264 SVC and raptor FEC. *IEEE Trans. Consum. Electron.* **2011**, *57*.
52. Canadian Assistive Technologies Ltd.. What to Know before You Buy a Braille Display. 2018. Available online: https://canasstech.com/blogs/news/what-to-know-before-you-buy-a-braille-display (accessed on 10 January 2019).

53. Consortium, T.D. Daisy Consortium Homepage. 2018. Available online: http://www.daisy.org/home (accessed on 6 October 2018).
54. IDPF (International Digital Publishing Forum). EPUB Official Homepage. 2018. Available online: http://idpf.org/epub (accessed on 6 October 2018).
55. Google Developers. Documentation for Android Developers—Message. 2018. Available online: https://developer.android.com/reference/android/os/Message (accessed on 7 October 2018).
56. Google Developers. Documentation for Android Developers—Activity. 2018. Available online: https://developer.android.com/reference/android/app/Activity (accessed on 7 October 2018).
57. Bray, T.; Paoli, J.; Sperberg-McQueen, C.M.; Maler, E.; Yergeau, F. Extensible Markup Language (XML) 1.0. Available online: http://ctt.sbras.ru/docs/rfc/rec-xml.htm (accessed on 3 December 2019).
58. Daisy Consortium. DAISY 2.02 Specification. 2018. Available online: http://www.daisy.org/z3986/specifications/daisy_202.html (accessed on 7 October 2018).
59. Daisy Consortium. Part I: Introduction to Structured Markup—Daisy 3 Structure Guidelines. 2018. Available online: http://www.daisy.org/z3986/structure/SG-DAISY3/part1.html (accessed on 7 October 2018).
60. Paul Siegmann. EPUBLIB—A Java EPUB Library. 2018. Available online: http://www.siegmann.nl/epublib (accessed on 18 October 2018).
61. Siegmann, P. Epublib for Android OS. 2018. Available online: http://www.siegmann.nl/epublib/android (accessed on 6 October 2018).
62. Hedley, J. Jsoup: Java HTML Parser. 2018. Available online: https://jsoup.org/ (accessed on 7 October 2018).
63. Jung, J.; Kim, H.G.; Cho, J.S. Design and implementation of a real-time education assistive technology system based on haptic display to improve education environment of total blindness people. *J. Korea Contents Assoc.* **2011**, *11*, 94–102.
64. Jung, J.; Hongchan, Y.; Hyelim, L.; Jinsoo, C. Graphic haptic electronic board-based education assistive technology system for blind people. In Proceedings of the 2015 IEEE International Conference on Consumer Electronics (ICCE), Las Vegas, NV, USA, 9–12 January 2015; pp. 364–365.
65. Shi, J.; Tomasi, C. *Good Features to Track*; Technical Report; Cornell University: Ithaca, NY, USA, 1993.
66. Daisy Consortium. Daisy Sample Books. 2018. Available online: http://www.daisy.org/sample-content (accessed on 10 October 2018).
67. PRESSBOOKS. EPUB Sample Books. 2018. Available online: https://pressbooks.com/sample-books/ (accessed on 8 October 2018).
68. Jeong, I.; Ahn, E.; Seo, Y.; Lee, S.; Jung, J.; Cho, J. Design of Electronic Braille Learning Tool System for low vision people and Blind People. In Proceedings of the 2018 Summer Conference of the Korean Institute of Information Scientists and Engineers, Seoul, Korea, 20–22 June 2018; pp. 1502–1503.
69. Seo, Y.S.; Joo, H.J.; Jung, J.I.; Cho, J.S. Implementation of Improved Functional Router Using Embedded Linux System. In Proceedings of the 2016 IEIE Summer Conference, Seoul, Korea, 22–24 June 2016; pp. 831–832.
70. Park, J.; Sung, K.K.; Cho, J.; Choi, J. Layout Design and Implementation for Information Output of Mobile Devices based on Multi-array Braille Terminal. In Proceedings of the 2016 Winter Korean Institute of Information Scientists and Engineers, Seoul, Korea, 21–23 December 2016; pp. 66–68.
71. Goyang-City. Cat character illustration by Goyang City. 2014. Available online: http://www.goyang.go.kr/www/user/bbs/BD_selectBbsList.do?q_bbsCode=1054 (accessed on 16 October 2019).
72. Repo, P. Like PNG Icon. 2019. Available online: https://www.pngrepo.com/svg/111221/like (accessed on 16 October 2019).
73. FLATICON. Chat Icon. 2019. Available online: https://www.flaticon.com/free-icon/chat_126500#term=chat&page=1&position=9 (accessed on 16 October 2019).
74. NV Access. About NVDA. 2019. Available online: https://www.nvaccess.org/ (accessed on 12 March 2019).

Article

Using 3D Convolutional Neural Networks for Tactile Object Recognition with Robotic Palpation

Francisco Pastor *, Juan M. Gandarias, Alfonso J. García-Cerezo and Jesús M. Gómez-de-Gabriel

Robotics and Mechatronics Group, University of Málaga, 29071 Málaga, Spain; jmgandarias@uma.es (J.M.G.); ajgarcia@uma.es (A.J.G.-C.); jesus.gomez@uma.es (J.M.G.-d.-G.)

* Correspondence: fpastor@uma.es

Received: 31 October 2019; Accepted: 2 December 2019; Published: 5 December 2019

Abstract: In this paper, a novel method of active tactile perception based on 3D neural networks and a high-resolution tactile sensor installed on a robot gripper is presented. A haptic exploratory procedure based on robotic palpation is performed to get pressure images at different grasping forces that provide information not only about the external shape of the object, but also about its internal features. The gripper consists of two underactuated fingers with a tactile sensor array in the thumb. A new representation of tactile information as 3D tactile tensors is described. During a squeeze-and-release process, the pressure images read from the tactile sensor are concatenated forming a tensor that contains information about the variation of pressure matrices along with the grasping forces. These tensors are used to feed a 3D Convolutional Neural Network (3D CNN) called 3D TactNet, which is able to classify the grasped object through active interaction. Results show that 3D CNN performs better, and provide better recognition rates with a lower number of training data.

Keywords: tactile perception; robotic palpation; underactuated grippers; deep learning

1. Introduction

Recent advances in Artificial Intelligence (AI) have brought the possibility of improving robotic perception capabilities. Although most of them are focused on visual perception [1], existing solutions can also be applied to tactile data [2–4]. Tactile sensors measure contact pressure from other physical magnitudes, depending on the nature of the transducer. Different types of tactile sensors [5–9] have been used in robotic manipulation [10,11] for multiple applications such as slippage detection [12,13], tactile object recognition [14,15], or surface classification [16,17], among others.

Robotic tactile perception consists of the integration of mechanisms that allow a robot to sense tactile properties from physical contact with the environment along with intelligent capacities to extract high-level information from the contact. The sense of touch is essential for robots the same way as for human beings for performing both simple and complex tasks such as object recognition or dexterous manipulation [18–20]. Recent studies focused on the development of robotic systems that behaves similar to humans, including the implementation of tactile perception capabilities [21,22]. However, tactile perception is still a fundamental problem in robotics that has not been solved so far [23]. In addition, there are multiple applications, not limited to classic robotic manipulation problems that can benefit from tactile perception such as medicine [24], food industry [3], or search-and-rescue [4], among others.

Many works related to tactile perception use pressure images after the interaction [25], which means that the interaction is considered static or passive. However, tactile perception in the real world is intrinsically active [26]. A natural or bio-inspired haptic Exploratory Procedure (EP) for perceiving pressure or stiffness of an object must consider dynamic information [27]. According to [28], the haptic attributes that can be perceived depends on the EP.

A survey on the concept of active tactile perception considering biological and psychological terms is presented in [29]. In this chapter, and according to [30], two approaches for tactile perception in robots are possible: perception for action, which means that the perceived information is used to guide the robot (i.e., dexterous manipulation and grasp control), and action for perception, which means that the robot explores the environment to collect data (i.e., active perception and haptic exploration). Hence, an active tactile perception approach can be defined as one in which the data are collected during an active EP using an active sensing approach (e.g., tactile sensing). This means that action and perception are not separated, and the robot collects dynamic data depending on the action, while this action is occurring. Therefore, although both static and dynamic tactile data are useful for many robotic applications, it can be considered that active perception is more faithful to the real sense of touch, and the information acquired using active tactile sensing reflects the attributes of the grasped objects better. Static pressure images only contain information about stiffness and shape of the object when a certain force is applied [14], while the changes of the pressure distribution over force contain information about the variation of shape and stiffness during the whole EP [31]. This dynamic information allows us to distinguish both rigid and deformable objects [32].

This paper addresses the shortcomings mentioned above and is focused on the active tactile perception problem in robotics. A robotic palpation process with active tactile sensing, based on a squeeze-and-release motion for distinguishing grasped objects, both rigid and deformable, is presented (see Figure 1). The robotic EP conceives a novel representation of dynamic tactile information based on sequences of pressure images and an AI method based on 3D Convolutional Neural Networks (3D CNNs) for active tactile perception. A tactile sensor array is integrated into the thumb of a gripper with two underactuated fingers to get sequences of tactile images. These sequences are represented as 3D tensors similar to Magnetic Resonance Imaging (MRI). However, in this case, 3D tactile tensors represent the variation of pressure distribution over applied force, whereas MRI contains information about cross-sectional images of internal structures and organs over distance. Although the type of information contained in MRIs and 3D tactile tensors is different, methods such as 3D CNNs used to process MRI information [33,34] might be used for tactile data with good results in this application as we explored in our previous work [35]. In this work, our preliminary study is expanded: a high-resolution tactile sensor has been integrated into a new gripper where the palpation process (e.g., the EP) is fully autonomous, so the robot controls the grasping force. As a result, not only objects with different elasticity are compared and classified, but also objects that contain internal inclusions and bags of objects which provide different pressure images each time, have been tested. In particular, 24 objects have been used: rigid, deformable, and in-bag; and the results are compared against 2D CNN-based methods. Altogether, the main contribution of this paper relates to the entire process of active tactile perception, considering the use of an underactuated, sensorized gripper to carry out the EP, and a 3D CNN for tactile perception.

The relevance of this contribution relies on different factors. First, the presented method achieves better performance in the discrimination problem for all kinds of objects, and, in case the number of classes increases, a lower number of training data are needed to obtain higher accuracy rates than classic 2D networks. Second, it is also shown that, in case of misclassification, the resulting object class has almost indistinguishable physical features (e.g., soda cans of different capacities), where 2D CNNs, in the event of failure, give disparate output classes unrelated to the class of the grasped object.

This paper is structured as follows: In Section 2, the current state-of-the-art related to this topic is introduced. In Section 3, the underactuated gripper and the 3D CNN-based method used for tactile perception are described. The experimental protocol and results are explained in Section 4, and a thorough and detailed discussion of our results in comparison with related works is presented in Section 5. Finally, the conclusion and future research lines are exposed in Section 6.

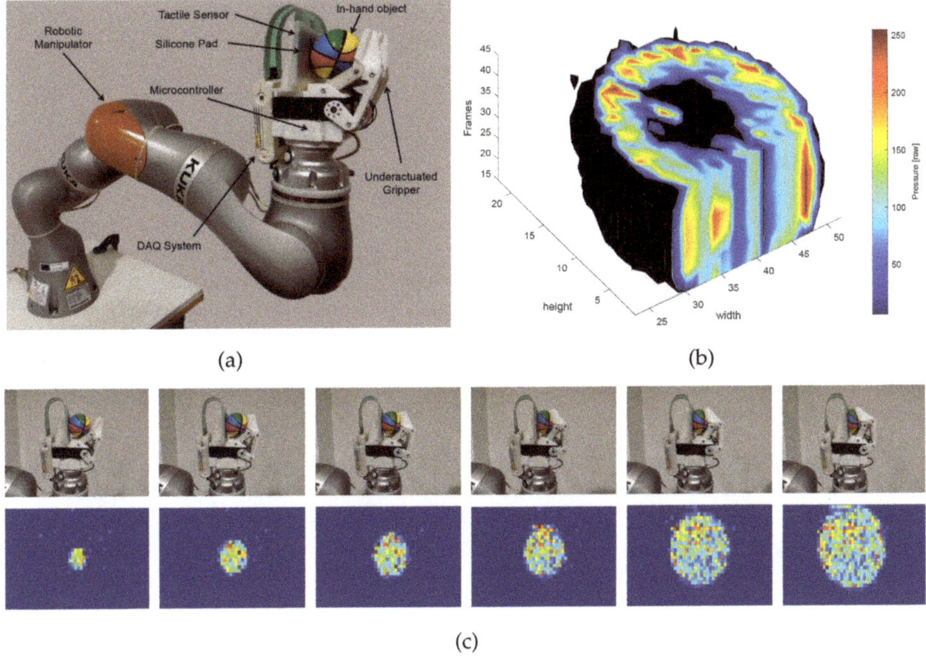

Figure 1. The full experimental system formed by a robotic manipulator, an underactuated gripper with a tactile sensor, and the control electronics (**a**), a 3D tensor representation of active tactile information when the gripper is grasping a squeezable ball (**b**), and a subset of pictures and their respective tactile images of a grasping sequence of another squeezable ball (**c**). In (**b**), the tensor is sectioned to show the intrinsic attributes and pressure variations of the grasped object.

2. Related Work

Related works within the scope of tactile perception in robotics focus on tactile object-recognition from pressure-images, deep-learning methods based on CNNs, and active tactile perception.

2.1. Tactile Object Recognition

Two main approaches for tactile object recognition may be considered depending on the nature of the EP: On one hand, perceiving attributes from the material composition, which are typically related to superficial properties like roughness, texture, or thermal conductivity [36–38]. On the other hand, other properties related to stiffness and shape may also be considered for object discrimination [39–41]. Most of these works are based on the use of novel machine learning-based techniques. That way, different approaches can be followed, such as Gaussian Processes [42], k-Nearest Neighbour (kNN) [25], Bayesian approaches [43], k-mean and Support Vector Machines (SVM) [44], or Convolutional Neural Networks (CNNs) [45], among others. Multi-modal techniques have also been considered in [46], where they demonstrated that considering both haptic and visual information generally gives better results.

2.2. Tactile Perception Based on Pressure Images

Concerning the latter approach, most of the existing solutions in literature acquire data from tactile sensors, in the form of matrices of pressure values, analog to common video images [47]. In this respect, multiple strategies and methodologies can be followed. In [25], a method, based on *Scale Invariant Feature Transform* (SIFT) descriptors, is used as a feature extractor, and the kNN algorithm is used to classify objects by their shape. In [15], Luo et al. proposed a novel multi-modal algorithm that mixes kinesthetic and tactile data to classify objects from a 4D point cloud where each point is represented by the 3D position of the point and the pressure acquired by a tactile sensor.

2.3. CNNs-Based Tactile Perception

One recent approach for tactile object discrimination consists of the incorporation of modern deep learning-based techniques [48,49]. In this respect, the advantages of Convolutional Neural Networks (CNNs) such as translational and rotational invariant property enable the recognition in any pose [50]. A CNN-based method to recognize human hands in contact with an artificial skin has been presented in [44]. The proposed method benefits from the CNN's translation-invariant properties and is able to identify whether the contact is made with the right or the left hand. Apart from that, the integration of the dropout technique in deep learning-based tactile perception has been considered in [49], where the benefits of fusing kinesthetic and tactile information for object classification are also described, as well as the differences of using planar and curved tactile sensors.

2.4. Active Tactile Perception

In spite of the good results obtained by existing solutions in tactile object recognition, one of the main weaknesses is that most of these solutions only consider static or passive tactile data [25]. As explained, static tactile perception is not a natural EP to perceive attributes like pressure or stiffness [27]. Pressure images only have information about the shape and pressure distribution when a certain force is applied [14]. On the other hand, sequences of tactile images also contain information about the variation of shape (in the case of deformable objects [32]), stiffness, and pressure distribution over time [31].

Time-series or sequential data are important to identify some properties. This approach has been followed in some works for material discrimination [51,52]. In [53], an EP is carried out by a robotic manipulator to get dynamic data using a 2D force sensor. The control strategy of the actuator is critical to apply a constant pressure level and perceive trustworthy data. For this purpose, a multi-channel neural network was used achieving high accuracy levels.

Pressure images obtained from tactile sensors have also been used to form sequences of images. In [3], a flexible sensor was used to classify food textures. A CNN was trained with sequences of tactile images obtained during a food-biting experiment in which a sensorized press is used to crush food, simulating the behavior of a mouth biting. The authors found that the results when using the whole biting sequence or only the first and last tactile images were very similar because the food was crushed when a certain level of pressure was applied. Therefore, the images before and after the break point were significantly different. For other applications, as it was demonstrated in [54], Three-Dimensional Convolutional Neural Networks (3D CNNs) present better performance when dealing with sequences of images than common 2D CNNs.

3. Materials and Methods

The experimental setup is composed of a gripper with a tactile sensor. The gripper, the representation of 3D tactile information, and the 3D CNN are described next.

3.1. Underactuated Gripper

The active perception method has been implemented using a gripper with two parallel underactuated fingers and a fixed tactile-sensing surface (see Figure 2). The reason for using an underactuated gripper is that this kind of gripper allows us to apply evenly spread pressure to the grasped objects, and the fingers could adapt to their shape, which is especially useful when grasping deformable or in-bag objects. In our gripper, each underactuated finger has two phalanxes with two

(DOFs) θ_1 and θ_2, and a single actuator θ_a capable of providing different torque values τ_a. The values of the parameters of the kinematics are included in Table 1. A spring provides stiffness to the finger to recover the initial position when no contact is detected.

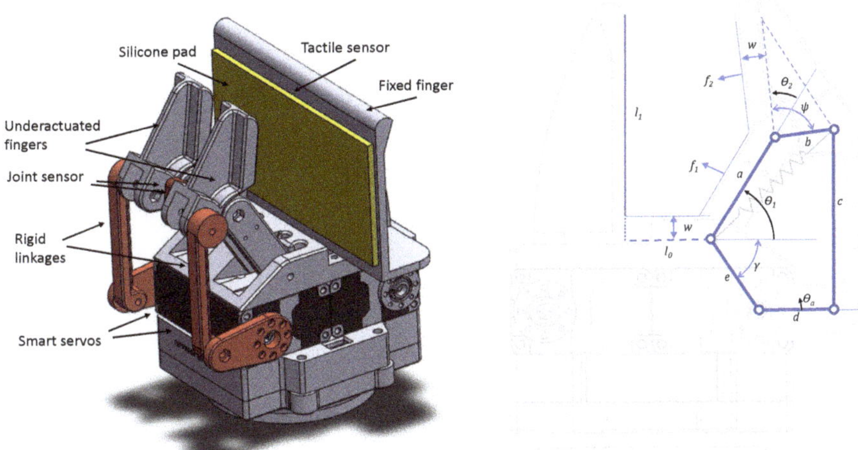

Figure 2. Gripper design (**left**) with two independent underactuated fingers and one fixed thumb with a tactile sensor covered with a silicone pad. The kinematic structure of the underactuated fingers (**right**) shows the five-bar structure with associated parameters and degrees of freedom (DOFs).

Two smart servos (Dynamixel XM-430 from ROBOTIS (Seoul, Korea)) have been used to provide different torques trough rigid-links, using a five-bar mechanical structure to place the servos away from the first joint. Thus, the relationship between τ_a and the joint torques (τ_1, τ_2) can be expressed as a transfer matrix T, and the computation of the Cartesian grasping forces (f_1, f_2) from the joint torques is defined by the Jacobian matrix $\mathbf{F} = \mathbf{J}(\boldsymbol{\theta})\boldsymbol{\tau}$.

The computation of those matrices requires knowledge of the actual values of the underactuated joints. For this reason, a joint sensor has been added to the second joint of each finger. The remaining joint can be computed as the actual value of the servo joint, which is obtained from the smart servos. Two miniature potentiometers (*muRata* SV01) have been used to create a special gripper with both passive adaptation and proprioceptive feedback.

The dynamic effects can be neglected when considering slow motions and lightweight fingers. This way, a kinetostatic model of the forces can be derived in Equation (1) as described in [55]:

$$\mathbf{F} = \mathbf{J}(\boldsymbol{\theta})^{-T}\,\mathbf{T}(\boldsymbol{\theta})^{-T}\,\boldsymbol{\tau}. \tag{1}$$

Although the actual Cartesian forces could be computed, each object with a different shape should require feedback control to apply the desired grasping forces. In order to simplify the experimental setup, an open-loop force control has been used for the grasping operations, where the actuation (pulse-width modulation - PWM) of the direct current (DC) motors of the smart servos follows a slow triangular trajectory from a minimum value (5%) to a maximum (90%) of the maximum torque of 1.4 N.m of each actuator. The resulting position of each finger depends on the actual PWM and the shape and impedance of each contact area.

Table 1. Parameter values for the kinematic model of the gripper with underactuated fingers.

Parameter	Value	Parameter	Value
a	40 mm	e	27.8 mm
b	20 mm	ψ	90°
c	60 mm	γ	56°
d	25 mm	w	10 mm
l_0	25–45 mm	l_1	70 mm

Finally, a microcontroller (Arduino Mega2560) has been used to acquire angles form the analog potentiometers and communicating with the smart servos in real time, with a 50 ms period.

3.2. Tactile Sensor

A Teskcan (South Boston, MA, USA) sensor model 6077 has been used. This high-resolution tactile-array has 1400 tactels (also called taxels o sensels), with 1.3 × 1.3 mm size each. The sensor presents a density of 27.6 tactels/cm^2 distributed in a 28 × 50 matrix. The main features of the sensor are presented in Table 2. The setup includes the data acquisition system (DAQ) (see Figure 1a), and the Tekscan real-time software development kit (SDK) (South Boston, MA, USA).

Table 2. Main features of the *Tekscan* 6077 tactile sensor.

Parameter	Value
Max. pressure	34 KPa
Number of tactels	1700
Tactels density	27.6 tactels/cm^2
Temperature range	−40 °C to +60 °C
Matrix height	53.3 mm
Matrix width	95.3 mm
Thickness	0.102 mm

A silicone pad of 3 mm has been added to the tactile sensor to enhance the grip and the image quality, especially when grabbing rigid objects. In particular, the Ecoflex$^{TM}00-30$ rubber silicone has been chosen due to its mechanical properties.

3.3. Representation of Active Tactile Information

As introduced in Section 1, a natural palpation EP to get information about the stiffness of an in-hand object is dynamic. In this respect, it seems evident that a robotic EP should also be dynamic so that the information acquired during the whole squeeze-and-release process describes the external and internal tactile attributes of an object.

The pressure information can be represented in multiple ways, commonly as sequences of tactile images. However, in this case, a more appropriate structure is in the form of 3D tactile tensors. An example of this type of representation is presented in Figure 1b, which is similar to MRI, except that, in this case, the cross-sectional images contain information about the pressure distribution at the contact surface for different grasping forces.

To show the advantages of 3D tactile tensors, sectioned tensors of the same sponge, with and without hard inclusions, are shown in Figure 3. The inclusions become perfectly visible as the grasping force increases.

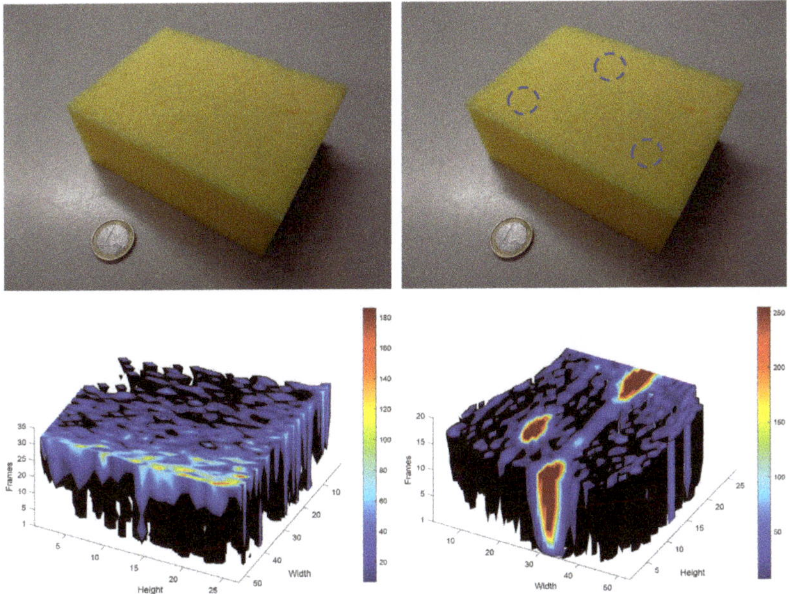

Figure 3. 3D tactile tensors (**bottom**) of the same sponge with and without hard inclusions (**top**). The inclusions become visible as grasping force increases but cannot be seen in the picture of the sponge.

3.4. 3D TactNet

When using 3D tactile information, it is necessary to control the applied forces to obtain a representative pressure-images from a certain object. For 3D CNNs, each tensor has information about the whole palpation process. On the other hand, when dealing with soft or shape-changing objects, this operation is more challenging using 2D CNNs, as a high amount of training data would be necessary, or selected data captured at optimal pressure levels, which also depends on the stiffness of each object.

In previous works, we trained and validated multiple 3D CNNs with different structures and hyperparameters to discriminate deformable objects in a fully-supervised collection and classification process [35]. Here, although the classification is still supervised, the grasping and data collection processes have been carried out autonomously by the robotic manipulator. According to the results of our previous work, the 3D CNN with the highest recognition rate, and compatible with the size of the 3D tensors read from our tactile sensor, was a neural network with four layers, where the first two were 3D convolutional, and the last two were fully connected layers. The network's parameters have been slightly modified to fit a higher number of classes and to adjust the new 3D tensor, which has a dimension of $[28 \times 50 \times 51]$.

The architecture of this network, called TactNet3D, is presented in Figure 4. This network has two 3D convolutional layers ($\mathscr{C} = [\text{3D conv}_1\ , \text{3D conv}_2]$) with kernels $16 \times [3 \times 5 \times 8]$ and $32 \times [3 \times 5 \times 8]$, respectively, and two fully connected layers ($\mathscr{F} = [\text{fc}_3, \text{fc}_4]$) with 64 and 24 neurons, respectively. Each convolutional layer also includes a Rectified Linear Unit (ReLU), batch normalization with $\epsilon = 10^{-5}$, and max-pooling with filters and stride equal to 1. In addition, fc_3 incorporated a dropout factor of 0.7 to prevent overfitting. Finally, a softmax layer is used to extract the probability distribution of belonging to each class. The implementation, training, and testing of this network has been done using the Deep Learning Toolbox in Matlab (R2019b, MathWorks,Natick, MA, USA).

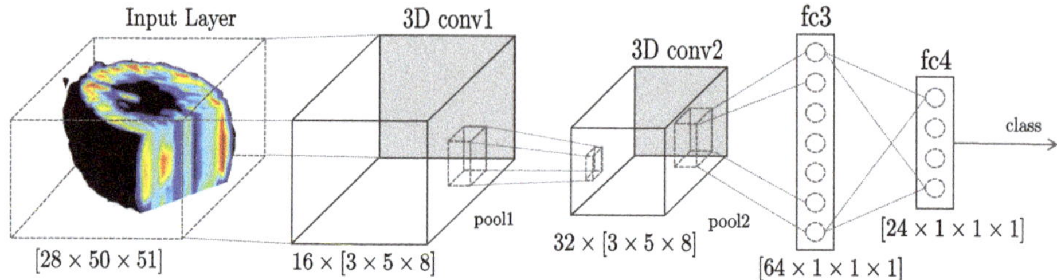

Figure 4. Architecture of TactNet3D, which is formed by four layers, the first two are 3D convolutional layers with kernel sizes $16 \times [5 \times 3 \times 8]$ and $32 \times [5 \times 3 \times 8]$, respectively, and two fully connected layers with 64 and 24 neurons, respectively.

4. Experimental Protocol and Results

This section presents the procedure for the dataset collection and the experiments. The dataset is conformed by three subsets of data: Rigid, deformable, and in-bag objects, which are described in more detailed below. Similarly, four experiments have been carried out to show the performance of the method and compare the results of dynamic and static methods: Experiment 1 for rigid objects, experiment 2 for deformable objects, experiment 3 for in-bag objects, and experiment 4 for the whole dataset.

4.1. Dataset

4.1.1. Collection Process

The dataset collection process consists of capturing sequences of tactile images and creating a 3D tactile tensor. For this purpose, the underactuated gripper holds an object and applies incremental forces while recording images over the whole palpation process. Each object, depending on its internal physical attributes, has a unique tactile frame for each amount of applied force. The dataset collection has been carried out by the gripper, recording 51 tactile frames per squeeze. This process is made by the two active fingers of the gripper, which are moved by the two smart servos in torque control mode with incremental torque references. Finally, 1440 3D tactile tensors have been obtained, for a total of 24 objects with 60 tactile tensors each. In Figure 1c, a grasping sequence is shown. The sequence at the top, from the left to the right, shows the grasping sequence due to the progressive forces applied by the underactuated gripper to the ball 2, and the sequence at the bottom, from the left to the right, shows the tactile images captured by the pressure sensor.

For machine learning methods, it is important to have the greatest possible variety in the dataset. In order to achieve this goal, the incremental torque is increased in random steps, so that the applied forces between two consecutive frames are different in each case. This randomness is also applied due to the intention to take a dataset that imitates the palpation procedure that could be carried out by a human, in which the exact forces are not known. Another fact that has been considered for the dataset collection process is that the force is applied to the object through the fingers of the gripper; therefore, non-homogeneous pressure is exerted on the whole surface of the object. Therefore, in order to obtain all of the internal features of the objects, multiple grasps with random positions and orientations of the objects have been obtained.

4.1.2. Rigid Objects

Eight objects of the dataset are considered as rigid because they barely change their shape when the gripper tightens them. The rigid dataset is composed of subsets of objects with similar features (e.g., the subset of bottles and the subset of cans) which are very different from each other. The subset of rigid objects is shown in Figure 5a.

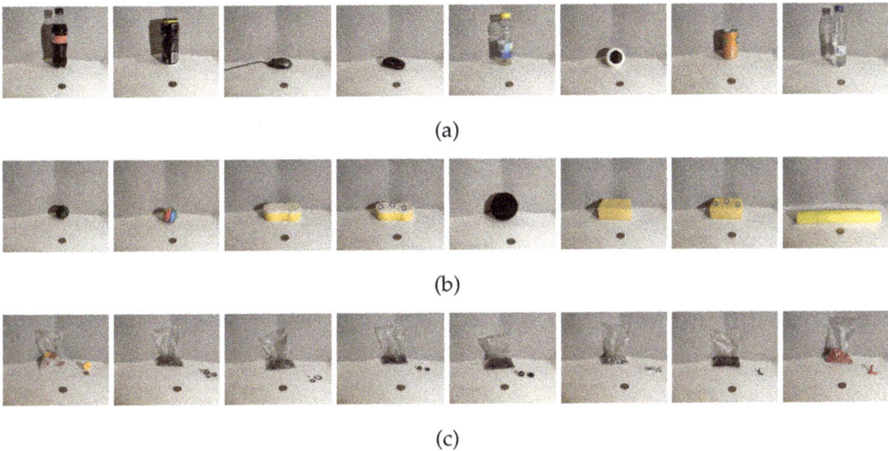

Figure 5. Pictures of the 24 objects used in experiments. Rigid objects (**a**), from left to right: bottle of coke, energy drink can, mouse 1, mouse 2, bottle of ice tea, skate wheel, soda can, and bottle of water. Deformable objects (**b**), from left to right: ball 1, ball 2, sponge rough, sponge rough with inclusions, sponge scrunchy, sponge soft, sponge soft with inclusions, and sponge pipe. In-bag objects (**c**), from left to right: gears, mixed nuts, mixed washer, M6 nuts, M8 nuts, M10 nuts, rivets, and rubber pipes.

4.1.3. Deformable Objects

Another subset of the dataset is the deformable objects. This subset consists of eight objects that change his initial shape substantially when a pressure is applied over it but recover its initial shape when the pressure ends. This subgroup also has objects with similar elasticity (e.g., balls and sponges). The set of deformable objects is shown in Figure 5b.

4.1.4. In-Bag Objects

The last subset of objects included in the dataset is composed of plastic bags with a number of small objects. Bags are shuffled before every grasp, so that the objects in the bag are placed in different positions and orientations. Hence, the tactile images are different depending on the position of the objects. Another characteristic of this group is that in-bag objects may change their position randomly during the grasping process. As in the other subgroups, bags with similar objects have been chosen (e.g., M6, M8, or M10 nuts). In-bag objects are shown in Figure 5c.

4.2. Experiments and Results

According to [45], three approaches can be followed to classify tactile data with 2D CNNs: training the network from scratch (method 1), using a pre-trained network with standard images and re-training the last classification layers (method 2), or changing the last layers by another estimator (method 3). The best results for each approach were obtained by TactNet6, ResNet50_NN, and VGG16_SVM, respectively. In this work, four experiments have been carried out to validate and compare the performance of TactNet3D against these 2D CNN structures considering only the subset of rigid objects, the subset of deformable objects, the subset of in-bag objects, and the whole dataset. The training, validation, and test sets to train the 2D CNN-based methods are formed using the individual images extracted from the 3D tactile tensors.

The performance of each method has been measured in terms of recognition accuracy. Each network has been trained 20 times with each subset, and the mean recognition rate and standard deviation for each set of 20 samples have been compared in Figure 6, where, for each experiment, the results of each method have been obtained using data from 1, 2, 5, 10, and 20 grasps of each object.

Moreover, representative confusion matrices for each method trained in subsets of rigid, deformable, and in-bag objects are presented in Figure 7. In contrast, the confusion matrices related to the whole dataset are presented in Figure 8. These confusion matrices have been obtained for the case in which each method is trained using data from two grasps to show the differences in classification performance.

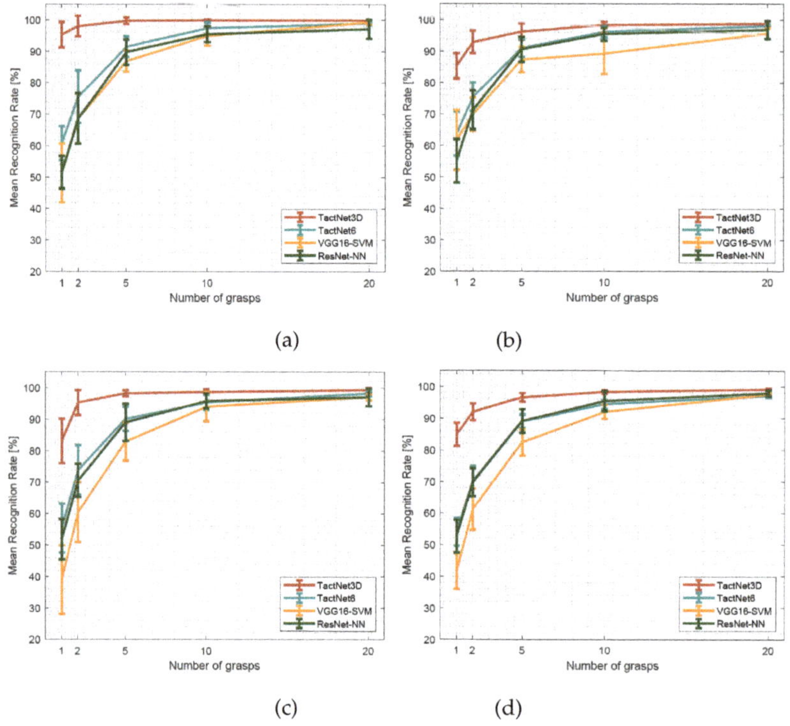

Figure 6. Experimental results of the experiment with rigid objects (**a**), deformable objects (**b**), in-bag objects (**c**), and all objects (**d**). Error bars represent the standard deviation σ of each recognition rate distribution over a 20-sample testing process.

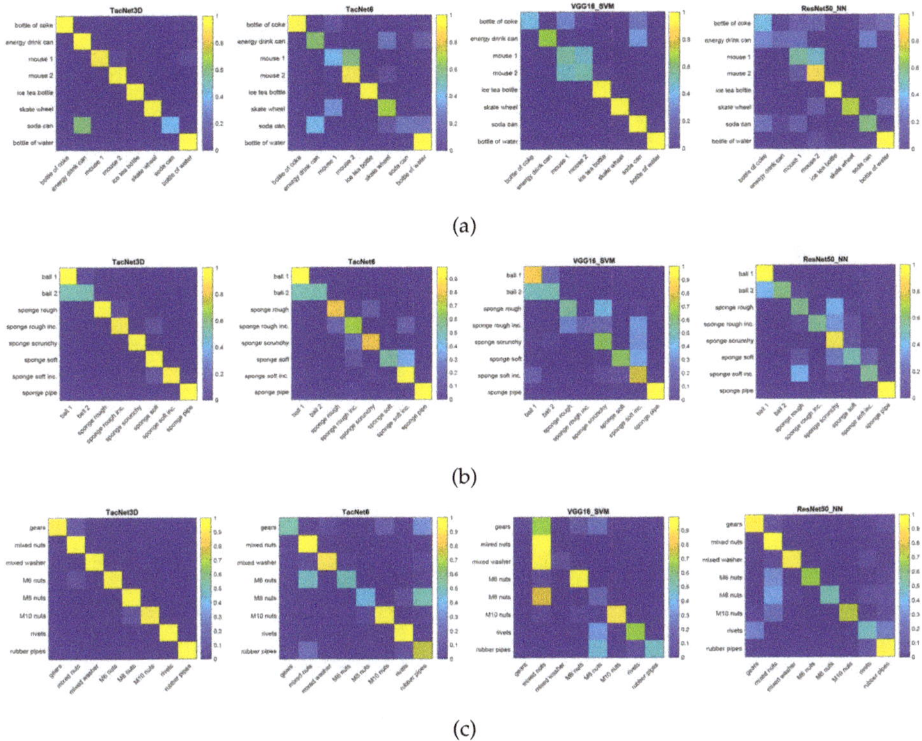

Figure 7. Confusion matrices of the methods, from left to right, TactNet3D, TactNet6, VGG16_SVM and ResNet50_NN, in experiments with rigid objects (**a**), deformable objects (**b**), and in-bag objects (**c**). All of the methods are trained using data from two grasps.

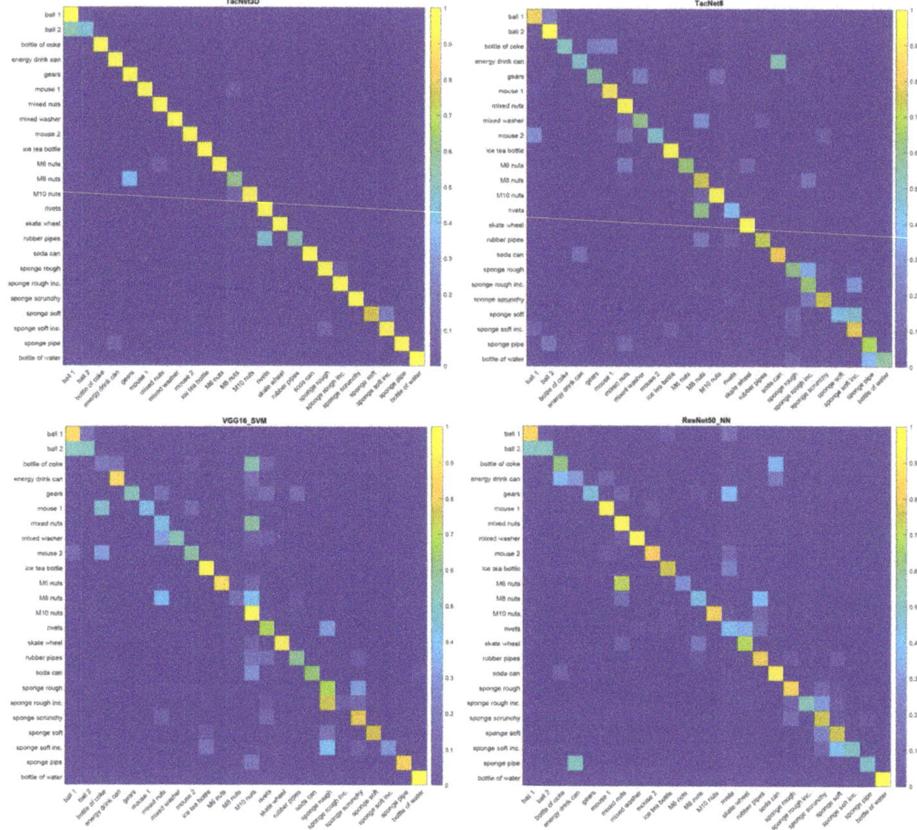

Figure 8. Confusion matrices of the methods, from left to right, TactNet3D, TactNet6, VGG16_SVM and ResNet50_NN, in the experiments with the whole dataset. All the methods are trained using data from two grasps.

5. Discussion

Regarding the performance of TactNet3D in comparison with 2D CNN-based methods, the results shown in Figure 6 prove that the recognition rate of the first one is better than the latter in all the studied cases. For all kinds of objects, rigid, deformable, or in-bag, and all the amount of grasps used as training data, TactNet3D outperforms 2D CNNs.

In addition, the differences in classification accuracy are higher when the number of training data are lower, getting better results when training TactNet3D with one or two grasps than 2D CNNs with five or ten grips in some cases. Therefore, it is not only shown that the performance is better, but also the adaptability of TactNet3D as the amount of data needed to train the network is lower, which is especially interesting for online-learning.

In addition, in the misclassification cases, the resulting object class given by TactNet3D has almost indistinguishable physical features to those of the grasped object, unlike 2D CNNs, which may provide disparate results, as can be seen in the confusion matrices presented in Figures 7 and 8. Looking at some object subsets with similar physical features such as the sponges, the different bag of nuts or the cans, it can be observed that the output given by TactNet3D corresponds to objects form the same subset, whereas 2D CNN output classes of objects with different features in some cases (e.g., bottle of coke and M10 nuts in Figure 8 bottom left). This phenomenon is interesting from the neurological point of view of an artificial touch sense as 3DTactNet behaves more similarly to human beings' sense of touch. However, a broad study of this aspect is out of the scope of this paper and will be considered in future works.

6. Conclusions

A novel method for the active tactile perception based on 3D CNN has been presented and used for an object recognition problem in a new robot gripper design. This gripper includes two underactuated fingers that accommodate to the shape of different objects, and have additional proprioceptive sensors to get its actual position. A tactile sensor has been integrated into the gripper, and a novel representation of sequences of tactile images as 3D tactile tensors has been described.

A new 3D CNN has been designed and tested with a set of 24 objects classified in three main categories that include rigid, deformable, and in-bag objects. There are very similar objects in the set, and objects that have changing and complex shapes such as sponges or bags of nuts, in order to assess the recognition capabilities. 3D CNN and classical CNN with 2D tensors have been tested for comparison. Both perform well with high recognition rates when the amount of training data are high. Nevertheless, 3D CNN gets higher performance even with a lower number of training samples, and misclassification is obtained just in very similar classes.

As future works, we propose the use of additional proprioceptive information to train multi-channel neural networks using the kinesthetic information about the shape of the grasped object, along with the tactile images for multi-modal tactile perception. In addition, the use of other dynamic approaches, such as temporal methods (e.g., LSTMs), for both tactile-based and multi-modal-based perception strategies, need to be addressed in more detail. Moreover, a comparison of new active tactile perception methods will be studied in depth.

Author Contributions: Conceptualization, F.P. and J.M.G.; software, F.P.; validation, F.P., J.M.G.-d.-G. and J.M.G.-d.-G.; investigation, F.P., J.M.G.-d.-G., and J.M.G.; data curation, F.P.; writing—original draft preparation, F.P., J.M.G.-d.-G., and J.M.G.; writing—review and editing, J.M.G.-d.-G. and J.M.G.; visualization, F.P. and J.M.G.; supervision, J.M.G.-d.-G.; project administration, A.J.G.-C. and J.M.G.-d.-G.; funding acquisition, A.J.G.-C.

Funding: This research was funded by the University of Málaga, the Ministerio de Ciencia, Innovación y Universidades, Gobierno de España, Grant No. DPI2015-65186-R and RTI2018-093421-B-I00, and the European Commission, Grant No. BES-2016-078237.

Conflicts of Interest: The authors declare no conflict of interest. The funders had no role in the design of the study; in the collection, analyses, or interpretation of data; in the writing of the manuscript, or in the decision to publish the results.

References

1. Krizhevsky, A.; Sutskever, I.; Hinton, G.E. ImageNet Classification with Deep Convolutional Neural Networks. In *Advances in Neural Information Processing Systems*; Neural Information Processing Systems Foundation, Inc.: Lake Tahoe, Nevada, 2012; pp. 1–9.
2. Cao, L.; Sun, F.; Liu, X.; Huang, W.; Kotagiri, R.; Li, H. End-to-End ConvNet for Tactile Recognition Using Residual Orthogonal Tiling and Pyramid Convolution Ensemble. *Cogn. Comput.* **2018**, *10*, 1–19. [CrossRef]
3. Shibata, A.; Ikegami, A.; Nakauma, M.; Higashimori, M. Convolutional Neural Network based Estimation of Gel-like Food Texture by a Robotic Sensing System. *Robotics* **2017**, *6*, 37. [CrossRef]
4. Gandarias, J.M.; Gómez-de Gabriel, J.M.; García-Cerezo, A.J. Tactile Sensing and Machine Learning for Human and Object Recognition in Disaster Scenarios. In *Third Iberian Robotics Conference*; Springer: Berlin, Germany, 2017.
5. Vidal-Verdú, F.; Oballe-Peinado, Ó.; Sánchez-Durán, J.A.; Castellanos-Ramos, J.; Navas-González, R. Three realizations and comparison of hardware for piezoresistive tactile sensors. *Sensors* **2011**, *11*, 3249–3266. [CrossRef] [PubMed]
6. Chathuranga, D.S.; Wang, Z.; Noh, Y.; Nanayakkara, T.; Hirai, S. Magnetic and Mechanical Modeling of a Soft Three-Axis Force Sensor. *IEEE Sens. J.* **2016**, *16*, 5298–5307. [CrossRef]

7. Ward-Cherrier, B.; Pestell, N.; Cramphorn, L.; Winstone, B.; Giannaccini, M.E.; Rossiter, J.; Lepora, N.F. The TacTip Family: Soft Optical Tactile Sensors with 3D-Printed Biomimetic Morphologies. *Soft Robot.* **2018**, *5*, 216–227. [CrossRef]
8. Gong, D.; He, R.; Yu, J.; Zuo, G. A pneumatic tactile sensor for co-operative robots. *Sensors* **2017**, *17*, 2592. [CrossRef]
9. Maiolino, P.; Maggiali, M.; Cannata, G.; Metta, G.; Natale, L. A Flexible and Robust Large Scale Capacitive Tactile System for Robots. *IEEE Sens. J.* **2013**, *13*, 3910–3917. [CrossRef]
10. Gandarias, J.M.; Gómez-de Gabriel, J.M.; García-Cerezo, A.J. Enhancing Perception with Tactile Object Recognition in Adaptive Grippers for Human–Robot Interaction. *Sensors* **2018**, *18*, 692. [CrossRef]
11. Chitta, S.; Sturm, J.; Piccoli, M.; Burgard, W. Tactile sensing for mobile manipulation. *IEEE Trans. Robot.* **2011**, *27*, 558–568. [CrossRef]
12. James, J.W.; Pestell, N.; Lepora, N.F. Slip Detection With a Biomimetic Tactile Sensor. *IEEE Robot. Autom. Lett.* **2018**, *3*, 3340–3346. [CrossRef]
13. Romeo, R.; Oddo, C.; Carrozza, M.; Guglielmelli, E.; Zollo, L. Slippage Detection with Piezoresistive Tactile Sensors. *Sensors* **2017**, *17*, 1844. [CrossRef] [PubMed]
14. Gandarias, J.M.; Gomez-de Gabriel, J.M.; Garcia-Cerezo, A. Human and object recognition with a high-resolution tactile sensor. In Proceedings of the IEEE Sensors Conference, Glasgow, UK, 29 October–1 November 2017.
15. Luo, S.; Mou, W.; Althoefer, K.; Liu, H. Iterative Closest Labeled Point for Tactile Object Shape Recognition. In Proceedings of the IEEE/RSJ International Conference on Intelligent Robots and Systems (IROS), Daejeon, Korea, 9–14 October 2016.
16. Yuan, Q.; Wang, J. Design and Experiment of the NAO Humanoid Robot's Plantar Tactile Sensor for Surface Classification. In Proceedings of the 4th International Conference on Information Science and Control Engineering (ICISCE), Changsha, China, 21–23 July 2017.
17. Hoelscher, J.; Peters, J.; Hermans, T. Evaluation of tactile feature extraction for interactive object recognition. In Proceedings of the IEEE-RAS International Conference on Humanoid Robots, Seoul, Korea, 3–5 November 2015.
18. Luo, S.; Bimbo, J.; Dahiya, R.; Liu, H. Robotic tactile perception of object properties: A review. *Mechatronics* **2017**, *48*, 54–67. [CrossRef]
19. Trujillo-Leon, A.; Bachta, W.; Vidal-Verdu, F. Tactile Sensor-Based Steering as a Substitute of the Attendant Joystick in Powered Wheelchairs. *IEEE Trans. Neural Syst. Rehabil. Eng.* **2018**, *26*, 1381–1390. [CrossRef] [PubMed]
20. Schiefer, M.A.; Graczyk, E.L.; Sidik, S.M.; Tan, D.W.; Tyler, D.J. Artificial tactile and proprioceptive feedback improves performance and confidence on object identification tasks. *PLoS ONE* **2018**, *13*. [CrossRef]
21. Bartolozzi, C.; Natale, L.; Nori, F.; Metta, G. Robots with a sense of touch. *Nat. Mater.* **2016**, *15*, 921–925. [CrossRef]
22. Jamone, L.; Natale, L.; Metta, G.; Sandini, G. Highly Sensitive Soft Tactile Sensors for an Anthropomorphic Robotic Hand. *IEEE Sens. J.* **2015**, *15*, 4226–4233. [CrossRef]
23. Roncone, A.; Hoffmann, M.; Pattacini, U.; Fadiga, L.; Metta, G. Peripersonal space and margin of safety around the body: Learning visuo-tactile associations in a humanoid robot with artificial skin. *PLoS ONE* **2016**, *11*, e0163713. [CrossRef]
24. Tanaka, Y.; Nagai, T.; Sakaguchi, M.; Fujiwara, M.; Sano, A. Tactile sensing system including bidirectionality and enhancement of haptic perception by tactile feedback to distant part. In Proceedings of the IEEE World Haptics Conference (WHC), Daejeon, Korea, 14–17 April 2013; pp. 145–150.
25. Luo, S.; Mou, W.; Althoefer, K.; Liu, H. Novel Tactile-SIFT Descriptor for Object Shape Recognition. *IEEE Sens. J.* **2015**, *15*, 5001–5009. [CrossRef]
26. Lee, H.; Wallraven, C. Exploiting object constancy: Effects of active exploration and shape morphing on similarity judgments of novel objects. *Exp. Brain Res.* **2013**, *225*, 277–289. [CrossRef]
27. Lepora, N.F. Biomimetic Active Touch with Fingertips and Whiskers. *IEEE Trans. Haptics* **2016**, *9*, 170–183. [CrossRef]
28. Okamura, A.M. Feature Detection for Haptic Exploration with Robotic Fingers. *Int. J. Robot. Res.* **2001**, *20*, 925–938. [CrossRef]
29. Lepora, N. Active Tactile Perception. In *Scholarpedia of Touch*; Atlantis Press: Paris, France 2016; pp. 151–159.

30. Dahiya, R.S.; Metta, G.; Valle, M.; Sandini, G. Tactile sensing-from humans to humanoids. *IEEE Trans. Robot.* **2010**, *26*, 1–20. [CrossRef]
31. Zapata-Impata, B.; Gil, P.; Torres, F.; Zapata-Impata, B.S.; Gil, P.; Torres, F. Learning Spatio Temporal Tactile Features with a ConvLSTM for the Direction Of Slip Detection. *Sensors* **2019**, *19*, 523. [CrossRef] [PubMed]
32. Drimus, A.; Kootstra, G.; Bilberg, A.; Kragic, D. Design of a flexible tactile sensor for classification of rigid and deformable objects. *Robot. Auton. Syst.* **2014**, *62*, 3–15. [CrossRef]
33. Dolz, J.; Desrosiers, C.; Ayed, I.B. 3D fully convolutional networks for subcortical segmentation in MRI: A large-scale study. *NeuroImage* **2018**, *170*, 456–470. [CrossRef] [PubMed]
34. Chaddad, A.; Desrosiers, C.; Niazi, T. Deep radiomic analysis of MRI related to Alzheimer's Disease. *IEEE Access* **2018**, *6*, 58213–58221. [CrossRef]
35. Gandarias, J.M.; Pastor, F.; García-Cerezo, A.J.; Gómez-de Gabriel, J.M. Active Tactile Recognition of Deformable Objects with 3D Convolutional Neural Networks. In Proceedings of the IEEE World Haptics Conference (WHC), Tokyo, Japan, 9–12 July 2019; pp. 551–555.
36. Feng, D.; Kaboli, M.; Cheng, G. Active Prior Tactile Knowledge Transfer for Learning Tactual Properties of New Objects. *Sensors* **2018**, *18*, 634. [CrossRef]
37. Kaboli, M.; Cheng, G. Robust Tactile Descriptors for Discriminating Objects From Textural Properties via Artificial Robotic Skin. *IEEE Trans. Robot.* **2018**, *34*, 1–19. [CrossRef]
38. Baishya, S.S.; Bauml, B. Robust material classification with a tactile skin using deep learning. In Proceedings of the IEEE/RSJ International Conference on Intelligent Robots and Systems (IROS), Daejeon, Korea, 9–14 October 2016.
39. Jamali, N.; Sammut, C. Majority voting: Material classification by tactile sensing using surface texture. *IEEE Trans. Robot.* **2011**, *27*, 508–521. [CrossRef]
40. Liu, H.; Song, X.; Nanayakkara, T.; Seneviratne, L.D.; Althoefer, K. A computationally fast algorithm for local contact shape and pose classification using a tactile array sensor. In Proceedings of the IEEE International Conference on Robotics and Automation (ICRA), Saint Paul, MN, USA, 14–18 May 2012.
41. Martinez-Hernandez, U.; Dodd, T.J.; Prescott, T.J. Feeling the Shape: Active Exploration Behaviors for Object Recognition With a Robotic Hand. *IEEE Trans. Syst. Man Cybern. Syst.* **2017**, pp. 1–10. [CrossRef]
42. Yi, Z.; Calandra, R.; Veiga, F.; van Hoof, H.; Hermans, T.; Zhang, Y.; Peters, J. Active tactile object exploration with gaussian processes. In Proceedings of the IEEE/RSJ International Conference on Intelligent Robots and Systems (IROS), Daejeon, Korea, 9–14 October 2016.
43. Corradi, T.; Hall, P.; Iravani, P. Bayesian tactile object recognition: Learning and recognising objects using a new inexpensive tactile sensor. In Proceedings of the IEEE International Conference on Robotics and Automation (ICRA), Seattle, WA, USA, 26–30 May 2015.
44. Albini, A.; Denei, S.; Cannata, G. Human Hand Recognition From Robotic Skin Measurements in Human-Robot Physical Interactions. In Proceedings of the IEEE/RSJ International Conference on Intelligent Robots and Systems (IROS), Vancouver, BC, Canada, 24–28 September 2017.
45. Gandarias, J.M.; García-Cerezo, A.J.; Gómez-de Gabriel, J.M. CNN-based Methods for Object Recognition with High-Resolution Tactile Sensors. *IEEE Sens. J.* **2019**. [CrossRef]
46. Falco, P.; Lu, S.; Cirillo, A.; Natale, C.; Pirozzi, S.; Lee, D. Cross-modal visuo-tactile object recognition using robotic active exploration. In Proceedings of the IEEE International Conference on Robotics and Automation (ICRA), Singapore, 29 May–3 June 2017.
47. Luo, S.; Liu, X.; Althoefer, K.; Liu, H. Tactile object recognition with semi-supervised learning. In Proceedings of the International Conference on Intelligent Robotics and Applications (ICIRA); Portsmouth, UK, 24–27 August 2015.
48. Khasnobish, A.; Jati, A.; Singh, G.; Bhattacharyya, S.; Konar, A.; Tibarewala, D.; Kim, E.; Nagar, A.K. Object-shape recognition from tactile images using a feed-forward neural network. In Proceedings of the International Joint Conference on Neural Networks (IJCNN), Brisbane, QLD, Australia, 10–15 June 2012.
49. Schmitz, A.; Bansho, Y.; Noda, K.; Iwata, H.; Ogata, T.; Sugano, S. Tactile object recognition using deep learning and dropout. In Proceedings of the IEEE-RAS International Conference on Humanoid Robots, Madrid, Spain, 18–20 November 2014.
50. Lawrence, S.; Giles, C.L.; Tsoi, A.C.; Back, A.D. Face recognition: A convolutional neural-network approach. *IEEE Trans. Neural Netw.* **1997**, *8*, 98–113. [CrossRef] [PubMed]

51. Madry, M.; Bo, L.; Kragic, D.; Fox, D. ST-HMP: Unsupervised Spatio-Temporal feature learning for tactile data. In Proceedings of the IEEE International Conference on Robotics and Automation (ICRA), Hong Kong, China, 31 May–7 June 2014.
52. Liu, H.; Guo, D.; Sun, F. Object Recognition Using Tactile Measurements: Kernel Sparse Coding Methods. *IEEE Trans. Instrum. Meas.* **2016**, *65*, 656–665. [CrossRef]
53. Kerzel, M.; Ali, M.; Ng, H.G.; Wermter, S. Haptic material classification with a multi-channel neural network. In Proceedings of the International Joint Conference on Neural Networks (IJCNN), Anchorage, AK, USA, 14–19 May 2017.
54. Tran, D.; Bourdev, L.; Fergus, R.; Torresani, L.; Paluri, M. Learning spatiotemporal features with 3d convolutional networks. In Proceedings of the IEEE International Conference on Computer Vision (ICCV), Santiago, Chile, 11–18 December 2015; pp. 4489–4497.
55. Birglen, L.; Laliberté, T.; Gosselin, C.M. *Underactuated Robotic Hands*; Springer: Berlin, Germany, 2007; Volume 40.

Article

An Embedded, Multi-Modal Sensor System for Scalable Robotic and Prosthetic Hand Fingers

Pascal Weiner [1,*], Caterina Neef [1,†], Yoshihisa Shibata [2], Yoshihiko Nakamura [2] and Tamim Asfour [1,*]

1 KIT Department of Informatics, Karlsruhe Institute of Technology, 76137 Karlsruhe, Germany; caterina.neef@th-koeln.de
2 Department of Mechano-Informatics, University of Tokyo, Tokyo 113-8656, Japan; shibata@ynl.t.u-tokyo.ac.jp (Y.S.); nakamura@ynl.t.u-tokyo.ac.jp (Y.N.)
* Correspondence: pascal.weiner@kit.edu (P.W.); asfour@kit.edu (T.A.); Tel.: +49-721-608-43546 (P.W.); +49-721-608-47379 (T.A.)
† Current address: Cologne Cobots Lab, TH Köln—University of Applied Sciences, 50679 Köln, Germany.

Received: 31 October 2019; Accepted: 20 December 2019; Published: 23 December 2019

Abstract: Grasping and manipulation with anthropomorphic robotic and prosthetic hands presents a scientific challenge regarding mechanical design, sensor system, and control. Apart from the mechanical design of such hands, embedding sensors needed for closed-loop control of grasping tasks remains a hard problem due to limited space and required high level of integration of different components. In this paper we present a scalable design model of artificial fingers, which combines mechanical design and embedded electronics with a sophisticated multi-modal sensor system consisting of sensors for sensing normal and shear force, distance, acceleration, temperature, and joint angles. The design is fully parametric, allowing automated scaling of the fingers to arbitrary dimensions in the human hand spectrum. To this end, the electronic parts are composed of interchangeable modules that facilitate the mechanical scaling of the fingers and are fully enclosed by the mechanical parts of the finger. The resulting design model allows deriving freely scalable and multimodally sensorised fingers for robotic and prosthetic hands. Four physical demonstrators are assembled and tested to evaluate the approach.

Keywords: sensorised fingers; tactile sensors; parametric model; robotic fingers; prosthetic fingers; hand prostheses; anthropomorphic robotic hands

---

1. Introduction and Related Work

Humans possess a very sophisticated sense of touch that enables intelligent physical interactions with the environment [1]. If the sense of touch is missing, otherwise trivial manipulation tasks are difficult or impossible to achieve despite visual feedback [2]. Therefore, efforts have been made to replicate the sense of touch in artificial systems such as robotic and prosthetic hands. An overview of different technologies that were explored to create such tactile sensors can be found in [3–6], a review on the use of tactile information for robotic manipulation is found in [7].

While contact and force sensing play a key role for the control of artificial hands, other modalities contribute to the understanding of the interaction between an artificial hand and its environment. Joint angle sensors can provide information regarding the shape of a grasped object, temperature sensors measure thermal conductivity of an object, accelerometers can detect slip and distance sensors are able to detect the presence of an object before it makes contact with the hand.

Integration of such multi-modal sensor technologies into robotic and prosthetic hands is a complex and challenging task as sensing electronics and electric connections need to be distributed throughout the already complex mechanical structure of an artificial hand and its fingers. An additional challenge is given by the tight space constraints inside fingers and joints. A high degree of integration of the signal conditioning electronics is therefore desirable in order to save space, ensure low noise of the sensor signals and keep the overall system complexity manageable. A low wire count is beneficial as well for reducing system complexity, especially when routing cables through narrow finger joints. Additionally, a simple production and application process eases prototyping and integration into artificial hands.

Prosthetic hands, in particular, need to be scalable to match the size of the able hand of an amputee [8]. However, even for a humanoid robotic hand of any given size, scalability of the fingers is an issue, since the four fingers require different dimensions in order to ensure an anthropomorphic appearance. To ensure the ability to build individually sized hands, the overhead stemming from designing different versions of fingers should be kept to a minimum.

In this work, we propose a solution to implement freely scalable fingers based on a parametrised, skeleton-based computer aided design (CAD) model. The fingers are equipped with a multi-modal sensor setup leveraging recent developments in miniaturised commercial sensors for smartphones and consumer electronics. Four physical demonstrators, derived from the presented model, have been assembled and are shown in Figure 1a. We plan to utilise the presented development in differently sized hands as shown in Figure 1b.

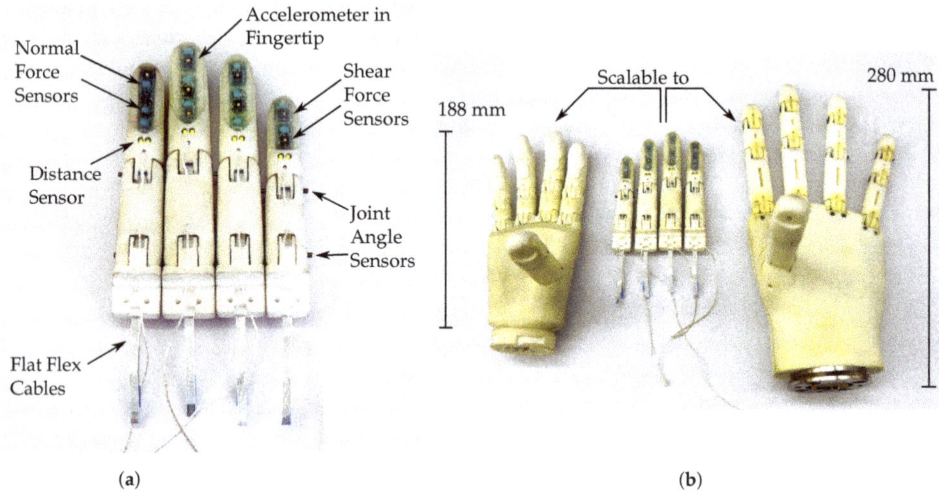

Figure 1. (**a**) The four manufactured demonstrators derived from the scalable model. (**b**) A comparison of the physical demonstrators with our most recent KIT Prosthetic Hand (left) and our robotic KIT ARMAR-6 Hand (right).

There exist a number of approaches that make use of diverse commercially available sensors with integrated signal processing electronics as well as a minimal bus interface to reduce system complexity and size. In many cases a soft material is used to transduce the applied force, protect the electronics from mechanical damage and ensure compliant interaction with the environment.

Tomo et al. [9,10] use commercially available digital three-axis Hall effect sensors together with a magnet embedded into flexible material above the sensor to measure normal and shear forces through the displacement of the magnet. These modular sensors were integrated into the Allegro [11] and the iCub hands and fingers [12,13]. An interesting variation of this sensing method is presented in [14], where a single sensor is used to measure elastic deformation of the finger structure induced by applied forces.

Tenzer et al. [15] use miniature barometric pressure sensors to design tactile normal force sensors. The sensors are completely covered and filled with polyurethane, which acts as the force transmitting medium. When a force is exerted on the soft material, it is transmitted to the sensor and is measured as a change in pressure. These sensors have been integrated into the iHY hand [16]. The design is adapted by Jentoft et al. [17] to allow for a flexible and stretchable array of sensors.

A multi-modal sensor setup is used in [18] to assemble small hexagonal printed circuit boards (PCBs) for the use as artificial skin. Each hexagon houses a microcontroller, an accelerometer, proximity sensors, and temperature sensors. In an extension of the work, force sensing cells are added [19]. A silicone cover protects the skin from damage. Individual cells can communicate with adjacent cells through an event-driven protocol that allows efficient data transfer over a small number of wires as well as offering redundancy in case of failures [20]. The skin is used to cover an entire dual arm manipulation platform [21].

For the humanoid robot iCub, capacitive tactile sensors were developed [22] and integrated into the hand. One plate of the capacitor is formed by a flexible PCB, the other by a deformable conductor. Both are separated by silicone foam. The sensor signal is digitalised using a commercially available capacitance-to-digital converter. In an extension of the work, the production process is optimised in order to make the sensors easier to manufacture and more robust [23]. The capacitive sensing principle is also used for skin spanning larger parts of the iCub robot [24]. Alagi et al. [25] expand on the principle of capacitive tactile sensors by incorporating capacitance-based distance sensing. This sensor is also incorporated into an anthropomorphic hand [26].

In [27], Wörn and Weiss introduce a resistive sensing method using conductive foam and interlaced traces on PCBs. When the foam is compressed onto the traces, resistance between the traces is lowered. Kõiva et al. implement a tactile fingertip based on this measurement method using a PCB lasered onto a three-dimensional structure for the Shadow hand [28]. Recently, this fingertip was extended by a fingernail [29] measuring static forces by combining methods by Tenzer et al. [15] and Tomo et al. [10]. In addition, dynamic events are measured using a three-axis accelerometer. A similar method for force measurement is used in [30] to construct curved tactile fingertips using a flexible PCB and conductive rubber.

Three tactile systems using commercially available distance sensors have recently been introduced [31–33]. The sensors are covered by a layer of a translucent elastomer, allowing the sensor to measure distance even through the material. When an object comes into contact with the elastic layer, the layer width is reduced, which can also be measured by the distance sensors. The measured distance can then be translated into a force measurement as the material properties are known. All three systems were evaluated on robotic platforms.

There exist a number of works dedicated to the engineering challenges associated with the design of artificial anthropomorphic fingers. Laurentis et al. [34] present an approach based on a 3D-printed model where all parts of the finger are already assembled during printing, thus reducing production time. Only the sensors have to be added after the printing process. Liu et al. present a prosthetic hand with fingers able to withstand deformations and impacts, while also embedding a three-axis force sensor [35]. The fingers are made of individual segments connected by ligaments and rolling contact joints. The fingertips are equipped with three-axis force sensors based on an optical measurement principle [36]. Cheng et al. [37] incorporate a tactile sensor fingertip and joint angle encoders in an anthropomorphic prosthetic finger fabricated from metal. The finger is actuated by a motor directly attached to the finger with a rigid transmission mechanism, all encapsulated inside of the finger. The design of a prosthetic finger with integrated force sensing is presented by Imbinto et al. [38]. The included sensor is based on the work by Jentoft et al. [17]. The finger includes a motor in the proximal phalanx as well as a non-back driveable mechanism that locks the finger while grasping.

The multimodal BioTac tactile fingertip (SynTouch Inc., Montrose, CA, USA) incorporates 19 electrodes embedded into a fluid-filled soft fingertip.When pressure is applied to the fingertip, the distribution of the liquid inside the finger changes and the resistance between the different electrodes changes with it. A hydro-acoustic pressure sensor measures high frequency vibrations through the liquid and additional temperature sensors in combination with heating elements enable temperature flux measurements [39]. The rich set of information generated by the multimodal sensor setup used in the BioTac sensor is well suited for feature extraction and classification based on machine learning approaches [40].

With the development of sensorised fingers we set out to provide a rich multi-modal sensory feedback while addressing the challenge of tight system integration, aiming at a practical and useable yet versatile solution. We intend to use these fingers in both our prosthetic hand developments [41] as well as the hands for our humanoid robot ARMAR-6 [42]. As with our first version [43], we make use of commercial components and proven sensing methods in order to limit system complexity while introducing scalability, improving robustness, the anthropomorphic design, and extending the sensor suit. The developed sensor system allows sensing of normal and shear forces, distance, vibration, joint angles, as well as temperature. The finger's mechanics and its electronics are freely scalable to the dimensions of any human finger, which to the authors' knowledge, is the first solution to address this problem in literature. We consider the integration of a rich sensor suit into a scalable and robust robotic finger to be the major contribution of this work.

In the remainder of this work we give a detailed overview over the mechanical and electrical design of the fingers. We describe how the central requirement of scalability influences the design and how the multi-modal sensor system is integrated into the finger. Finally, extensive tests are conducted to characterise the different sensor modalities individually and as a system to show the validity of our approach.

2. Design of Anthropomorphic Robotic Fingers

With our development of robotic fingers with a multi-modal sensor system we strive to enhance grasping and manipulation skills of both prosthetic and robotic hands as well as to enable haptic feedback for prosthetic applications. When used as robotic fingers, the sensor system should provide sufficient information for closed-loop control throughout all phases of the grasping process, namely approach/pregrasp, contact with the object, lift, placing and breaking contact. In a prosthetic setting the fingers should enable haptic feedback to the user. The information provided by the sensor system can also be used for semi-autonomous control schemes where the prosthesis automates part of the grasping process. This relieves the user from the high cognitive burden of manually controlling a prosthetic device.

As especially prosthetic hands should match the dimensions of the human hand, it is important to be able to scale the dimensions of the model and manufacture new fingers easily. This free scalability is particularly challenging for the embedded electronics. Application of such fingers in robotic, as well as prosthetic, settings also means that the requirements regarding the sensor setup vary in terms of sensor position, quantity and type of sensors. The mechanical design and electronics must be flexible enough to allow for such reconfiguration.

Different sensor modalities have been proven to be useful in the robotic and prosthetic domain. Apart from normal and shear force sensing, accelerometers enabling slip detection, as well as distance sensors have been used. For both robotic and prosthetic hands, under-actuation is an important principle for building lightweight hands with simplified control strategies, where multiple joints are coupled and actuated jointly. Such under-actuated hands rely heavily on exploiting interaction and contact with the objects and environment to perform successful grasping. For such hands, contact and force information is essential and joint angle encoders are needed for estimating the current kinematic state of the hand.

Our approach to implementing a multi-modal sensor system into an anthropomorphic finger combines a completely parametrised CAD-model with a modular electronic system consisting of commercially available sensors and standard PCBs. For force sensing we combine and extend methods described by Tenzer et al. [15] and Tomo et al. [10]. Additionally, we incorporate distance, vibration and joint angle sensing into the system. Mechanical parts are realised using 3D-printing. In the remainder of this section we describe the interplay of mechanical scalability and electrical modularity and how these two central concepts are implemented in detail.

2.1. Finger Skeleton

For the mechanical structure of the fingers we created a single scalable CAD-model of the finger. It is based on a skeleton that contains all sketches for all important features of the finger, as can be seen in Figure 2. Based on this skeleton, the three individual parts for knuckle, proximal and distal phalanx are derived by referencing the sketches in the skeleton model. The skeleton can be parametrised using seven high-level parameters that can be set individually and independently of each other. These parameters define the width and height of the proximal interphalangeal (PIP) joint (PIP_w and PIP_h) as well as the lengths of the proximal, intermediate and distal phalanx ($proximal_l$, $intermediate_l$ and $distal_l$). Two additional parameters s_w and s_h represent scaling factors that are used to determine the width and height of the distal interphalangeal (DIP) and metacarpophalangeal (MCP) joints as follows:

$$\begin{aligned} MCP_w &= PIP_w \cdot s_w & \qquad DIP_w &= PIP_w \cdot (1 - (s_w - 1)) \\ MCP_h &= PIP_h \cdot s_h & \qquad DIP_h &= PIP_h \cdot (1 - (s_h - 1)) \end{aligned}$$

The scaling factors s_w and s_h are calculated as the average ratios of joint height/width as measured by Vergara et al. [44]. If desired, the height/width of all joints can also be changed individually and independently by replacing the above calculations for MCP and DIP in the root sketch with independent scalar values.

The seven parameters (PIP_w, PIP_h, $proximal_l$, $intermediate_l$, $distal_l$, s_w, s_h) define lengths and radii in a root sketch, together with basic definitions of useful axes and planes. This root sketch is referenced by further sketches in the model that define further features like loft guides or faces later used for extrusions. The model makes use of splines which allow for round and smooth shapes. To keep the number of spline parameters small, we define the whole shape of the finger with as few splines as possible and use a small number of via points for each spline. Once the whole skeleton is defined as a set of sketches, the three individual parts of the fingers are designed by deriving relevant sketches from the skeleton and building the geometry based on these sketches.

In addition to the seven high-level parameters, the model contains 257 other parameters, which are either derived from the high-level parameters or are constant like fittings, so that a change in high-level parameter values results in a change of the dimensions of the entire finger.

The parameters can, for example, be set to match the measured dimensions of the able hand of an amputee. We successfully tested the model using the 5th percentile female hand dimensions as well as the 95th percentile male hand dimensions. Figure 3 shows two specimens of the developed CAD-model scaled to the dimensions of the little and middle finger sized according to a median female hand.

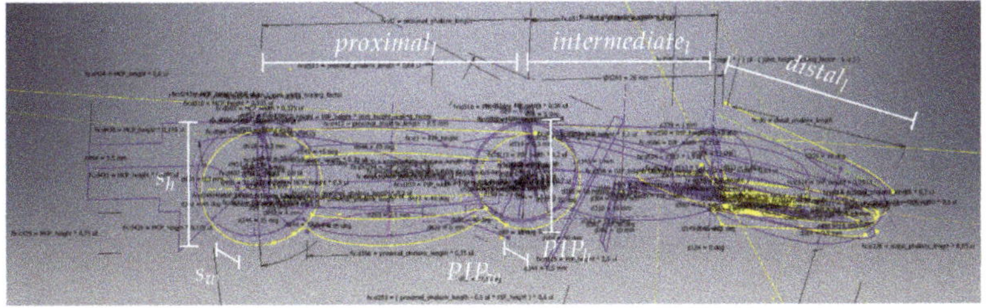

Figure 2. Skeleton sketches used for the individual phalanges. The seven high-level parameters are used in these sketches to derive all dependent dimensions of the finger.

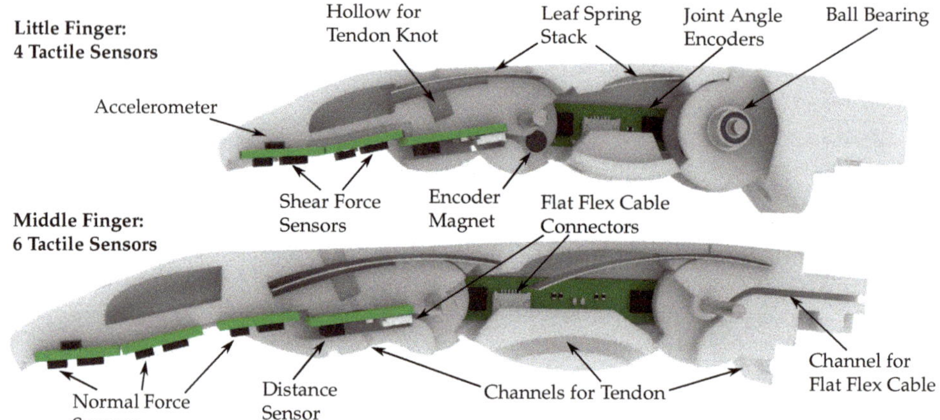

Figure 3. Section view of the little and middle finger with the model scaled to the 50th percentile female dimensions. The section view of the little finger shows the joints of the finger while the deeper cut into the middle finger shows the paths for both the flat flex cables (FFCs) and the tendon.

In addition to human sizing, special attention was given to the anthropomorphic shape of the fingers which is an important factor especially for the acceptance in prosthetic applications [45]. In Figure 4, the profile of a physical demonstrator is shown.

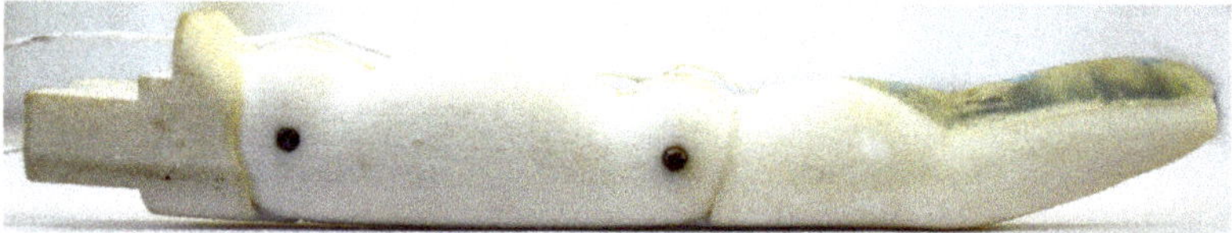

Figure 4. Profile of the ring finger showing a curved design for the individual phalanges.

We consciously made the decision to fuse the distal and intermediate phalanx into one part. Each finger hence has two joints, the MCP joint and the PIP joint. Omitting the DIP joint reduces the complexity of the assembly and allows for sufficient space in the distal part to house the sensor system. This is a common design choice for both commercially available prostheses [46] as well research prostheses and anthropomorphic robotic hands [47].

2.2. Physical Demonstrators

The physical demonstrators in this work were sized according to the 50th percentile female dimensions as described by the German standard specification (DIN 33402-2) in finger length and additional dimensions as identified by Vergara et al. [44]. The dimensions for all fingers and segments are shown in Table 1. As the range for prosthetic and robotic hand sizes varies greatly, we have chosen dimensions at the smaller end of the range to ensure that the model and electronics can be used in even small hands and can consequently also be easily modified and extended for larger hands.

The individual components for the fingers are manufactured from polyamide (PA 2200) using 3D-printing methods (selective laser sintering) to enable individualised sizing as needed. Through the use of additive manufacturing for mechanical parts, commercially available sensors and standard manufacturing techniques for the PCBs the price for an individual finger can be kept lower than 70€.

Table 1. Dimensions in [mm] of the finger demonstrators.

	Index Finger	Middle Finger	Ring Finger	Little Finger
Finger length	80.64	90.02	83.98	66.17
Proximal phalanx length	34.15	38.15	34.45	27.05
Distal phalanx length	46.49	51.87	49.53	39.12
Proximal phalanx height	17.28	17.28	17.28	17.28
Distal phalanx height	14.4	14.4	14.4	14.4
Proximal phalanx width	18.7	20.13	18.81	16.5
Distal phalanx width	17	18.3	17.1	15

2.3. Joint Structure and Actuation

Each finger consists of three individual parts, the distal/intermediate phalanx, the proximal phalanx and the knuckle, as well as two joints, the metacarpophalangeal (MCP) joint and the proximal interphalangeal (PIP) joint, that connect the phalanges. Each joint is supported by two miniature metal ball bearings to reduce friction to a minimum. The joints are actuated in flexion direction by a tendon that is located on the palmar side of the fingers. Actuation via tendons was chosen as it does not require levers inside of the finger, hence leaving necessary space for the electronics and cables.

Each joint is extended by a stack of leaf springs, as opposed to the torsion springs, that are used in the previous version of the KIT prosthetic hand [41], in order to free up space in the joints for sensors and their cables. The leaf springs are installed completely inside each finger and are not visible from the outside. The pockets for holding the leaf springs are slightly curved to minimise friction and simultaneously create pretension in the springs. One end of the leaf springs is glued into the phalanx, the other slides in and out of the pocket. In order to decrease the friction between the finger material and leaf springs, we created a hollow space inside the finger. Both the distal/intermediate as well as the proximal phalanx are hollow for the most part to allow the springs to extend without friction and in order to save weight. We used spring steel with 0.1 mm thickness as the leaf spring material. Each joint is equipped with a stack of three leaf springs to reach a sufficiently high torque.

2.4. Embedded Sensor System Overview

As shown in Figures 1 and 3, a customizable number of sensors can be integrated into a finger, depending on the finger pad area available for the integration of tactile sensors. The middle, and thus largest, finger used in this work, shown in the bottom of Figure 3, contains a total number of ten sensors, which include two joint angle encoders, a distance sensor, three normal force sensors, three shear force sensors and one accelerometer. The sensor PCBs can be connected to a controller using FFCs via connectors on the joint angle encoder and distance sensor PCBs (see Figure 3), while the tactile sensor PCBs are connected through magnet wires.

2.5. Sensor Placement Experiments

In order to find the optimal configuration and placement positions of the tactile sensors in each finger, we conducted tests to determine which surfaces of each finger are in contact when grasping different objects. These experiments were carried out prior to the definition of the tactile sensor layout and their results were used to define the area on the finger that should be covered with tactile sensors. We used the 50th percentile female version of the KIT Prosthetic Hand [41], as well as five objects (banana, baseball, bowl, drill, spam) from the YCB object set [48] and two objects (cola, green cup) from the KIT object set [49], in order to have a variation of shapes and sizes. The outside surface of each object was painted green, after which the object was immediately grasped using either a top or a side grasp (approaching the object from the top or the side, respectively). From the experiments it is evident that the finger area that is in contact with most objects is the finger pad, especially that of the index finger. Examples of the experiment results can be seen in Figure 5. Thus, contact forces and slip can be measured most accurately by placing sensors in these areas.

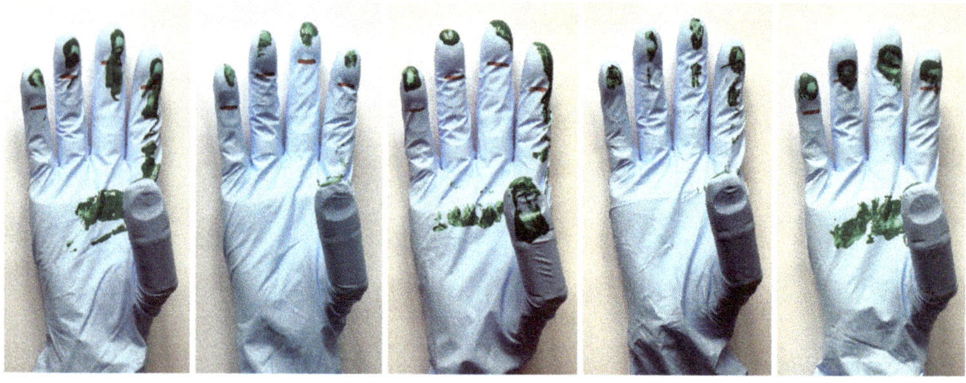

(a) cup, side grasp (b) Spam, top grasp (c) Bowl, top grasp (d) Bottle, side grasp (e) Drill, side grasp

Figure 5. Examples of results from the finger contact surface experiment.

2.6. Sensor System Configuration

We use a combination of three types of sensors to acquire a wide range of tactile information in the available finger pad area, shown in Figure 6. The normal force sensors (NPA201, Amphenol, Wallingford, USA) are based on ideas presented by Tenzer et al. [15]. In contrast to the design used by Tenzer et al., they are constructed by leaving a small pressure chamber in the silicone above the barometric pressure sensor as explained in [43]. When a force is exerted on the silicone in vicinity to the sensor, this force compresses the pressure chamber which in turn is measured as an increase in pressure by the sensor. These sensors will be called barometer-based sensors in the rest of this publication. While estimating forces only in one direction, they are very sensitive and offer a high resolution, as shown in Section 3.1. The shear force sensors (MLX90393, Melexis, Ypres, Belgium) are based on work presented by Tomo et al. [9,10]. They can be used to estimate both normal and shear forces and offer a larger measurement range than the normal force sensors. These sensors will be called Hall effect-based sensors in the rest of this publication. The normal force estimation is hence performed using different measurement principles for each sensor, enabling a more accurate overall measurement and event detection through sensor fusion.

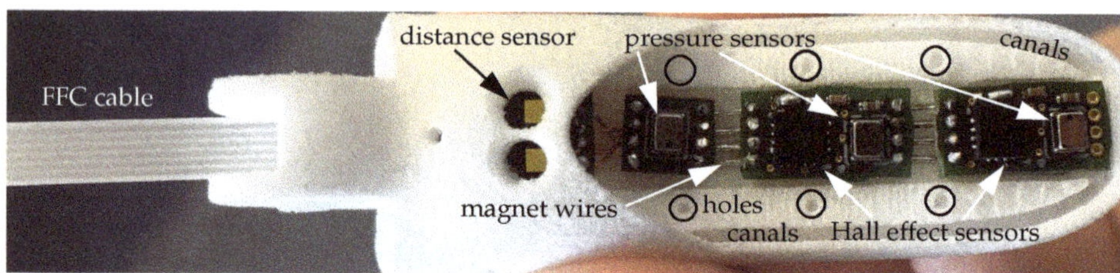

Figure 6. The individual tactile sensor printed circuit boards (PCBs), containing normal and shear force sensors and an accelerometer, as well as the distance sensor and design details are shown.

An accelerometer (BMA456, Bosch Sensortec, Reutlingen, Germany) with a sample rate of 1.6 kHz is mounted at the back of the most distal sensor PCB. It can be used to detect slip of grasped objects as it is, unlike the barometer- and Hall effect-based sensors, able to achieve the necessary sampling rates for this task.

To gain additional information even before contact is made with an object, we implemented a distance sensor into the finger, shown in Figure 6. The sensor is a state of the art time of flight (ToF) device that is able to measure the distance of objects independent of their reflectance (VL53L1X,

STMicroelectronics, Geneva, Switzerland). The distance information acquired with this sensor could for example be used to control an artificial hand to form a pre-grasp when approaching the object, as it can determine the distance between the finger and the object which in turn can then be used to adjust the pre-grasp accordingly.

All sensors in the sensor system are commercially available sensors that integrate all signal conditioning and digitalisation circuits. All sensors also include a temperature sensing element that can be sampled together with the main sensor signal. This way no additional electronics are needed for signal processing. All sensors, except for the accelerometer, are sampled with 50 Hz, while the accelerometer is sampled at 1.6 kHz. All sensors communicate using the two-wire I_2C communication bus. Together with the supply voltage lines, only four wires are needed to operate and read out all sensors. As there are no commercially available flat flex cables and connectors with four terminals, the smallest available configuration with six terminals is chosen to connect the sensor system to a central processor, typically inside of the palm of the robotic hand.

2.7. Manufacturing Process

The manufacturing process of the tactile sensors is described in the following. First the tactile sensor PCBs for each finger are glued into place. We then use the same methodology described in our previous work [43] to cast the barometer- and Hall effect-based sensors in silicone rubber.

For the Hall effect-based sensors a thin pad of silicone with Shore hardness A (ShA) 13 is glued to the top of the sensor using silicone glue. A magnet is placed on the centre above the sensor and a drop of silicone is used to fix the magnet to the silicone pad.

For the barometer-based sensors a small mould is placed around the sensor. A small magnet is placed directly above the opening in the casing of the barometric pressure sensor and the mould is filled to the top of the magnet with silicone (ShA 45). Since the housing of the pressure sensor is magnetic, the magnet is tightly held in place. As soon as the silicone is hardened, the magnet is removed. The resulting hole forms the walls of the pressure chamber above the sensor. The hole is then covered with a thin sheet of silicone (ShA 22) and fixated by a drop of silicone. Both moulds and placeholder magnets are shown in Figure 7a. The 3D printed moulds (as shown in Figure 7a) can be placed directly onto the PCBs glued into the finger. The individual sensors are cast in rubber, as shown in Figure 7a, the result of which can be seen in Figure 7b.

The entire area of the finger pad is then cast in an additional layer of silicone rubber (ShA 13), shown in Figure 7c. In order to ensure sufficient stability of the silicone layer, holes and canals with undercuts, as shown in Figure 6, are integrated into the finger tip. This design allows for increased adhesion of the silicone to the 3D printed material. Compared to the previous work, a larger part of the fingertip is cast in silicone rubber to enable a configuration with a larger number of sensors. This also contributes to an improved grasping behaviour as the silicone rubber is more elastic and has a higher friction coefficient compared to the 3D printed material [50].

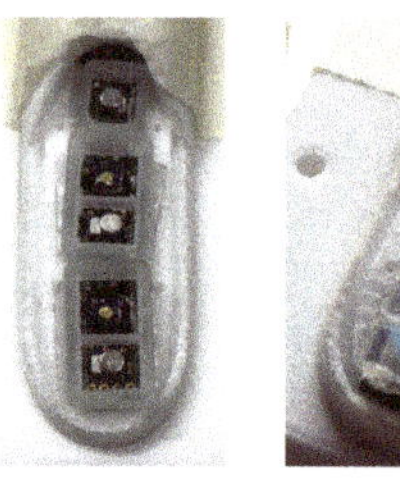

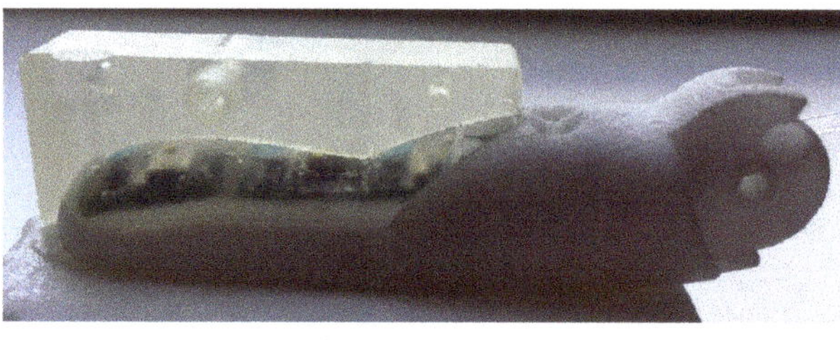

(a) (b) (c)

Figure 7. (**a**) The moulds for the silicone rubber casting process are placed directly into the finger. (**b**) The individual sensors have been cast in rubber. The next step is to cast the remaining finger pad area in rubber. (**c**) The result of the casting process of the finger tip and one half of the used mould is shown. A silicone canal is used to inject silicone rubber into the moulds.

2.8. Joint Angle Measurement

Additionally required for the forming of pre-grasps are joint angle encoders that determine the rotation angle of each joint and can be used to control and adjust the finger flexion. One possibility to measure joint rotation angles is to place a diametrically polarised magnet directly on the joint axis that rotates with the joint. A Hall effect sensor that is placed directly above the magnet is then able to measure the change in magnetic field strength induced by the rotation of the magnet. As the Hall effect sensor outputs the magnetic field strength in x, y and z direction, these values can be used to derive the rotation angle of each joint.

To calculate the rotation angle α_z around the z axis, the following equation is used, where x_{Mag} and y_{Mag} are the magnetic field strengths in x and y direction, respectively:

$$\alpha_z = \arctan 2(y_{Mag}, x_{Mag}) \cdot \frac{180^\circ}{\pi} \quad (1)$$

In this work, however, we perform this measurement off-axis both for the magnet and the sensor (MLX90393, Melexis), due to space constraints in the joints, which is shown in Figure 8. The magnets are glued into the distal/intermediate phalanx (blue) and the knuckle (green), and rotate around their respective joints when these are rotated. At 45 degrees rotation the magnets are positioned directly above the sensors, so that they have the same distance between each other at 0 and 90 degrees, which corresponds to the minimum and maximum angle, respectively, of all joints. The above stated Equation (1) can nevertheless be used to provide an approximation of the joint rotation angle using this off-axis measurement. However, when the placement of the magnet above the sensor changes in x or y direction, the magnetic field strength and therefore sensor output of the Hall effect sensor changes. Therefore, an experiment is necessary to determine the correlation between the sensor output and the actual rotation angle, which is described in Section 3.3.

As the PCB for the joint angle encoders contains just two sensors and a connector, it is fairly easy to adjust the distance between the two sensors for different finger sizes directly in the PCB layout. We hence produced PCBs in three different sizes for the little finger, the middle finger, and the other two fingers which can accommodate the same PCB size due to their similar dimension.

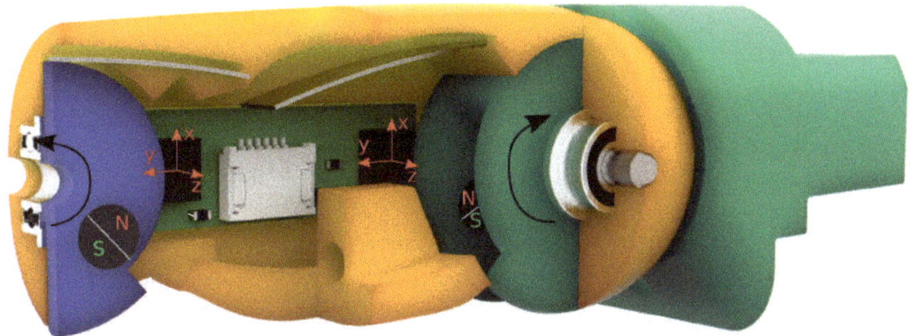

Figure 8. Section view of the little finger, with the distal/intermediate phalanx in blue, the proximal phalanx in orange and the knuckle in green, showing the joint angle encoders and magnets used to determine the joint rotational angles.

3. Experimental Results

A series of experiments were conducted on the physical demonstrators to assess the performance of the sensors individually and as a coherent system. For the experiments regarding normal and shear forces, as well as a spatial mapping, a two-axis linear table was used as depicted in Figure 9. Each axis was a precision linear stage (PT4808, MM Engineering GmbH, Brackenheim, Germany) with 0.5 mm displacement per turn attached to a stepper motor with 200 steps per turn. A force/torque sensor (Mini 40, ATI Industrial Automation, Apex, NC, USA) was mounted on one axis. The sensor could be equipped with a cylindrical probe with a diameter of 5.3 mm, which was small enough to allow applying loads to individual sensors. A sensorised finger could be attached to the other axis, enabling the probe to apply normal forces to different parts of the finger along one axis, as well as shear forces when the finger was moved while normal forces were applied.

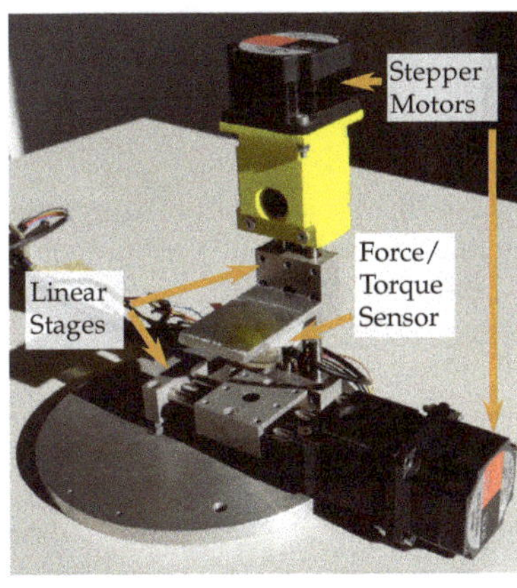

(a)

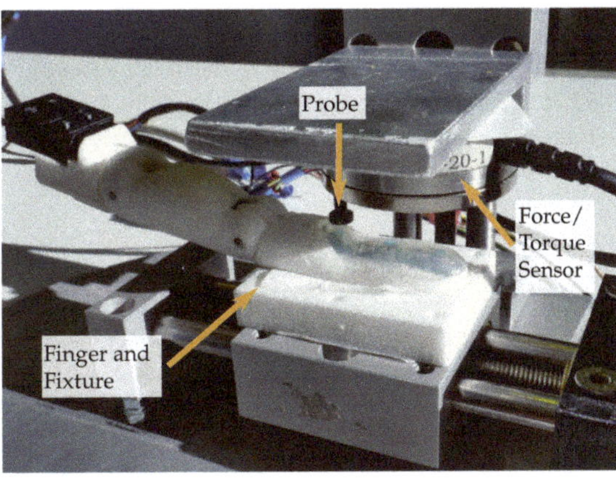

(b)

Figure 9. Linear table for normal force, shear force and spatial mapping experiments; (**a**) depicts an overview of the linear table; (**b**) shows a close up of the f/t sensor and finger attached to the axes

The communication with the sensors during the experiments was implemented on the embedded system also used in the KIT prosthetic hand [41]. This embedded system was based on a microcontroller (STM32H7 series, STMicroelectronics, Geneva, Switzerland) and provides four I^2C ports.

In general, we tried to incorporate sensors from different fingers into the experiments to examine if they exhibited similar characteristics. The following experiments for the tactile sensors were intended to determine that the methods described by Tomo et al. [9,10], Tenzer et al. [15] and Weiner et al. [43]

could be successfully adapted despite differences in design like smaller magnets and the curved shape of the finger. A thorough characterisation of the tactile sensor technologies used in this work can be found in the works above. The experiments primarily give an overview of the quality, correlation and coherence of the signals generated by the different sensor modalities.

To be able to identify the individual tactile sensors for the experiments we adopted the following naming scheme: the sensor names started with the beginning letter of the finger they were included in—I for index finger, M for middle finger, R for ring finger, and L for little finger. The second letter denoted the position inside the finger—D for distal, I for intermediate and P for proximal. Note that the little finger did not have intermediate sensors. The third letter distinguishes the type of sensor—H for Hall effect-based sensors and B for barometer-based sensors. The Hall effect-based sensor at the tip of the index finger would hence be IDH. An overview over the positions of all tactile sensors and their corresponding designators in all physical demonstrators can be seen in Figure 10.

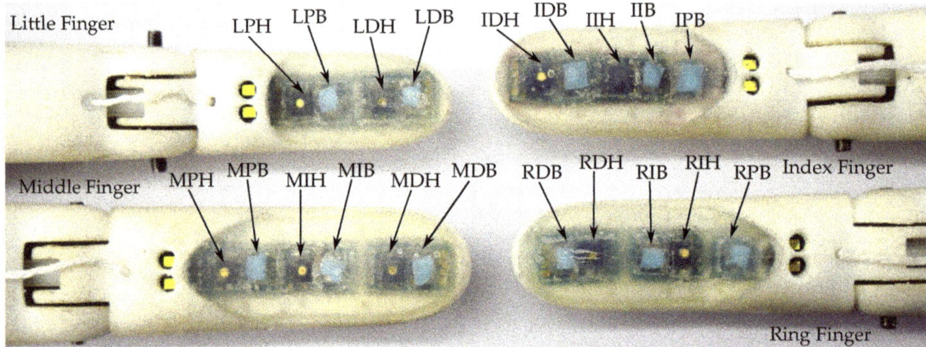

Figure 10. Identifiers for all sensors on all physical demonstrators. The identifiers are named after the first letter of the finger name, their position inside the finger and the sensor type.

3.1. Normal Force Sensor Characterization

Two types of normal force sensors were built into the fingers. The barometer-based normal force sensors were able to resolve small forces but also saturate at comparatively low forces. The Hall effect-based sensors did not offer the same level of resolution but were able to measure magnitudes higher forces before saturation sets in. For the normal force experiments we used the aforementioned linear table (see Figure 9) to allow applying and measuring well-defined forces. In Figure 11 two measurements for Hall effect-based sensors (a,b) and two measurements for barometer-based sensors (c,g) are presented in green. The ground truth measurement of the force/torque sensor is plotted in orange (labelled F_n). These two measurements together are combined in the hysteresis plots (d–f,h) corresponding to the four sensor measurements (a–c,g).

To show the difference in resolution we carried out an additional experiment where a small metal plate was placed on the adjacent sensors RIB and RIH on the ring finger. The plate distributes the load of any weight placed in its centre evenly on the two sensors. For the experiment consecutive weights with an increasing mass of 0.4 g, 0.85 g, 1.1 g, 2.2 g, 4.65 g and 10.75 g were placed on the plate. The resulting sensor readings can be seen in Figure 11i.

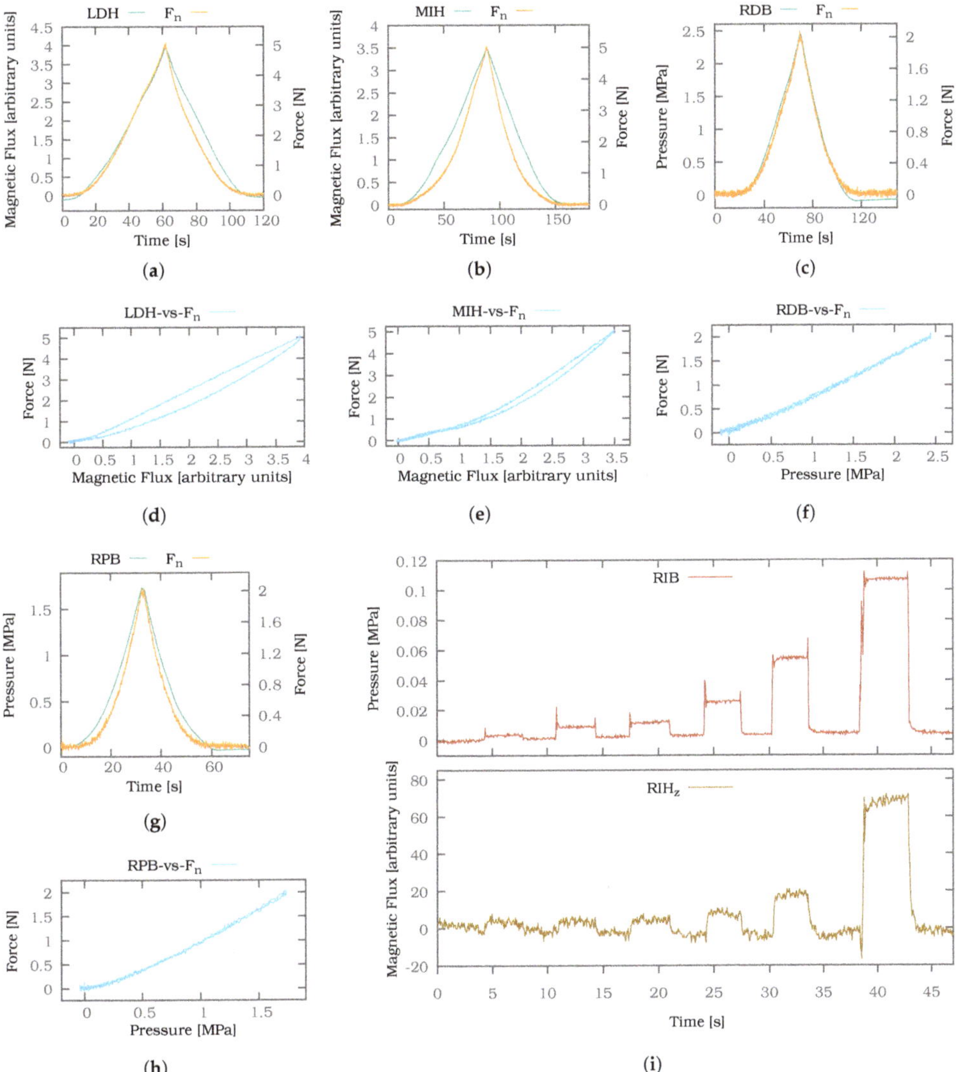

Figure 11. (**a**,**b**) Normal force measurements for Hall effect-based sensors. (**d**,**e**) Corresponding hysteresis plots. (**c**,**g**) Normal force measurements for the barometer-based normal force sensors. (**f**,**h**) Corresponding hysteresis plots for the barometer-based sensors. (**i**) Weights distributed on a Hall effect- and barometer-based sensor.

Both types of sensors were able to track the applied normal forces. Differences were visible in the hysteresis behaviour as the barometer-based normal force sensors RDB and RPB showed a more linear correspondence between their signals and the normal forces measured by the force/torque sensor. Furthermore, the hysteresis was directed in different directions for both sensor types while unloading the sensor. While the Hall effect-based sensors LDH and MIH showed a notable lag when returning to the unloaded state compared to the ground truth, the barometer-based sensors overshot the unloaded state.

In terms of sensitivity the barometer-based sensors had a clear advantage over the Hall effect-based sensor as can be seen in Figure 11i. The barometer-based sensor RIB showed a discernible response even to the smallest weight of 0.4 g, whereas the noise in the signal of the Hall effect-based sensor RIH only allowed the detection of the fourth 2.2 g weight with sufficient confidence. The barometer-based sensors saturated by design at 2.6 MPa which was only slightly above the maximum sensor readings observed during the above experiments at 2 N. The Hall effect-based sensors on the other hand showed a clear signal at forces up to 5 N.

Overall the barometer-based sensors offered a good performance for low forces coupled with a comparatively low hysteresis. The Hall effect-based sensors offered a far wider sensing range at the expense of a more nonlinear behaviour and a stronger hysteresis effect, which could arguably also be caused by the applied forces being higher.

3.2. Shear Force Sensor Characterization

To reliably allow applying shear forces, the force sensor without a probe was used to first apply a normal force of 5 N to the fingertip. The larger sensor surface compared to the probe then allowed to evenly shear the soft silicone material whereas a small probe would only cause a local and undefined distortion. As soon as the normal force threshold was reached, an increasing shear force was applied by the second axis of the linear table up to a limit of 2 N. After the limit was reached the shear force was lowered again until it reached a value close to zero. The direction of the exerted shear forces was chosen to correspond to one of the two measurement axes of the shear force sensors in the fingers. For the experiments, shear force sensors in the ring finger (x-axis), middle finger (x-axis) and little finger (y-axis) were chosen. The resulting measurements can be seen in figure 12. For each measurement the shear force sensor signal, as well as force/torque sensor values, are plotted in diagrams (a–c) and corresponding hysteresis plots are provided in (d,e).

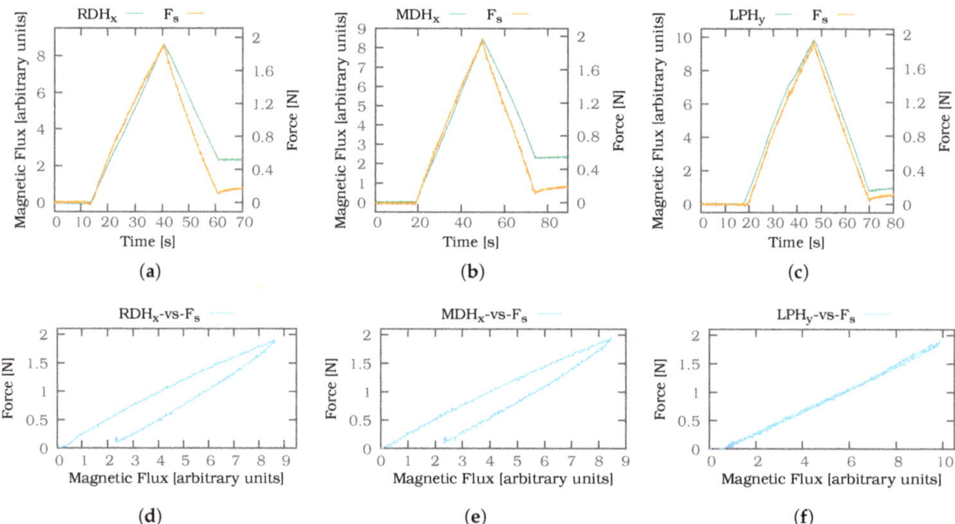

Figure 12. Exemplary shear force measurements for sensor signals RDH_x (**a**), MDH_x (**b**) and LPH_y (**c**) as well as the corresponding hysteresis plots (**d–f**).

In general, the shear force sensors are able to correctly track the direction and rate of change of the applied shear forces. The amplitude of the signal is similar for all sensors, although not identical. Due to the anthropomorphic shape of the finger the silicone is not evenly distributed onto the sensors but follows the curved shape of the human finger. Hence different sensors are covered by silicone of different heights as shown in Figure 4 and the amount of transduced pressure changes accordingly.

In addition, the force/torque sensor is in almost all cases not perfectly aligned with the sensor plane since the PCBs for the sensors are mounted at a slight angle.

From the three hysteresis plots a significant hysteresis is noticeable for the sensors RDH_y and MDH_x. This is also evident at the end of the plots (a) and (b) as the signal of the shear force sensor remains notably higher than that of the force/torque sensor. For the shear force sensor LPH_y in the little finger the hysteresis is much less noticeable. As the little finger is the smallest, the silicone layer on top of the sensor, as well as the overall amount of silicone, is smaller than for the middle and ring finger. Hence the effect of hysteresis should also be reduced for this finger. During the experiments we found no sign of crosstalk between the sensors, meaning the magnet on one Hall-effect-based sensor did not affect the other Hall effect-based sensors. There was also no noticeable crosstalk between the Hall effect-based tactile sensors and the joint angle encoders.

It can be concluded from the above observations that the shear force sensor signals are able to track direction and dynamic of shear forces well. The sensors exhibit notable hysteresis for more static forces. The shape of the finger does not seem to influence the sensor performance too much.

3.3. Joint Angle Sensor Characterization

As mentioned, due to space constraints the measurement of the joint rotation angles is performed off-axis in this work (see Section 2.4). Therefore, an experiment, shown in Figure 13a for the MCP joint of the index finger, was necessary to determine the correlation between the calculated sensor output α_z (using Equation (1), based on the magnetic field strengths x_{Mag} and y_{Mag} in x and y direction) and the actual rotation angle of the joint. To determine this correlation, we moved each joint of each finger incrementally in steps of 5°, starting at 0° and ending at 90°, which corresponds to the minimum and maximum rotation angle of each joint, respectively. To ensure that only the correct joint was rotated, we fixed the other joint during the measurements. At each step the sensor output (α_z) and rotation angle were recorded, after which the joint was bent five degrees further. The resulting correlation between rotation angle and sensor output was then used to obtain a 3rd order polynomial fit for each joint, shown for the MCP joint of the index finger in Figure 13b. This fit can be used for the real time control of the finger to directly calculate the rotation angle of each joint during the data processing step.

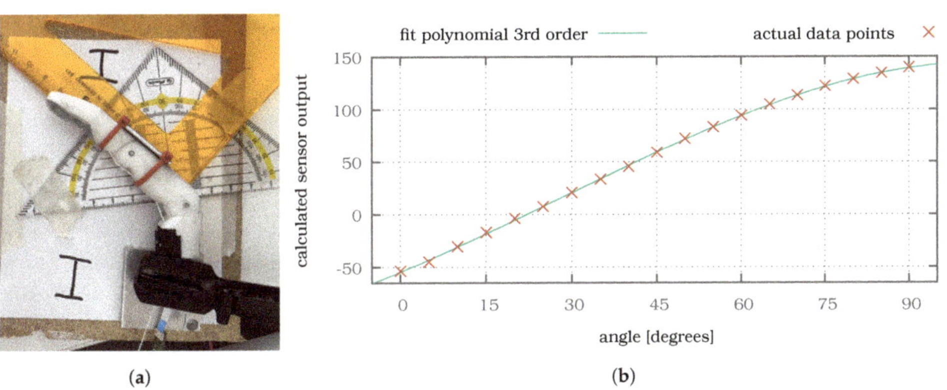

Figure 13. (**a**) The PIP joint is fixed and the MCP joint is incrementally rotated by 5° while the sensor output data are acquired, to determine the correlation between sensor output and actual rotation angle. (**b**) The resulting data points and 3rd order polynomial fit.

As the rotational orientation of the diametrally polarised magnet can not be exactly controlled during assembly, this polynomial fit needs to be experimentally determined for each joint individually if accurate joint angle measurements are needed. The position of the PCBs with the Hall effect sensors inside the proximal phalanx can also vary slightly. Alternatively, the curve can be linearly interpolated using the lowest and highest measured value for increased calibration speed at the cost of angular resolution.

In addition to calibration, we investigated the influence of crosstalk between the magnet of one joint and the Hall effect sensor of the other joint. For this experiment the distal joint was fixated and the proximal joint actuated across the full range while recording the values of the distal sensor. For the little finger, with a minimal distance of 16.7 mm between distal Hall sensor and proximal magnet, we measured a maximum of 1.1° of crosstalk. The other fingers did not show significant crosstalk as the distances between sensor and magnet are larger (23.3 mm for index/ring finger and 27.3 mm for the middle finger).

3.4. Object Grasping and Slip Detection

To evaluate the performance of the multimodal sensor system, we devised a grasping experiment where an object is grasped, held and released using two sensorised fingers. During the holding phase slip is induced. For this experiment the little and ring finger are fixed in direct opposition to each other, meaning both sensor surfaces are roughly facing each other. The tendon of the little finger can be actuated manually so that an object can be grasped in a pinch grasp configuration.

Using this setup, a wooden block of $4 \times 4 \times 20$ cm and a mass of 215 g is grasped firmly. The holding force is then lowered until slip occurs, after which the grasp is quickly fastened again twice. Afterwards, the grasp is released. All normal-force, shear-force, accelerometer and joint angle sensors for both fingers are recorded simultaneously. Figure 14 shows the signals of the different sensors throughout the experiment. For clarity only changing sensor values are plotted. To make the characteristic frequencies generated by incipient and gross slip visible, the accelerometer values are transformed using a short-time Fourier transformation (STFT).

At the beginning of the experiment the wooden block is grasped just above the centre of mass, as can be seen in Figure 14(1). The distal barometer-based sensor of the little finger LDB and the distal Hall effect-based sensors LDH_z and RDH_z are loaded. This means that the point of contact on the little finger is located between the LDB and LDH sensors, whereas the contact point on the ring finger is close to the RDH sensor. Both shear force components LDH_x and RDH_x show a signal proportional to how near they are to the contact point, according to their normal force component. Since the shear force sensor in the ring finger is rotated by 180°, its values are negative, whereas the values of the shear force sensor in the little finger are positive.

After around 5 seconds the first slip event occurs, marked by box a). Just prior to the event, grip strength is reduced as indicated by all sensors LDB, LDH_z and RDH_z. The reduction of grip strength also results in a slight reduction of the joint angle in the distal joint of the ring finger (fourth plot). The gross slip is detected by the accelerometer y-axis, as can be seen in the STFT of the signal at box a) (fifth plot). As soon as the slip occurs, the grip is manually strengthened again. During the slip event the contact point of the block on the fingers changes, as can be seen when comparing Figure 14(1) and Figure 14(2). On the ring finger it moves between the sensors RDB and RDH, whereas on the little finger it moves away from LDB towards LDH. Hence the normal force sensor RDB gets loaded while RDH_z gets partly released. The opposite is true for the little finger. The slip event also induces a small pendulum motion on the wooden block around the two contact points which can be seen in the small waves in all loaded sensors after the first slip event.

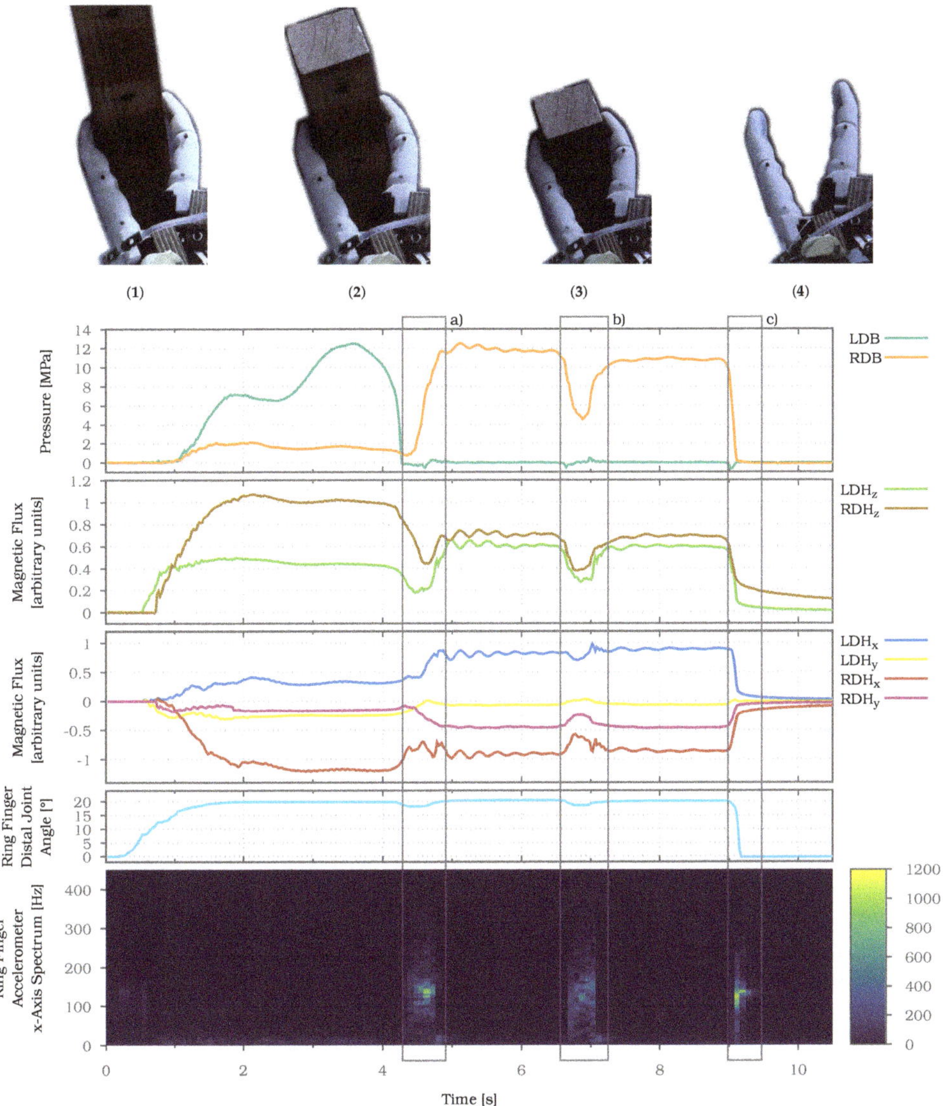

Figure 14. Sensor signals recorded while grasping a block with two fingers in a pinch grasp configuration, letting the object slide twice and releasing the grasp. The top row images (1)–(4) show the static states between the slip events (a)–(c).

The second slip event occurs at around 7 s and is again induced by reducing the grip force, as can be seen in the signals of LDH_z, RDB and RDH_z. The joint angle also changes slightly as the grip is released. Again, the slip itself is clearly visible in the signal of the accelerometer in the ring finger. After the slip event the block is grasped near the top, as can be seen in Figure 14(3). At around 10 s the grip is released, causing a very short but intensive slip event. The fingers are fully opened again, as can be seen in Figure 14(4), and is also visible in the joint angle measurement. As can be seen in the last two seconds of the plot, the shear force sensors LDH and RDH exhibit hysteresis after unloading, whereas the normal force sensor RDB returns to zero immediately. The same holds true for the normal force sensor LDB at the time it is unloaded.

The experiment shows that distinct events during grasping, such as making or breaking contact, as well as gross slip, can be detected by not only one single sensor modality but multiple different modalities. This allows for the fusion of sensor data from different modalities in order to gain more confidence for the detection of events during grasping.

3.5. Spatial Resolution and Sensitivity

The following experiment determines how the different sensors and sensor types in the fingertip, namely normal force and shear force sensors, respond to a fixed normal force applied at varying locations along the fingertip. The linear table is used to apply a normal force to the finger using the probe on the force/torque sensor. As soon as 2 N of normal force are reached, a measurement of the finger's sensors is taken. The probe is then lifted again and moved by 0.25 mm along the long axis of the fingertip. The probe is lowered again to apply force and read the resulting sensor outputs. This process is repeated incrementally, starting from the proximal end of the sensorised surface of the fingertip and ending at the distal end. Measurements were taken at an interval of 30 s to limit the influence of hysteresis on the experiment results. The result for the ring finger can be seen in Figure 15.

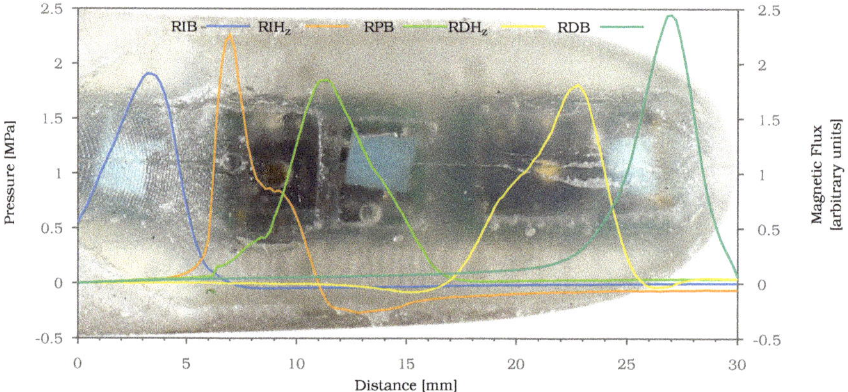

Figure 15. Normal force signals for sensors of the ring finger while probed with 2 N of normal force along the axis from the proximal end of the fingertip to the distal end. The background image shows the approximate position of the probe on the finger at the time of each measurement.

As can be seen, even a small probe of 5.3 mm could be detected almost everywhere along the fingertip. Only between 16 mm and 18 mm the probe remained hard to detect. Normally, the sensor response should be highest above the sensor itself, so in the case of the barometer-based sensors at the position of the blue pad and in case of the Hall effect based sensors around the golden magnet. As can be seen in the plot this was not the case. The spatial shift in sensor response can be explained by the uneven surface of the fingertip, which can be seen in Figure 4. Since the surface was not even, not all parts of the probe made contact with the finger at all positions. At the curved parts the contact area was smaller and more to the edge of the probe. This in turn shifted the positions of the signals perceived by the sensors.

Together with the observations from Section 3.4, it can be concluded that the spatial resolution of the finger should suffice for use cases concerned with grasping and lifting objects of daily life, while for fine-grained manipulation tasks a higher sensor density can be desirable.

4. Discussion and Future Work

In this paper we introduced a concept and implementation of complete scalable robotic fingers with a sophisticated multi-modal sensor system. The fingers are modelled using a skeleton-based parametric model that allows adaptation of all relevant finger dimensions. The embedded electronics are based on readily available sensors and rely on standard design and production techniques. Different sensor modalities have been included in the finger, namely normal and shear force sensors, a distance sensor, an accelerometer as well as joint angle encoders. In addition, each sensor chip includes a temperature sensing element. The sensor system is realised as a number of interchangeable modules that reflect the scalability of the model and allow easy adaptation of the sensor suit to different applications and finger sizes. All tactile sensors are encased in soft silicone while cables and other sensors are encapsulated in the finger itself to increase mechanical robustness. Conceptually, the sensor system is not limited to the presented sensors but can be completely exchanged with any sensor(s) that interface to an electronic bus.

In the experiments presented, we evaluated the tactile sensors, allowing for an informed comparison of two promising tactile sensing methods from literature and show how the detection of distinct events during grasping can benefit from a multi-modal sensor setup. The experiments regarding normal (see Section 3.1) and shear force measurements (see Section 3.2) for the tactile sensors have shown that these are susceptible to hysteresis induced by the silicone. Evaluation of multiple sensors in different fingers shows that this hysteresis, as well as the magnitude of response to forces, is similar for all sensors of each type, indicating that the influence of different shapes of the fingers is minor. The large range of tested sensors also shows that the production process is reliable, as well as repeatable. The density of sensors in the finger is sufficient for the location of the point of contact with an object without larger blind spots (see Section 3.5). Detection of distinct events during grasping and manipulation is not only dependent on tactile sensors but can be realised through sensor fusion of all available sensor data from distance sensors, accelerometers and joint angle encoders (see Section 3.4). The accelerometers have also proven to be a valuable tool for gross slip detection despite being damped by soft material.

In the future we will integrate the presented fingers into our ongoing work on hand prostheses, as well as our humanoid robotic hand development. This will give us the opportunity to further test the robustness of the proposed design as well as validate the usefulness of all parts of the multi-modal sensor setup.

Integration of the fingers into an artificial hand will also make it possible to evaluate different sensor fusion approaches to extract semantic information from the high dimensional sensor information of four fingers. The intention is to utilise the generated information in a similar way to the human, where individual events during grasping like making or breaking contact, lifting and slip seem to define sub-goals during the grasping process [1]. Detection of such events allows breaking down and controlling the grasping process in small steps.

Further experiments are planned regarding the examination of the ability for incipient slip detection based on the accelerometer signals and potential changes of the mounting position of this sensor will be considered.

The addition of further electronics and sensing modalities will also be considered, taking advantage of the modularity of the proposed system. The inclusion of regulated heating elements in the fingers would, for example, enable measurement of temperature flux to objects in contact with the fingers using the temperature sensing elements included in the already present sensor chips.

Author Contributions: Conceptualization, P.W. and T.A.; methodology, P.W. and T.A.; software, P.W.; validation, P.W., C.N. and Y.S.; formal analysis, C.N., P.W. and Y.S.; investigation, P.W., C.N. and Y.S.; resources, T.A.; data curation, P.W.; writing—original draft preparation, P.W., C.N. and Y.S.; writing—review and editing, T.A.,P.W. and Y.N.; visualization, P.W. and C.N.; supervision, T.A. and Y.N.; project administration, T.A. and Y.N.; funding acquisition, T.A. and Y.N. All authors have read and agreed to the published version of the manuscript.

Funding: This work has been supported by the German Federal Ministry of Education and Research (BMBF) under the project INOPRO (16SV7665) and has received funding from the European Union's Horizon 2020 Research and Innovation programme under grant agreement No 643950 (SecondHands).

Acknowledgments: The authors would like to thank Hans Haubert for his valuable input on how to optimise the assembly procedure.

Conflicts of Interest: The authors declare no conflict of interest. The funders had no role in the design of the study; in the collection, analyses, or interpretation of data; in the writing of the manuscript, or in the decision to publish the results.

References

1. Johansson, R.S.; Flanagan, J.R. Coding and use of tactile signals from the fingertips in object manipulation tasks. *Nat. Rev. Neurosci.* **2009**, *10*, 345. [CrossRef]
2. Johansson, R.S.; Westling, G. Influences of Cutaneous Sensory Input on the Motor Coordination during Precision Manipulation. In *Somatosensory Mechanisms: Proceedings of an International Symposium held at The Wenner-Gren Center, Stockholm, 8–10 June 1983*; von Euler, C.; Franzén, O., Lindblom, U., Ottoson, D., Eds.; Springer US: Boston, MA, USA, 1984; pp. 249–260. [CrossRef]
3. Kappassov, Z.; Corrales, J.A.; Perdereau, V. Tactile sensing in dexterous robot hands—Review. *Rob. Autom. Syst.* **2015**, *74*. [CrossRef]
4. Saudabayev, A.; Varol, H.A. Sensors for Robotic Hands: A Survey of State of the Art. *IEEE Access* **2015**, *3*. [CrossRef]
5. Lucarotti, C.; Oddo, C.M.; Vitiello, N.; Carrozza, M.C. Synthetic and Bio-Artificial Tactile Sensing: A Review. *Sensors* **2013**, *13*, 1435–1466. [CrossRef]
6. Dahiya, R.S.; Metta, G.; Valle, M.; Sandini, G. Tactile Sensing - from Humans to Humanoids. *IEEE Trans. Rob.* **2010**, *26*. [CrossRef]
7. Luo, S.; Bimbo, J.; Dahiya, R.; Liu, H. Robotic tactile perception of object properties: A review. *Mechatronics* **2017**, *48*, 54 – 67. [CrossRef]
8. Wijk, U.; Carlsson, I. Forearm amputees' views of prosthesis use and sensory feedback. *J. Hand Ther.* **2015**, *28*, 269 – 278. [CrossRef]
9. Tomo, T.P.; Somlor, S.; Schmitz, A.; Jamone, L.; Huang, W.; Kristanto, H.; Sugano, S. Design and Characterization of a Three-Axis Hall Effect-Based Soft Skin Sensor. *Sensors* **2016**, *16*, 491. [CrossRef]
10. Tomo, T.P.; Wong, W.K.; Schmitz, A.; Kristanto, H.; Sarazin, A.; Jamone, L.; Somlor, S.; Sugano, S. A modular, distributed, soft, 3-axis sensor system for robot hands. In Proceedings of the 16th International Conference on Humanoid Robots (Humanoids) (2016 IEEE-RAS), Cancun, Mexico, 15–17 November 2016; pp. 454–460. [CrossRef]
11. Tomo, T.P.; Schmitz, A.; Wong, W.K.; Kristanto, H.; Somlor, S.; Hwang, J.; Jamone, L.; Sugano, S. Covering a Robot Fingertip with uSkin: A Soft Electronic Skin With Distributed 3-Axis Force Sensitive Elements for Robot Hands. *IEEE Rob. Autom. Lett.* **2017**, *3*. [CrossRef]
12. Tomo, T.P.; Regoli, M.; Schmitz, A.; Natale, L.; Kristanto, H.; Somlor, S.; Jamone, L.; Metta, G.; Sugano, S. A New Silicone Structure for uSkin—A Soft, Distributed, Digital 3-Axis Skin Sensor and Its Integration on the Humanoid Robot iCub. *IEEE Rob. Autom. Lett.* **2018**, *3*, 2584–2591. [CrossRef]
13. Holgado, A.C.; Piga, N.; Tomo, T.P.; Vezzani, G.; Schmitz, A.; Lorenzo, N.; Sugano, S. Magnetic 3-axis Soft and Sensitive Fingertip Sensors Integration for the iCub Humanoid Robot. In Proceedings of the 2019 IEEE-RAS 19th International Conference on Humanoid Robots (Humanoids), Toronto, ON, Canada, 15–17 October 2019; pp. 344–351.
14. Votta, A.M.; Günay, S.Y.; Erdoğmus, D.; Önal, C. Force-Sensitive Prosthetic Hand with 3-axis Magnetic Force Sensors. In Proceedings of the 2019 IEEE International Conference on Cyborg and Bionic Systems, Munich, Germany, 18–20 September 2019.
15. Tenzer, Y.; Jentoft, L.P.; Howe, R.D. The Feel of MEMS Barometers: Inexpensive and Easily Customized Tactile Array Sensors. *IEEE Rob. Autom. Mag.* **2014**, *21*. [CrossRef]
16. Odhner, L.U.; Jentoft, L.P.; Claffee, M.R.; Corson, N.; Tenzer, Y.; Ma, R.R.; Buehler, M.; Kohout, R.; Howe, R.D.; Dollar, A.M. A compliant, underactuated hand for robust manipulation. *Int. J. Rob. Res.* **2014**, *33*. [CrossRef]
17. Jentoft, L.P.; Tenzer, Y.; Vogt, D.; Liu, J.; Wood, R.J.; Howe, R.D. Flexible, stretchable tactile arrays from MEMS barometers. In Proceedings of the 2013 16th International Conference on Advanced Robotics (ICAR), Karlsruhe, Germany, 6–10 May 2013; pp. 1–6. [CrossRef]
18. Mittendorfer, P.; Cheng, G. Humanoid Multimodal Tactile-Sensing Modules. *IEEE Trans. Rob.* **2011**, *27*. [CrossRef]

19. Mittendorfer, P.; Cheng, G. Integrating discrete force cells into multi-modal artificial skin. In Proceedings of the 2012 12th IEEE-RAS International Conference on Humanoid Robots (Humanoids 2012), Osaka, Japan, 29 November–1 December 2012; pp. 847–852. [CrossRef]
20. Bergner, F.; Dean-Leon, E.; Cheng, G. Event-based signaling for large-scale artificial robotic skin—Realization and performance evaluation. In Proceedings of the 2016 IEEE/RSJ International Conference on Intelligent Robots and Systems (IROS), Daejeon, Korea, 9–14 October 2016; pp. 4918–4924. [CrossRef]
21. Dean-Leon, E.; Pierce, B.; Bergner, F.; Mittendorfer, P.; Ramirez-Amaro, K.; Burger, W.; Cheng, G. TOMM: Tactile omnidirectional mobile manipulator. In Proceedings of the 2017 IEEE International Conference on Robotics and Automation (ICRA), Singapore, 29 May–3 June 2017; pp. 2441–2447. [CrossRef]
22. Schmitz, A.; Maggiali, M.; Natale, L.; Bonino, B.; Metta, G. A tactile sensor for the fingertips of the humanoid robot iCub. In Proceedings of the 2010 IEEE/RSJ International Conference on Intelligent Robots and Systems, Taipei, Taiwan, 18–22 October 2010; pp. 2212–2217. [CrossRef]
23. Jamali, N.; Maggiali, M.; Giovannini, F.; Metta, G.; Natale, L. A new design of a fingertip for the iCub hand. In Proceedings of the 2015 IEEE/RSJ International Conference on Intelligent Robots and Systems (IROS), Hamburg, Germany, 29 September–2 October 2015; pp. 2705–2710. [CrossRef]
24. Maiolino, P.; Maggiali, M.; Cannata, G.; Metta, G.; Natale, L. A Flexible and Robust Large Scale Capacitive Tactile System for Robots. *IEEE Sens. J.* **2013**, *13*, 3910–3917. [CrossRef]
25. Alagi, H.; Navarro, S.E.; Mende, M.; Hein, B. A versatile and modular capacitive tactile proximity sensor. In Proceedings of the 2016 IEEE Haptics Symposium (HAPTICS), Philadelphia, PA, USA, 8–11 April 2016; pp. 290–296. [CrossRef]
26. Göger, D.; Alagi, H.; Wörn, H. Tactile proximity sensors for robotic applications. In Proceedings of the 2013 IEEE International Conference on Industrial Technology (ICIT), Cape Town, South Africa, 25–28 February 2013; pp. 978–983. [CrossRef]
27. Weiss, K.; Wörn, H. The working principle of resistive tactile sensor cells. In Proceedings of the IEEE International Conference Mechatronics and Automation, 2005, Niagara Falls, ON, Canada, 29 July–1 August 2005; pp. 471–476. [CrossRef]
28. Kõiva, R.; Zenker, M.; Schürmann, C.; Haschke, R.; Ritter, H.J. A highly sensitive 3D-shaped tactile sensor. In Proceedings of the 2013 IEEE/ASME International Conference on Advanced Intelligent Mechatronics, Wollongong, Australia, 9–12 July 2013; pp. 1084–1089. [CrossRef]
29. Kõiva, R.; Schwank, T.; Walck, G.; Haschke, R.; Ritter, H.J. Mechatronic fingernail with static and dynamic force sensing. In Proceedings of the 2018 IEEE/RSJ International Conference on Intelligent Robots and Systems (IROS), Madrid, Spain, 1–5 October 2018; pp. 2114–2119. [CrossRef]
30. Suzuki, Y. Multilayered Center-of-Pressure Sensors for Robot Fingertips and Adaptive Feedback Control. *IEEE Rob. Autom. Lett.* **2017**, *2*, 2180–2187. [CrossRef]
31. Patel, R.; Cox, R.; Correll, N. Integrated proximity, contact and force sensing using elastomer-embedded commodity proximity sensors. *Auton. Rob.* **2018**, *42*, 1443–1458. [CrossRef]
32. Lancaster, P.E.; Smith, J.R.; Srinivasa, S.S. Improved Proximity, Contact, and Force Sensing via Optimization of Elastomer-Air Interface Geometry. In Proceedings of the 2019 International Conference on Robotics and Automation (ICRA), Montreal, QC, Canada, 20–24 May 2019; pp. 3797–3803. [CrossRef]
33. Yamaguchi, N.; Hasegawa, S.; Okada, K.; Inaba, M. A Gripper for Object Search and Grasp Through Proximity Sensing. In Proceedings of the 2018 IEEE/RSJ International Conference on Intelligent Robots and Systems (IROS), Madrid, Spain, 1–5 October 2018; pp. 1–9. [CrossRef]
34. De Laurentis, K.J.; Mavroidis, C. Rapid fabrication of a non-assembly robotic hand with embedded components. *Assembly Autom.* **2004**, *24*, 394–405. [CrossRef]
35. Liu, H.; Ferrentino, P.; Pirozzi, S.; Siciliano, B.; Ficuciello, F. The PRISMA Hand II: A Sensorized Robust Hand for Adaptive Grasp and In-Hand Manipulation. In Proceedings of the 2019 International Symposium on Robotics Research, Hanoi, Vietnam, 6–10 October 2019.
36. Maria, G.D.; Natale, C.; Pirozzi, S. Force/tactile sensor for robotic applications. *Sens. Actuators A* **2012**, *175*, 60–72. [CrossRef]
37. Cheng, M.; Jiang, L.; Fenglei Ni, F.; Fan, S.; Liu, Y.; Liu, H. Design of a highly integrated underactuated finger towards prosthetic hand. In Proceedings of the 2017 IEEE International Conference on Advanced Intelligent Mechatronics (AIM), Munich, Germany, 3–7 July 2017; pp. 1035–1040. [CrossRef]

38. Imbinto, I.; Montagnani, F.; Bacchereti, M.; Cipriani, C.; Davalli, A.; Sacchetti, R.; Gruppioni, E.; Castellano, S.; Controzzi, M. The S-Finger: A Synergetic Externally Powered Digit With Tactile Sensing and Feedback. *IEEE Trans. Neural Syst. Rehabil. Eng.* **2018**, *26*, 1264–1271. [CrossRef]
39. Wettels, N.; Fishel, J.A.; Loeb, G.E. Multimodal Tactile Sensor. In *The Human Hand as an Inspiration for Robot Hand Development*; Balasubramanian, R., Santos, V.J., Eds.; Springer International Publishing: Cham, Switzerland, 2014; pp. 405–429. [CrossRef]
40. Wettels, N.; Loeb, G.E. Haptic feature extraction from a biomimetic tactile sensor: Force, contact location and curvature. In Proceedings of the 2011 IEEE International Conference on Robotics and Biomimetics, Phuket Island, Thailand, 7–11 December 2011; pp. 2471–2478. [CrossRef]
41. Weiner, P.; Starke, J.; Hundhausen, F.; Beil, J.; Asfour, T. The KIT Prosthetic Hand: Design and Control. In Proceedings of the IEEE/RSJ International Conference on Intelligent Robots and Systems (IROS), Madrid, Spain, 1–5 October 2018; pp. 3328–3334.
42. Asfour, T.; Wächter, M.; Kaul, L.; Rader, S.; Weiner, P.; Ottenhaus, S.; Grimm, R.; Zhou, Y.; Grotz, M.; Paus, F. ARMAR-6: A High-Performance Humanoid for Human-Robot Collaboration in Real World Scenarios. *IEEE Rob. Autom. Mag.* **2019**. to be published. [CrossRef]
43. Weiner, P.; Neef, C.; Asfour, T. A Multimodal Embedded Sensor System for Scalable Robotic and Prosthetic Fingers. In Proceedings of the IEEE/RAS International Conference on Humanoid Robots (Humanoids); Beijing, China, 6–9 November 2018; pp. 286–292.
44. Vergara, M.; Agost, M.J.; Gracia-Ibáñez, V. Dorsal and palmar aspect dimensions of hand anthropometry for designing hand tools and protections. *Hum. Factors Ergon. Manuf. Serv. Ind.* **2018**, *28*, 17–28. [CrossRef]
45. Cordella, F.; Ciancio, A.L.; Sacchetti, R.; Davalli, A.; Cutti, A.G.; Guglielmelli, E.; Zollo, L. Literature Review on Needs of Upper Limb Prosthesis Users. *Front. Neurosci.* **2016**, *10*, 209. [CrossRef] [PubMed]
46. Belter, J.T.; Segil, J.L.; Dollar, A.M.; Weir, R.F. Mechanical design and performance specifications of anthropomorphic prosthetic hands: A review. *J. Rehabil. Res. Dev.* **2013**, *50*. [CrossRef] [PubMed]
47. Piazza, C.; Grioli, G.; Catalano, M.; Bicchi, A. A Century of Robotic Hands. *Annu. Rev. Control Rob. Autom. Syst.* **2019**, *2*, 1–32. [CrossRef]
48. Calli, B.; Walsman, A.; Singh, A.; Srinivasa, S.; Abbeel, P.; Dollar, A.M. Benchmarking in Manipulation Research: Using the Yale-CMU-Berkeley Object and Model Set. *IEEE Rob. Autom. Mag.* **2015**, *22*, 36–52. [CrossRef]
49. Kasper, A.; Xue, Z.; Dillmann, R. The KIT object models database: An object model database for object recognition, localization and manipulation in service robotics. *Int. J. Rob. Res.* **2012**, *31*, 927–934. [CrossRef]
50. Or, K.; Morikuni, S.; Ogasa, S.; Funabashi, S.; Schmitz, A.; Sugano, S. A study on fingertip designs and their influences on performing stable prehension for robot hands. In Proceedings of the 2016 IEEE-RAS 16th International Conference on Humanoid Robots (Humanoids), Cancun, Mexico, 15–17 November 2016; pp. 772–777. [CrossRef]

Article

Effects of Multi-Point Contacts during Object Contour Scanning Using a Biologically-Inspired Tactile Sensor

Lukas Merker [1,*], Sebastian J. Fischer Calderon [2], Moritz Scharff [1,3], Jorge H. Alencastre Miranda [3] and Carsten Behn [4,*]

1 Technical Mechanics Group, Technische Universität Ilmenau, 98693 Ilmenau, Germany
2 Institute for Process Measurement and Sensor Technology, Technische Universität Ilmenau, 98693 Ilmenau, Germany
3 Section of Mechanical Engineering, Pontificial Catholic University of Peru, Lima 15088, Peru
4 Faculty of Mechanical Engineering, Schmalkalden University of Applied Sciences, 98574 Schmalkalden, Germany
* Correspondence: lukas.merker@tu-ilmenau.de (L.M.); c.behn@hs-sm.de (C.B.)

Received: 18 March 2020; Accepted: 4 April 2020; Published: 7 April 2020

Abstract: Vibrissae are an important tactile sense organ of many mammals, in particular rodents like rats and mice. For instance, these animals use them in order to detect different object features, e.g., object-distances and -shapes. In engineering, vibrissae have long been established as a natural paragon for developing tactile sensors. So far, having object shape scanning and reconstruction in mind, almost all mechanical vibrissa models are restricted to contact scenarios with a single discrete contact force. Here, we deal with the effect of multi-point contacts in a specific scanning scenario, where an artificial vibrissa is swept along partly concave object contours. The vibrissa is modeled as a cylindrical, one-sided clamped Euler-Bernoulli bending rod undergoing large deflections. The elasticae and the support reactions during scanning are theoretically calculated and measured in experiments, using a spring steel wire, attached to a force/torque-sensor. The experiments validate the simulation results and show that the assumption of a quasi-static scanning displacement is a satisfying approach. Beyond single- and two-point contacts, a distinction is made between tip and tangential contacts. It is shown that, in theory, these contact phases can be identified solely based on the support reactions, what is new in literature. In this way, multipoint contacts are reliably detected and filtered in order to discard incorrectly reconstructed contact points.

Keywords: vibrissa; bio-inspired sensor; contour scanning; multi-point contact

1. Introduction

Rats and mice use their vibrissae for tactile determination of object features, e.g., shapes and textures [1,2]. In doing so, mechanical stimuli are transmitted along the hair-shaft to the follicle-sinus complex (support) of each vibrissa, where they are transduced into action potentials for further processing in the brain [3,4].

1.1. Vibrissa-Based Sensors

The biological principle of vibrissae frequently serves as a paragon developing tactile sensors, e.g., for object shape recognition. Even though vibrissa-inspired sensors are nowhere near from the accuracy of conventional tactile sensors, such as coordinate measuring machines, they offer a number of benefits: Firstly, the universal applicability for the detection of different object features shows an important potential. For example, the same experimental setup was used in [5] to characterize surface textures and in [6] to reconstruct object contours. Beyond that, the same setup has shown the basic suitability for flow detection in [7]. Secondly, vibrissa-sensors benefit in many ways from their highly flexible structure. Since most conventional sensors take advantage of stiff probes as force transmitters, allowing for an easy localization of objects in space, they are vulnerable to damage when accidentally colliding with an obstacle. In contrast, a flexible transmitter is advantageous in some fields of application, e.g., mobile robotics, due to its robustness against collisions. Moreover, a rigid edge probe, e.g., must be moved over an object of interest, making and releasing contact at various points in order to scan a greater part of the object. Repeatedly sensing different points does not only extend the process time, but might also require a quite sophisticated trajectory of the sensing device (co-ordinated movements of several axes). In contrast, sweeping a long and flexible rod over an object's surface by a single continuous movement of its base relative to the object, directly provides a large number of sensed points.

Dealing with the problem of object shape scanning and reconstruction using vibrissa-inspired sensors, most publications in the field of biomimetics focus on a single artificial vibrissa. Frequently, it is considered as a continuum using linear [8–12] or nonlinear bending theories [5,7,13–15] or discretized using multi-body systems or FE-models [6,16]. In these approaches, the sophisticated kinematics of the animal's head and vibrissae during whisking are greatly simplified, e.g., by changing either the rod's base position or angle in order to make contact with an object. As a consequence of object contact, the rod undergoes a deformation, resulting in measurable mechanical signals at its support, which in turn can be used to draw conclusions about the contact point position. Since the force transmission from the contact point to the support is often neglected, one focus of the paper at hand is setting up a model for theoretically generating the support reactions during scanning at the base of the rod. This allows to gain a deeper understanding of the scanning process and to carry out parameter studies without the necessity of performing a wide range of experiments.

1.2. Multi-Point Contacts

One major limitation, which can be found in all publications mentioned above, is the restriction to single-point contact (SPC) scenarios. This means that at each time, there is only one point along the axis of the rod in contact with the object. Some authors merely hypothesize, that multi-point contacts (MPCs) at discrete points of a rod are unlikely to occur and that deflecting a rod against an object typically generates SPCs [12,16]. Other publications connect the occurrence of MPCs with the geometry of the scanned object. In [13] the authors mention, that MPCs might occur if the curvature of the scanned object approaches zero. Similarly, it is stated in [10] that the contact scenario (SPC or MPC, see ([17] Figure 9) depends on the surface slope of the scanned object. There, it was observed that SPCs primarily occurred scanning a square object, but it happened that the rod made contact with two edges of the object for special arrangements. This situation requires a closer look to refine the definition of MPC as it is understood in the present paper. The square object was scanned using a straight rod. If some point along the axis of the rod (not the tip) makes contact with the object, there may be a situation, where the straight end of the rod (between the contact position and the tip) aligns with the straight part of the object. That situation might give the impression of a line-contact or MPC, where, e.g., both edges of the square are in contact with the rod (as stated in [10]). In the presence of multiple contact forces or a distributed load, however, the straight end of the rod would also have to deform, which is not the case (see ([10] Figure 5)). Thus, there is only one actual contact point at the first position, where the rod meets the object. At this point, a single contact force acts on the rod and its straight end aligns stress-free with the object.

In [14] object contours are assumed to be strictly convex in order to exclude the occurrence of MPCs. Instead of excluding the occurrence of MPCs, some publications briefly discuss their effects. In [18] two-point contacts (TPCs) occur while the rod crosses an object-gap in context of texture determination. It is assumed, that the first contact releases abruptly when a second contact at a discrete rod position is made. This assumption may be sufficient in this particular case, where possible MPCs are very close to each other - but it is not a suitable in general. The authors in [11] relate to the principle of superposition and conclude that any load distribution along the rod can be modeled as a resultant force acting at a single point. In other words, a single contact force might cause the same mechanical signals at the base of a rod as any load distribution. This raises the question of whether it is possible to distinguish between MPCs and SPCs if the only quantities at hand are mechanical signals at the base of the rod. The authors [13,14] avoid this problem in context of object contour reconstruction by presupposing SPC scenarios. The reconstruction algorithms presented in [13,14] use the support reactions of a rod to calculate the contact force and its orientation. The clamping moment and position are used to formulate an initial-value problem, which is integrated numerically, exploiting the termination condition that the bending moment at the contact point is zero. However, this condition is only valid for SPCs. In [15] the effect of MPCs is briefly discussed using an FEA model to calculate the support reactions. Nevertheless, the investigation is limited to an example, where two discrete contact points are comparatively close to each other. Therefore, the TPC phase during scanning is limited to a negligible interval and does not allow a deeper insight into the deformation of the rod. In addition, the calculated support reactions seem to be strongly affected by the discretization of the rod (see ([15] Figure 9.22)). Finally, it is shown in ([15] Figure 9.23) that the maximum reconstruction error occurs during TPC.

For an actual sensor application, it is important that the scanned objects do not have to meet any specific requirements. If the assumption of strictly convex objectsin [15] is dropped, MPCs may occur, which influence the mechanical signals at the base of a rod in an unknown way. Using these signals to localize the contact point, the object reconstruction is affected as well, what might lead to an incorrect estimation of the scanned contour. Although, multi-point boundary-value problems (BVPs) for the rod equation were treated in some publications, e.g., [19], to date, no investigation has been carried out to clarify how MPCs affect the scanning sweep of a rod along specific objects.

1.3. Objective of the Present Work

The present paper contributes to the general overall goal of scanning and reconstructing arbitrary shaped object contours by means of vibrissa-inspired sensors. Here, we deal with scanning different partly concave object contours including TPC scenarios, as a first step. A single vibrissa is modeled as a rod using nonlinear bending theory. The paper particularly addresses the following goals: In Section 2.1. we derive the deformation equations of a rod with a finite number of applied forces. Secondly, the model is refined considering the rod in contact with a partly concave object contour, which is composed of two parabolas. In doing so, we identify different possible contact phases to derive the underlying BVPs. An experimental setup, consisting of a spring steel wire attached to a force/torque-sensor is used to validate the model is presented in Section 2.2. In the first part of Section 3, the deformation behavior of the rod is investigated during both contact phases, SPC and TPC. The support reactions, generated using a Matlab-algorithm, are compared with measured data. The second part of Section 3. reveals the effect of TPCs when using the reconstruction algorithm from [14]. With the goal to identify concave parts of different objects, we present an approach exploiting only the support reactions of a single rod. In doing so, we show, that it is theoretically possible, to distinguish SPC and TPC based on the support reactions exploiting a contact phase diagram. Experimental results clarify the limitations of this approach: the resolution of the used force- and torque sensor as well as frictional effects. Therefore, we suggest another way to identify concave parts of objects by repeatedly scanning at different distances.

Although the paragon in the present work is a biological one, our focus is neither on copying a vibrissa as exactly as possible nor on explaining the morphological characteristics of vibrissae. Instead, the paper aims to develop technical sensor principles in which some of the sensor properties are inspired by vibrissae. In particular, the presented approaches might be used for environmental exploration in mobile robotics. In this way, a robot equipped with vibrissa-like sensors might be adapted to deal with complex scanning situations, including large deflections in combination with TPCs.

2. Materials and Methods

In this section we present a mechanical model and a method of simulating scanning sweeps of an artificial vibrissa along partly convex object contours. Following, we introduce an experimental setup, which is used to validate the model as well as the simulation algorithm.

2.1. Modeling & Simulation

The general model, shown in Figure 1, consists of a single vibrissa modeled as an Euler-Bernoulli bending rod of length L, which is cylindrically shaped (causing constant characteristics of its cross section: area $A = const.$, second moment of area $I_z = const.$). The rod is one-sided clamped to a horizontally movable measurement unit, see Figure 1 and consists of a homogeneous, isotropic material with a constant Young's modulus E.

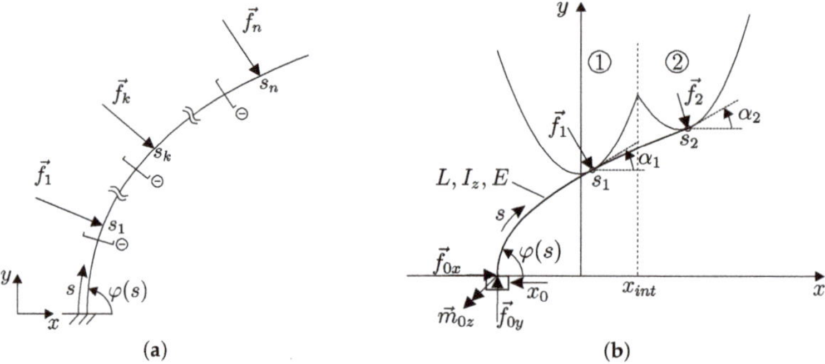

Figure 1. Mechanical model of a rod: (**a**) with a finite number of acting forces; (**b**) in contact with a piecewise-defined object contour function according to (7).

Remark. *From the outset, the following agreement is made regarding the units of measure for a dimensionless representation of the problem:*

$$[length] := L, \qquad [forces] := \frac{EI_z}{L^2}, \qquad [moments] := \frac{EI_z}{L} \tag{1}$$

In this way, the modeling equations within this subsection are simplified and may be used in arbitrary scalings. Afterwards, from Section 2.2 on, dimensional quantities are used in order to give the reader a more intuitive idea of the actual physical quantities. Based on this agreement, the introduction of a new notation for the dimensional quantities will be waived.

The parametrization of the rod axis by means of its slope angle $\varphi(s)$, where $s \in [0,1]$ is the natural coordinate arc length of the rod, yields:

$$\frac{x(s)}{ds} = \cos(\varphi(s)) \tag{2}$$

$$\frac{y(s)}{ds} = \sin(\varphi(s)) \tag{3}$$

$$\frac{\varphi(s)}{ds} = \kappa(s) \tag{4}$$

The curvature $\kappa(s)$ in (4) is substituted using Euler's constitutive law, in dimensionless representation:

$$\kappa(s) = m_z(s) \tag{5}$$

where $m_z(s)$ is the dimensionless bending moment of the rod's sections. Considering the rod in Figure 1a with a finite number of n forces $\vec{f}_j$, acting under the angles $\alpha_j \in [-\frac{\pi}{2}, \frac{\pi}{2}]$ at the contact positions $s_j \in (0,1]$, each force is represented by:

$$\vec{f}_j = f_j\Big(\sin(\alpha_j)\vec{e}_x - \cos(\alpha_j)\vec{e}_y\Big)$$

Thus, the bending moment $m_z(s)$ for the general case of an arbitrary number of applied forces writes:

$$m_z(s) = \begin{cases} \sum\limits_{j=1}^{n} f_j\Big(\big(y(s_j) - y(s)\big)\sin(\alpha_j) + \big(x(s_j) - x(s)\big)\cos(\alpha_j)\Big), & s \in (0, s_1) \\ \vdots & \\ \sum\limits_{j=k}^{n} f_j\Big(\big(y(s_j) - y(s)\big)\sin(\alpha_j) + \big(x(s_j) - x(s)\big)\cos(\alpha_j)\Big), & s \in (s_{k-1}, s_k) \\ \vdots & \\ f_n\Big(\big(y(s_n) - y(s)\big)\sin(\alpha_n) + \big(x(s_n) - x(s)\big)\cos(\alpha_n)\Big), & s \in (s_{n-1}, s_n) \\ 0. & s \in (s_n, 1) \end{cases} \tag{6}$$

with $1 < k < n$.

In order to realize the scanning sweep, the clamping position x_0 (input variable) of the rod is shifted incrementally, translationally relative to the object contour. This process is assumed to be slow enough to treat the problem as a quasi-static one. During scanning, contacts between the rod and the object may occur at either one or even at multiple points along the rod at the same time. As a first approach within the present paper, the problem of object scanning is limited to a special object type, which is composed of two parabolas (see Figure 1b). The object is assumed to be a rigid body, whose contour $g : x \mapsto g(x)$ is given by a continuous, piecewise-defined function, where each sub-function applying to a certain interval of the main function is strictly convex. The arbitrary combination of two strictly convex sub-functions results in a concave area in the overall function $g(x)$:

$$g : x \mapsto g(x) = \begin{cases} g_1(x) = 5x^2 + y_1 & x \in (-\infty, x_{int}) \\ g_2(x) = 5(x - x_2)^2 + y_2 & x \in [x_{int}, +\infty) \end{cases} \tag{7}$$

where $(x_1 = 0, y_1)$ and (x_2, y_2) are the vertex positions of g_1 and g_2, respectively, and x_{int} is the position of intersection of g_1 and g_2 (see Figure 1b). Each strictly convex sub-function is parameterized by means of its slope angle:

$$\alpha_1 \mapsto (\xi_1(\alpha_1), \eta_1(\alpha_1)) \qquad \text{and} \qquad \alpha_2 \mapsto (\xi_2(\alpha_2), \eta_2(\alpha_2))$$

Ignoring frictional effects, the contact force is always perpendicular to the object profile tangent. The deformation of the rod results in a set of support reactions corresponding to a certain clamping position x_0. Due to the chosen object contour (7), the number of contact points is always either one or two. While TPCs always include both object parts 1 and 2 (α_1 and α_2), SPCs might occur either at object part 1 or 2 (α_1 or α_2), see Figure 2. For the sake of brevity, the index i is used as an abbreviation for "1 or 2".

Thus, (6) simplifies to:

$$m_{bz}(s) = \begin{cases} f_1\Big(\big(y(s_1) - y(s)\big)\sin(\alpha_1) + \big(x(s_1) - x(s)\big)\cos(\alpha_1)\Big) \\ \qquad + f_2\Big(\big(y(s_2) - y(s)\big)\sin(\alpha_2) + \big(x(s_2) - x(s)\big)\cos(\alpha_2)\Big), & s \in (0, s_1) \\ f_2\Big(\big(y(s_2) - y(s)\big)\sin(\alpha_2) + \big(x(s_2) - x(s)\big)\cos(\alpha_2)\Big), & s \in (s_1, s_2) \\ 0. & s \in (s_2, 1) \end{cases} \tag{8}$$

for TPC, and for SPC:

$$m_{bz}(s) = \begin{cases} f_i\Big(\big(y(s_i) - y(s)\big)\sin(\alpha_i) + \big(x(s_i) - x(s)\big)\cos(\alpha_i)\Big), & s \in (0, s_1) \\ 0. & s \in (s_1, 1) \end{cases} \tag{9}$$

Substituting (8) and (9) in (5) and differentiating the curvature using (2) and (3) in order to get rid of the constants, we end up in the following differential equations describing the change of the curvature:

$$\text{TPC:} \qquad \frac{\kappa(s)}{ds} = \begin{cases} f_1\cos(\varphi(s) - \alpha_1) + f_2\cos(\varphi(s) - \alpha_2), & s \in (0, s_1) \\ f_2\cos(\varphi(s) - \alpha_2), & s \in (s_1, s_2) \\ 0. & s \in (s_2, 1) \end{cases} \tag{10}$$

$$\text{SPC:} \qquad \frac{\kappa(s)}{ds} = \begin{cases} f_i\cos(\varphi(s) - \alpha_i), & s \in (0, s_i) \\ 0. & s \in (s_i, 1) \end{cases} \tag{11}$$

Together with (2)–(4) the deformation of the rod is represented by the following system of ordinary differential equations (ODE) of first order:

$$\text{TPC:} \qquad \left.\begin{aligned} x'(s) &= \cos\big(\varphi(s)\big) \\ y'(s) &= \sin\big(\varphi(s)\big) \\ \varphi'(s) &= \kappa(s) \\ \kappa'(s) &= \begin{cases} f_1\cos(\varphi(s) - \alpha_1) + f_2\cos(\varphi(s) - \alpha_2), & s \in (0, s_1) \\ f_2\cos(\varphi(s) - \alpha_2), & s \in (s_1, s_2) \\ 0 & s \in (s_2, 1) \end{cases} \end{aligned}\right\} \tag{12}$$

and

$$\text{SPC:} \qquad \left.\begin{aligned} x'(s) &= \cos\big(\varphi(s)\big) \\ y'(s) &= \sin\big(\varphi(s)\big) \\ \varphi'(s) &= \kappa(s) \\ \kappa'(s) &= \begin{cases} f_i\cos(\varphi(s) - \alpha_i), & s \in (0, s_i) \\ 0 & s \in (s_i, 1) \end{cases} \end{aligned}\right\} \tag{13}$$

As Figure 2 shows, four different contact phases have to be distinguished during scanning:

(a) Phase A: SPC at the tip ($s_i = 1$) with unknown contact angle $\varphi(1)$;

(b) Phase B: tangential SPC at some unknown position $0 < s_i < 1$ with the contact angle condition $\varphi(s_i) = \alpha_i$;

(c) Phase BA: TPC with one tangential contact at $0 < s_1 < 1$ with $\varphi(s_1) = \alpha_1$ and one tip contact at $s_2 = 1$ with unknown contact angle $\varphi(1)$;

(d) Phase BB: TPC with two tangential contacts at $0 < s_1 < s_2 < 1$ with $\varphi(s_1) = \alpha_1$ and $\varphi(s_2) = \alpha_2$.

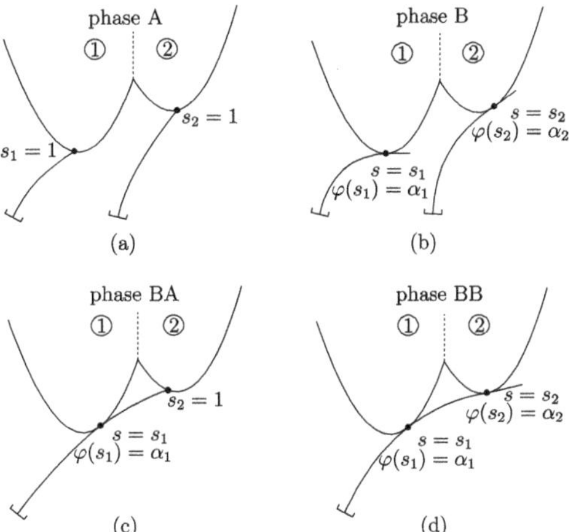

Figure 2. Contact phases–Phase A (**a**) and B (**b**) with single-point contact (SPC), phase BA (**c**) and BB (**d**) with two-point contact (TPC).

Note that a single scanning sweep along an object might contain various consequent contact phases. This sequence of contact phases during one scanning sweep is referred to as phase cycle in this paper.

The four contact phases in Figure 2 are characterized by the following boundary conditions (BCs):

Phase A:

$$\boxed{\begin{array}{ll} x(0) = x_0 & x(1) = \xi_i(\alpha_i) \\ y(0) = y_0 & y(1) = \eta_i(\alpha_i) \\ \varphi(0) = \pi/2 & \\ & \kappa(1) = 0. \end{array}} \tag{14}$$

Phase B:

$$\boxed{\begin{array}{ll} x(0) = x_0 & x(s_i) = \xi_i(\alpha_i) \\ y(0) = y_0 & y(s_i) = \eta_i(\alpha_i) \\ \varphi(0) = \pi/2 & \varphi(s_i) = \alpha_i \\ & \kappa(s_i) = 0. \end{array}} \tag{15}$$

Phase BA:

$$\boxed{\begin{array}{lll} x(0) = x_0 & x(s_1) = \xi_1(\alpha_1) & x(1) = \xi_2(\alpha_2) \\ y(0) = y_0 & y(s_1) = \eta_1(\alpha_1) & y(1) = \eta_2(\alpha_2) \\ \varphi(0) = \pi/2 & \varphi(s_1) = \alpha_1 & \\ & & \kappa(1) = 0. \end{array}} \tag{16}$$

Phase BB:

$$\boxed{\begin{array}{lll} x(0) = x_0 & x(s_1) = \xi_1(\alpha_1) & x(s_2) = \xi_2(\alpha_2) \\ y(0) = y_0 & y(s_1) = \eta_1(\alpha_1) & y(s_2) = \eta_2(\alpha_2) \\ \varphi(0) = \pi/2 & \varphi(s_1) = \alpha_1 & \varphi(s_2) = \alpha_2 \\ & & \kappa(s_2) = 0. \end{array}} \tag{17}$$

Simulating the scanning sweep, it is not a priori known, whether the contact resulting from a certain clamping position x_0 belongs to Phase A, B, BA or BB. Thus, one major problem is to determine which of the BVPs (13)&(14), (13)&(15), (12)&(16) and (12)&(17) has to be solved. One way to address this problem is to presuppose a certain contact phase examining the solution for contradictions afterwards. Such contradictions may include the exceeding of the rod length ($s_i > 1$) or intersections (permeation) of the calculated elastica and the object. In this way, a *Matlab*-algorithm is used to solve the unknown parameters f_i, s_i and α_i (s_i is known in phase A) in case of SPC, and f_1, f_2, s_1, s_2, α_1 and α_2 (s_2 is known in phase BA) in case of TPC. The support reactions are then determined in the following way:

$$\left.\begin{aligned} f_{0x} &= -f_i \sin(\alpha_i) \\ f_{0y} &= f_i \cos(\alpha_i) \\ m_{0z} &= f_i \ \cos(\alpha_i)\ \xi_i(\alpha_i) + f_i \ \sin(\alpha_i)\ \eta_i(\alpha_i) \end{aligned}\right\} \text{SPC} \tag{18}$$

and

$$\left.\begin{aligned} f_{0x} &= -f_1 \sin(\alpha_1) - f_2 \sin(\alpha_2) \\ f_{0y} &= f_1 \cos(\alpha_1) + f_2 \cos(\alpha_2) \\ m_{0z} &= f_1 \ \cos(\alpha_1)\ \xi_1(\alpha_1) + f_1 \ \sin(\alpha_1)\ \eta_1(\alpha_1) + f_2 \ \cos(\alpha_2)\ \xi_2(\alpha_2) + f_2 \ \sin(\alpha_2) \end{aligned}\right\} \text{TPC} \tag{19}$$

Once the support reactions are known either by simulation or measurement, they can be used to draw conclusions about the rod deformation and finally about the object's contour by using the reconstruction algorithm presented in [14]. Since this algorithm only provides correct reconstruction results if the support reactions result from SPCs, it is necessary to distinguish SPCs and TPCs solely based on the support reactions. In [15] the following criterion for the distinction of phase A and B (SPC) was derived:

$$m_{0z}^2 - 2f_{0y} = 0 \tag{20}$$

It can be shown, that this criterion remains valid for the case of TPC (see Appendix A). However, the criterion (20) does not allow a distinction between SPC and TPC so far. Instead, it is used for further investigations of contact phase determination, taking advantage of the simulation results.

2.2. Experiments

Figure 3 shows the experimental setup that is used to validate the simulation results. A straight spring steel wire according to DIN EN 10270-1:2017-09 with a diameter of $d = 0.5$ mm and a Young's modulus $E = 206$ GPa is used as transmitter. It is cut off from a long piece of wire and clamped by a miniature jaw chuck in a way that the free length of the wire is $L = 100$ mm. Both dimensions, diameter and length, are verified using a caliper. There is no after-treatment for the cutting edge (tip) of the wire. The small diameter of the steel wire is chosen for two reasons: firstly, with a view to the biological paragon, and secondly, because of the model assumption that object contacts are assumed to happen directly at the axis of the bending rod (geometric dimension of the diameter is neglected). The miniature jaw chuck is connected to a 3D force sensor of type K3D40 (ME-Meßsysteme), accuracy class 0.5, nominal load ± 2 N and a 1D torque sensor of type TD70 (ME-Meßsysteme), accuracy class 0.1, nominal load ± 50 Nm. The force and torque sensors are arranged in a way that the torque sensor measures signals with respect to the z-axis (see Figure 3). The signals are recorded using a GSV-1A4 M12/2 (ME-Meßsysteme) amplifier, a NI PXI 6221 M-Series multifunction data acquisition device and the software LabVIEW 2017 with a sampling rate of 500 Hz. Due to the small diameter of the wire, a material with a large Young's modulus is chosen in order to realize an adequate bending stiffness. In this way, the support reactions at the base of the wire are adapted for the measuring range of the available sensor equipment exploiting the well-known mechanical properties of the chosen steel wire. The entire assembly, consisting of transmitter, jaw chuck and sensors, is attached to two linear guides of type AMTEC Power Cube PLB 090 and 070 (position repeatability ± 0.005 mm) in serial arrangement.

Three object samples with contours according to (7) are 3D-printed from Tough PLA filament using a Ultimaker S5 3D-printer. Using (7), the object contours (see Figure 3) are given by:

- Object 1 (O1): $(x_1, y_1) = (0\,\text{mm}, 0\,\text{mm})$, $(x_2, y_2) = (30\,\text{mm}, 0\,\text{mm})$;
- Object 2 (O2): $(x_1, y_1) = (0\,\text{mm}, 0\,\text{mm})$, $(x_2, y_2) = (30\,\text{mm}, 10\,\text{mm})$;
- Object 3 (O3): $(x_1, y_1) = (0\,\text{mm}, 10\,\text{mm})$, $(x_2, y_2) = (30\,\text{mm}, 0\,\text{mm})$.

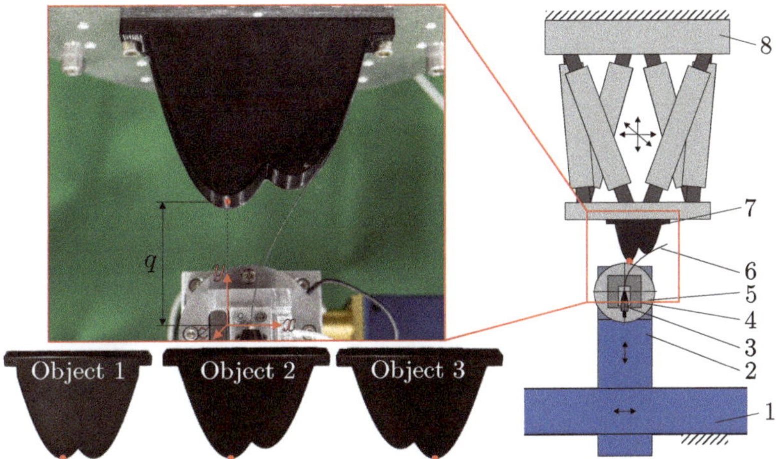

Figure 3. Experimental setup: 1, 2 – linear guides; 3 – jaw chuck; 4 – force sensor; 5 – torque sensor; 6 – spring steel wire; 7 – object, 8 – hexapod.

For each experiment, the corresponding object is mounted on the platform of a hexapod of type PI M-850.50 to position and align the object relative to the linear axis of the sensor. The scanning sweeps are performed in horizontal direction (negative x-direction) with a constant speed of 1 mm/s for three vertical object distances $q \in [20\text{ mm},\ 40\text{ mm},\ 60\text{ mm}]$ (vertical distance from the clamping to the closest parabola vertex, marked with a red point in Figure 3). The distance reference are determined by moving the undeformed wire vertically towards the objects closest vertex until the very first contact force is detected by the force sensor. Both, the objects O1, O2 and O3 and the selected object distances q were designed based on preliminary studies in order to realize well pronounced TPCs. The object distances are limited in a way that they must be greater than 20 mm to avoid plastic deformation of the wire and smaller than 60 mm to ensure the occurrence of TPCs (as a desired effect in the present paper).

Scanning sweeps are repeated three times for each of the three objects and three different object distances, starting from the very first contact between the wire and the object and terminating with a snap off of the wire from the object. This results in a total number of 27 experiments. In order to reduce dynamical effects during scanning, the experiments are performed in a way that the sensor is accelerated to the scanning velocity before the first contact with the object takes place. Due to the small scanning velocity, the impact of the first collision between the wire and the object is negligible.

3. Results and Discussion

The elasticae during a simulated scanning sweep (object distance $q = 40$ mm) of the rod along the objects O1, O2 and O3 are illustrated in Figure 4a–c. Note that all diagrams of Figure 4 are to be read from right to left, due to the fact that the scanning sweeps are performed in negative x-direction. The color of each deformation state (green, blue, orange or red) denotes the corresponding contact phase (A, B, BA or BB, respectively). Black colored deformation states represent the phase transitions. The simulation starts with the very first contact of the undeformed rod and the object and terminates at a specific position of the last equilibrium state found by the simulation algorithm.

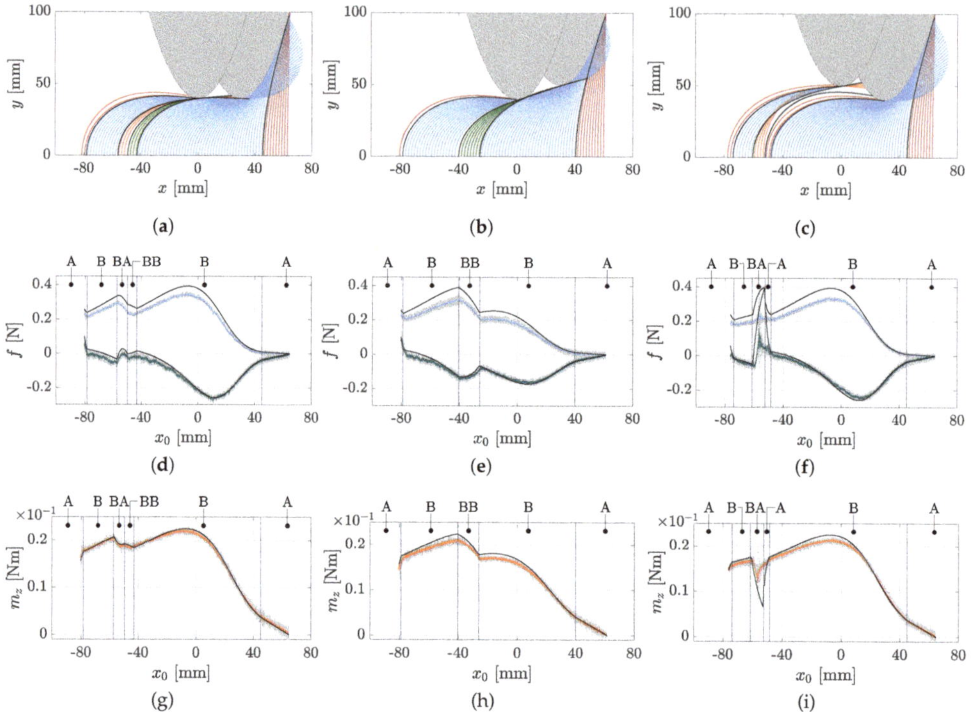

Figure 4. Scanning sweeps along object O1, O2 and O3: (**a–c**) elasticae (green: phase A, blue: phase B, orange: phase BA, red: phase BB, black: phase transitions); (**d–i**) simulated and measured support reactions (black: simulated data, green: measured data f_{0x}, blue: measured data f_{0y}, orange: measured data m_{0z}, grey: standard deviation, black vertical lines: phase transitions.

As clarified by the Figure 4a–c, considering concave objects and large bending deflections of the rod, two-point contacts are not unlikely to occur, as presupposed in some previous publications [12,16]. Instead, the fact that two-point contacts are limited to small intervals in proportion to the whole scanning sweep indicates that three-point or even higher multi-point contacts are statistically improbable and might only occur for very special object geometries or arrangements. Figure 4d–f and Figure 4g–i show the simulated and measured support reactions corresponding to the scanning sweeps in Figure 4a–c. The simulated support reactions are represented by black lines, while the experimental data is colored in blue (f_{0x}), green (f_{0y}) and orange (m_{0z}). The measured data are averaged from three repeated experiments and is shown with the standard deviation (grey tube: $\pm 3 \cdot$ *standard deviation*). The small standard deviation and the moderate noise of the averaged data indicate a good reproducibility of the experiment. The vertical black lines represent the phase transitions corresponding to the black elasticae in Figure 4a–c. The phase transitions are determined based on the simulation. Comparing Figure 4d,g with Figure 4f,i, it can be seen that the support reactions coincide to a large extend as a consequence of the similarity between the first part of the objects O1 and O3. They start to differ with the first TPC in Figure 4a. In general, it is obvious that there is a good correlation of the simulated and experimental data. Therefore, the assumption of a quasi-static displacement seems to be a satisfying approach. Only the reaction force f_{0y} shows a slight systematic deviation, with the simulated data exceeding the measured data. In order to aggregate the correlation into a single measure, we determined correlation factors using the *Matlab 2019a*-function *GoodnessOfFit* with the normalized mean square error. The correlation between simulated and experimental data of all objects and distances is shown in Figure 5.

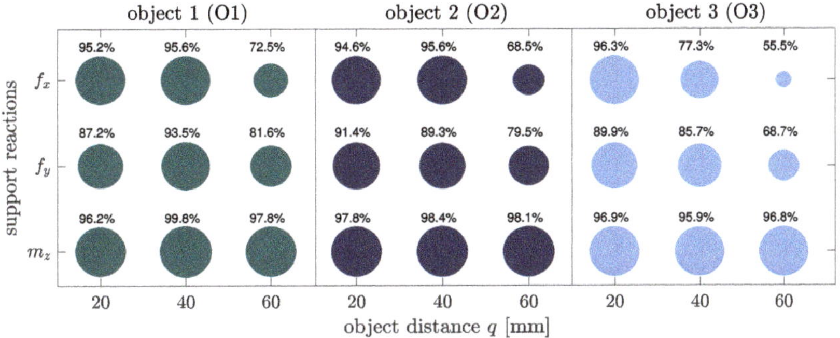

Figure 5. Goodness of fit between simulated and experimental data using the normalized mean square error (NMSE). The areas of the circles depend linearly on the correlation factors and are normalized to an interval $[0.1, 1]$, e.g., the smallest correlation factor corresponds to a value of 0.1, the largest one to a value of 1.

It becomes clear that in general the measured clamping moment m_{0z} correlates best with the simulated one. The reaction forces f_{0x} and f_{0y} also show a good correlation, except for large object distances. This is due to the fact that for larger object distances, the rod snaps from the first to the second part of the object. This dynamical transition can not be correctly described using a quasi-static model. A dynamic snap is also apparent scanning object O3: there is a small interval in Figure 4f,i, which is characterized by a large error. This deviation, even if limited to a small interval, strongly affects the correlation number in Figure 5, although there is a good overall correlation between the simulated and measured data in Figure 4f,i.

In order to demonstrate the contour reconstruction error, which arises as a consequence of TPCs, the reconstruction algorithm [14] is applied to the data in Figure 4f,i. Figure 6 shows the reconstructed contact point sequences based on the simulated data (SPC in orange and TPC in green) and experimental data (grey). The black dashed line denotes the original object contour. It becomes clear that during SPC, the object contour is well approximated. No filter or otherwise post-processing is applied to the experimental data. Even ignoring TPC, the objects O1, O2 and O3 can be clearly distinguished. Surprisingly, the dynamical snap of the rod between both parts of object O3, which caused a large error between the simulated and measured support reactions in Figure 4f,i did not cause any outliers in the reconstructed points. While the orange points coincide with the original object contour within numerical boundaries, the green contact points, based on TPC, are characterized by a large error (distance to the original contour), which is in the same order of magnitude as the experimental error. This error leads to an inaccurate estimation of the scanned object contour.

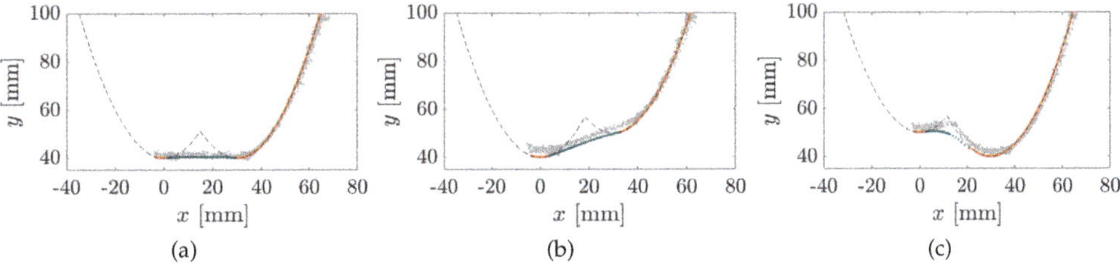

Figure 6. Contour reconstruction of objects O1 (**a**), O2 (**b**) and O3 (**c**) assuming SPC only: Contact points based on the simulated support reactions (SPC in orange and TPC in green) and the measured support reactions (gray).

Consequently, it would be advantageous to know, which reconstructed points result from TPCs, in order to exclude them from the contour reconstruction or to modify the reconstruction algorithm for TPC (discussed in Section 4). Therefore, we discuss possibilities of contact phase determination based on the support reactions of a single rod.

In Section 2.1, a criterion for distinguishing the phases A&BA from B&BB (20) is derived. The unit related representation of (20) is either zero in case of phase B&BB or non zero in case of phase A&BA. The criterion, corresponding to the scanning sweeps in Figure 4, is plotted against the support position x_0 in Figure 7, where simulated data is colored in red and experimental data in green.

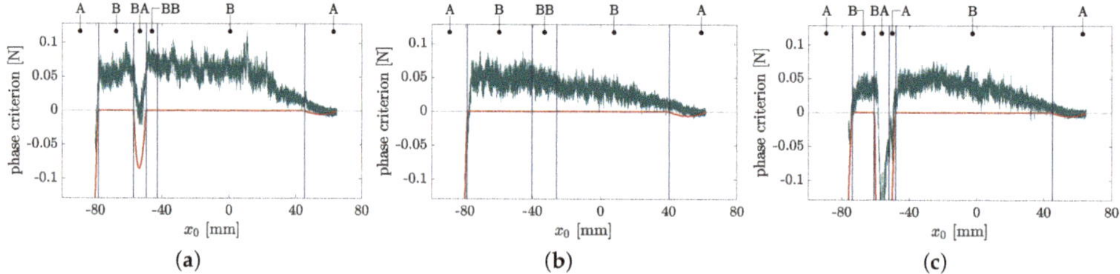

Figure 7. Unit related representation of phase criterion (20) for distinguishing phase A&BA from B&BB plotted against the support position x_0 (red: simulated data, green: experimental data, vertical black lines: phase transitions): scanning the objects O1 (**a**), O2 (**b**) and O3 (**c**).

Again, the phase transitions are represented by vertical black lines. The simulated data validates (20): it is zero during phase B&BB and non zero during phase A&BA (note the small deviation from zero during the first phase A). As Figure 4b shows, the scanning sweep along object O2 does not include phase BA. Instead, TPC only occurs in phase BB. For that reason, the phase criterion deviates from 0 only in phase A. The criterion based on the experimental data deviates from zero over the entire scan. This is probably due to frictional effects which were neglected in the model. In reality, friction causes a contact force component which is tangential to the scanned object contour. Nevertheless, especially phase BA and phase A at the end of the scanning sweep can be identified by peaks in the signals of Figure 7a–c. For example, the missing peak in Figure 7b shows, that even during experiments phase C only occurred scanning object O1 and O3. In order to distinguish between SPC and TPC based on the support reactions, the contact phases are examined in more detail. While the scanning sweep along object O1 (see Figure 4a,d,g) includes all contact phases, the sweeps along the objects O2 and O3 each include only three contact phases. A more comprehensive consideration including the contact phases of all simulations can be seen in Figure 8a,b, which need to be considered in context. Figure 8a results from evaluating all simulations with regard to the occurring contact phases. Each simulated scanning sweep consists of multiple consequent contact phases. It can be categorized into one of the three listed phase cycles C1, C2 or C3 (none but these three cycles occurred). For example, the scanning sweep shown in Figure 4a passes phase cycle C2, scanning sweep Figure 4b corresponds to C1 and scanning sweep Figure 4c to C3. The assignment of all simulations is shown in dependence on the scanned object and its distance q. A trend can be identified which shows that for each object the phase cycle changes from C1 over C2 to C3 with an increasing object distance. The phase cycles listed in Figure 8a are combined to a single phase diagram in Figure 8b. The symbol of each phase cycle in Figure 8a indicates the path through the phase diagram in Figure 8b. As the phase diagram shows, some phase transitions can not occur. For example, if the rod is in contact phase A (SPC at the tip) at a certain clamping position x_0, it can not change directly to phase BB (tangential TPC) in the next step ($x_0 + \Delta x_0$). Instead, the deformation state must change from phase A to phase B (transition A→B) before entering phase BB. The round arrows denote that the rod remains in the same contact phase as in the previous step.

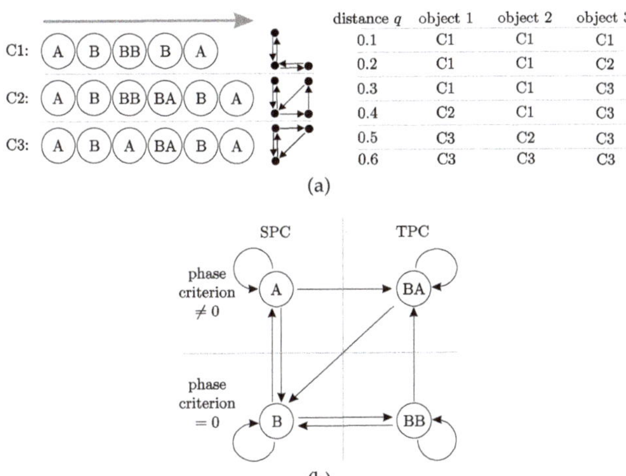

distance q	object 1	object 2	object 3
0.1	C1	C1	C1
0.2	C1	C1	C2
0.3	C1	C1	C3
0.4	C2	C1	C3
0.5	C3	C2	C3
0.6	C3	C3	C3

Figure 8. Phase transitions during object scanning: (**a**) phase cycle of each simulation in dependence on the object and its distance; (**b**) phase diagram combining all observed phase cycles.

Having a look at the theoretically generated support reactions in Figure 4d–i, all phase transitions (except transition A→B) caused kinks, which can be detected in the trends of the support reactions. The respective kinks can also be found in the experimental data, but can not be clearly localized due to the signal noise, occuring in every real-world measurement. However, theoretically, the final recognition of the contact phase might be made by combining the following aspects:

(a) the criterion (20) for distinguishing phase A&BA from B&BB (applicable in a restricted way for the experimental data, as Figure 7 shows);
(b) the phase diagram Figure 8 representing all possible phase cycles;
(c) the observation that all phase transitions (except transition A→B) lead to kinks in at least one of the support reactions.

In theory, the criterion (20) can be used to determine the initial contact phase right after detecting the first contact. Within the present paper, the initial contact is always in phase A. Then, according to Figure 8b, there are only three options: the contact either remains in phase A or it changes into phase B or BA. The transition A→B might not necessarily lead to a kink in the support reactions but either way it could be detected using the criterion (20). According to the simulation results, the transition A→BA causes a kink in at least one of the support reactions, which indicates a phase transition. In this case, the criterion (20) could be used in order to distinguish phase B from BA. The idea of contact phase identification is, that in each phase, there is a maximum of two subsequent phases, which can always be distinguished using the criterion (20) (see Figure 8b). Thus, knowing the current contact phase, it is always possible to determine the subsequent phase by evaluating kinks in the support reactions and exploiting (20). In this way, our reconstruction algorithm is adapted to identify and discard the orange points in Figure 6. It can be assumed that the phase diagram in Figure 8b is not only valid if the object is composed of two parabolas, as considered here, but might also be generalized for objects composed of any two strictly convex contours. However, so far the presented procedure applies only for the idealized model. It remains to be clarified whether the kinks in the experimental support reactions Figure 4d–i can be sufficiently accurately localized using appropriate signal processing or even artificial intelligence and whether the peaks in the experimental data in Figure 7 can be exploited to distinguish between phase A&BA and B&BB. Of course, this might be facilitated by the use of sensors with enhanced resolution.

Instead, another approach including multiple scanning sweeps with different object distances might be used in order to determine the concave interval of an object contour. Figure 9a–c show the reconstructed contact points of object O1 (a), O2 (b) and O3 (c) using the support reactions from object scans with $q = 20\ mm$ (green), $q = 40\ mm$ (orange) and $q = 60\ mm$ (blue). In Figure 9d–f the vertical offsets between the reconstruction results were eliminated.

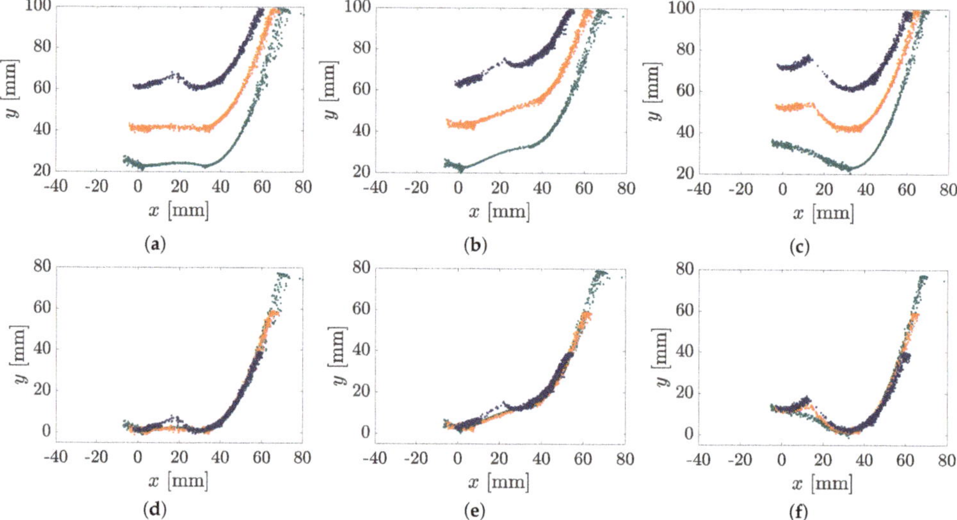

Figure 9. Contour reconstruction of objects O1 (**a**,**d**), O2 (**b**,**e**) and O3 (**c**,**f**) for different object distances $q = 20$ mm (green), $q = 40$ mm (orange), $q = 60$ mm (*blue*) (**a**–**c**) and excluding the distance offset (**d**–**f**).

It appears that in general, the closest object distance q leads to a maximum scanning range. At the same time, a larger object distance causes the rod to penetrate deeper into the concave part of the object. This suggests that the reconstruction of concave object parts might be improved, when the rod is in tip contact (phase A or BA). Similar findings were described in [20]. As Figure 9d–f reveal the reconstructed points based on different distances are consistent to a large extend, but differ in the area of the concave contour part. Thus, a concave object part could be localized checking multiple reconstruction results for inconsistency. Once a concave interval has been localized, the incorrectly reconstructed points can be discarded here. Of course, this results in reconstruction gaps with no information about the object. However, a reconstruction gap is preferred to undetected reconstruction errors. If a reconstruction gap would occur in a real world problem, it might just be scanned more precisely, e.g., varying the object distance or the scanning direction in order to realize SPCs. Returning to the biological paragon, it is also suspected in animals, that the first scan of an object might be used for a rough localization of the object in space and a second scan to detect more detailed information [21].

Future investigations are necessary to validate the kinds of conclusions that can be drawn from this study. This includes a more detailed investigation of the rod deformation during the experiment with regard to the contact phases by means of image processing. In this way, future research should further develop and confirm the findings of the present paper, especially the phase diagram Figure 8b. One problem of identifying two-point contacts was the detection of kinks in the measured support reactions and peaks in the phase criterion. In that context, future studies should aim to investigate the potential of further signal processing or neural networks. In addition it was hypothesized in [10], that edged objects, which are not strictly convex, would produce a more sudden change of deflection, what might also result in kinks in the support reactions. Therefore, the possibility of distinguishing between edges and two-point contacts should be examined in further work.

As the results have shown, the quality of the object contour reconstruction changes with the object distance in a way that the beam better reaches concave areas of the object for larger object distances. However, this observation seems to be strongly affected by the selection of the scanned object. Therefore, the range of objects should necessarily be extended in future works. In this way it should be clarified, whether there is a generally valid limit distance, which must not be undercut, in order to realize satisfactory reconstruction results. A more advanced way to handle the reconstruction of concave objects could be to modify the reconstruction algorithm for simultaneously localizing two contact points along the axis of a rod. Preliminary investigations have shown that in case of two-point contact, there are not enough boundary conditions to reconstruct the elastica of the rod. Therefore, the mechanical model could be changed to a statically indeterminate support arrangement. This might be realized by an additional roller support at some position along the rod, where an additional reaction force could be measured and used to formulate an additional boundary condition. Either way, frictional effects should be taken into account in future works. In this way, it might be possible to improve the phase criterion, presented in this paper, for the evaluation of experimental data.

Although the focus of the present work is on developing technical sensor principles, our findings are not limited to the large scale of the steel wire we used. Instead, the dimensionless parameters introduced in Section 2 allow a scaling of the modeling equations even to the smaller dimensions of a natural vibrissa. Returning to the biological paragon, animal's vibrissae show some geometrical features, which are neglected in the present paper. In particular, they are rather tapered and pre-curved than cylindrically shaped, as considered here [22,23]. It can be assumed, that MPCs along a tapered rod would cause more pronounced kinks, than observed in the present work. This hypothesis is based on the fact that a conical rod has a bending stiffness, which varies along its axis. Therefore, contacts at multiple discrete points along the rod would have varying degrees of influence on the support reactions. In addition, it is conceivable that certain types of pre-curvature might be suitable to allow deeper penetration of the rod into concave areas of objects. Therefore, the complexity of the model should be increased in future works taking more morphological properties of the biological paragon into account.

4. Conclusions

The present paper deals with the effect of multi-point contacts along the axis of a single bending rod in context of object contour scanning and reconstruction. For this purpose, the rod was one-sided clamped and swept along an object composed of two parabolas (restriction to a maximum of two contact points) translationally and quasi-statically. Deriving the modeling equations resulted in an ODE-system describing the deflection of the rod, where one equation depends on the contact scenario single- or two-point contact, respectively. The focus of investigation was on both the influence of two-point contacts on the support reactions at the clamping of the rod and on the object shape reconstruction. Beyond single- and two-point contact, further contact phases were identified: phase A—single-point contact at the tip; phase B—tangential single-point contact; phase BA—two-point contact with one contact at the tip; phase BB—two-point contact with two tangential contacts. Each contact phase is characterized by its own boundary-value problem. Simulating the scanning sweep, these four boundary-value problems were solved repeatedly using a *Matlab*-algorithm while changing the clamping position of the rod translationally. In this way, all unknown contact forces, their positions and orientations and finally the entire elastica of the rod as well as the support reactions were determined. As the results have shown, the calculated support reactions match well with the measured ones for the majority of the simulations, what validates the model and the simulation algorithm. One limitation became evident scanning objects with large distances, when dynamic transitions of the rod between both object parts occured in experiments. These dynamical effects resulted in large errors of the support reactions which, however, were limited to a small interval.

Using the support reactions in order to reconstruct the contours O1-O3, by means of an algorithm for single-point contacts, the error resulting from two-point contacts was highlighted. Since the reconstruction error might give an incorrect impression about the scanned object, we investigated, whether two-point contacts can be detected using the support reactions, in order to discard incorrectly reconstructed points. It was shown that all phase transitions, except the transition from phase A to phase B, can be identified by kinks in the simulated and experimental support reactions. In theory, a phase criterion allowed us to distinguish phase A&BA from B&BB using the simulated support reactions. In practice, frictional effects complicated the evaluation of the criterion, but, nevertheless, the presence of the phases BA and A was recognizable by peaks in the phase criterion based on the experimental data. Analyzing all simulations with regard to the occurring contact phases, only three different phase cycles (sequences of contact phases) were found. In this context, it was hypothesized that some phase transitions can not occur at all. The observed phase cycles were combined in a single phase diagram showing all possible phase transitions. Using this phase diagram, it was possible to determine the contact phase of each deformation state by evaluating kinks in the simulated support reactions and exploiting the phase criterion. Another approach of detecting concave parts of the object and thus the occurrence of two-point contacts, was realized by multiple object scans at different object distances. Finally, three different partly concave object contours were correctly reconstructed to a large extend and distinguished from each other.

Author Contributions: Conceptualization, L.M. and M.S.; Formal analysis, L.M. and S.J.F.C.; Investigation, L.M. and M.S.; Methodology, L.M. and S.J.F.C.; Project administration, C.B.; Software, L.M. and S.J.F.C.; Supervision, J.H.A.M. and C.B.; Visualization, L.M. and S.J.F.C.; Writing—original draft, L.M.; Writing—review & editing, M.S. and C.B. All authors have read and agreed to the published version of the manuscript.

Funding: We acknowledge financial support for the Article Processing Charge by the Thuringian Ministry for Economic Affairs, Science and Digital Society (Germany).

Conflicts of Interest: The authors declare no conflict of interest.

Appendix A

Deriving a criterion for the distinction between contact phase BA and BB using the support reactions we integrate (10):

$$\kappa^2(s) = \begin{cases} 2f_1 \sin(\varphi(s) - \alpha_1) + 2f_2 \sin(\varphi(s) - \alpha_2) + K_1, & s \in (0, s_1) \\ 2f_2 \sin(\varphi(s) - \alpha_2) + K_2, & s \in (s_1, s_2) \\ K_3. & s \in (s_2, 1) \end{cases}$$

The integration constants $K_1, ..., K_3$ are determined exploiting the boundary conditions $\kappa(0) = -m_{oz}$ and $\kappa(s_2) = 0$:

$$\begin{aligned}
s = 0: & \quad & m_{0z}^2 &= 2f_1 \cos(\alpha_1) + 2f_2 \cos(\alpha_2) + K_1 \\
& \Rightarrow & K_1 &= m_{0z}^2 - 2f_1 \cos(\alpha_1) - 2f_2 \cos(\alpha_2) \\
s = s_1: & & \kappa(s_1)^2 &= \begin{cases} 2f_1 \sin(\varphi_1 - \alpha_1) + 2f_2 \sin(\varphi_1 - \alpha_2) + K_1, s \in (0, s_1) \\ 2f_2 \sin(\varphi_1 - \alpha_2) + K_2, s \in (s_1, s_2) \end{cases} \\
& \Rightarrow & K_2 &= K_1 = m_{0z}^2 - 2f_1 \cos(\alpha_1) - 2f_2 \cos(\alpha_2) \\
s = s_2: & \Rightarrow & K_3 &= 0 \\
& & 0 &= 2f_2 \sin(\varphi_2 - \alpha_2) + m_{0z}^2 - 2f_1 \cos(\alpha_1) - 2f_2 \cos(\alpha_2) \\
& \Rightarrow & \varphi(s_2) &= \alpha_2 + \arcsin\left(\frac{m_{0z}^2 - 2f_1 \cos(\alpha_1) - 2f_2 \cos(\alpha_2)}{-2f_2}\right)
\end{aligned}$$

Obviously, the condition $\varphi(s_2) = \alpha_2$ characterizing phase BB is fulfilled if

$$m_{0z}^2 - 2f_1 \cos(\alpha_1) - 2f_2 \cos(\alpha_2) = 0$$

Substituting the support reaction f_{0y} (19) yields:

$$m_{0z}^2 - 2f_{0y} = 0$$

References

1. Brecht, M.; Preilowski, B.; Merzenich, M.M. Functional architecture of the mystacial vibrissae. *Behav. Brain Res.* **1997**, *84*, 81–97. [CrossRef]
2. Guić-Robles, E.; Valdivieso, C.; Guajardo, G. Rats can learn a roughness discrimination using only their vibrissal system. *Behav. Brain Res.* **1989**, *31*, 285–289. [CrossRef]
3. Ebara, S.; Furuta, T.; Kumamoto, K. Vibrissal mechanoreceptors. *Scholarpedia* **2017**, *12*, 32372. [CrossRef]
4. Zucker, E.; Welker, W.I. Coding of somatic sensory input by vibrissae neurons in the rat's trigeminal ganglion. *Brain Res.* **1969**, *12*, 138–156. [CrossRef]
5. Merker, L.; Scharff, M.; Zimmermann, K.; Behn, C. Signal tuning of observables at the support of a vibrissa-like tactile sensor in Different Scanning Scenarios. In Proceedings of the 7th IEEE International Conference on Biomedical Robotics and Biomechatronics (Biorob), Enschede, The Netherlands, 26–29 August 2018; pp. 1138–1143. [CrossRef]
6. Merker, L.; Scharff, M.; Behn, C. Approach to the dynamical scanning of object contours using tactile sensors. In Proceedings of the IEEE International Conference on Mechatronics (ICM), Ilmenau, Germany, 18–20 March 2019; pp. 364–369. [CrossRef]
7. Scharff, M.; Schorr, P.; Becker, T.; Resagk, C.; Alencastre, J.H.; Behn, C. An artificial vibrissa-like sensor for detection of flows. *J. Sens.* **2019**, *19*, 3892. [CrossRef] [PubMed]
8. Schultz, A.E.; Solomon, J.H.; Peshkin, M.A.; Hartmann, M.J.Z. Multifunctional whisker arrays for distance detection, terrain mapping, and object feature extraction. In Proceedings of the 2005 IEEE International Conference on Robotics and Automation, Barcelona, Spain, 18–22 April 2005; pp. 2588–2593. [CrossRef]
9. Kaneko, M.; Kanayama, N.; Tsuji, T. Active antenna for contact sensing. *IEEE Trans. Robot. Autom.* **1998**, *14*, 278–291. [CrossRef]
10. Kim, D.; Möller, R. Biomimetic whiskers for shape recognition. *Robot. Auton. Syst.* **2007**, *55*, 229–243. [CrossRef]
11. Birdwell, J.A.; Solomon, J.H.; Thajchayapong, M.; Taylor, M.A.; Cheely, M.; Towal, R.B.; Conrad, J.; Hartmann, M.J.Z. Biomechanical models for radial distance determination by the rat vibrissal system. *J. Neurophysiol.* **2007**, *98*, 2439–2455. [CrossRef] [PubMed]
12. Solomon, J.H.; Hartmann, M.J.Z. Artificial whiskers suitable for array implementation: Accounting for lateral slip and surface friction. *IEEE Trans. Robot.* **2008**, *24*, 1157–1167. [CrossRef]
13. Clements, T.N.; Rahn, C.D. Three-dimensional contact imaging with an actuated whisker. *IEEE Trans. Robot.* **2006**, *22*, 844–848. [CrossRef]
14. Will, C.; Behn, C.; Steigenberger, J. Object contour scanning using elastically supported technical vibrissae. *ZAMM J. Appl. Math. Mec.* **2018**, *98*, 289–305. [CrossRef]
15. Will, C. Continuum Models for Biologically Inspired Tactile Sensors: Theory, Numerics and Experiments. Ph.D. Thesis, Technische Universität Ilmenau, Ilmenau, Germany, 2018.
16. Huet, L.A.; Rudnicki, J.W.; Hartmann, M.J.Z. Tactile sensing with whiskers of various shapes: Determining the three-dimensional location of object contact based on mechanical signals at the whisker base. *Soft Robot.* **2017**, *4*, 88–102. [CrossRef] [PubMed]
17. Solomon, J.H.; Hartmann, M.J.Z. Extracting object contours with the sweep of a robotic whisker using torque information. *Int. J. Robot Res.* **2010**, *29*, 1233–1245. [CrossRef]
18. Boubenec, Y.; Claverie, L.N.; Shulz, D.E.; Debrégeas, G. An amplitude modulation/demodulation scheme for whisker-based texture perception. *Front. Neurosci.* **2014**, *6*. [CrossRef] [PubMed]
19. Graef, J.R.; Henderson, J.; Yang, B. Positive solutions to a fourth order three point boundary value problem. *Am. Ist. Math. Sci.* **2009**, 269–275. [CrossRef]
20. Scharff, M. Bio-inspired tactile sensing: Distinction of the overall object contour and macroscopic surface features. *Springer Proc. Math. Stat.* **2020**, accepted.

21. Grant, R.A.; Mitchinson, B.; Fox, C.W.; Prescott, T.J. Active touch sensing in the rat: Anticipatory and regulatory control of whisker movements during surface exploration. *J. Neurophysiol.* **2009**, *101*, 862–874. [CrossRef] [PubMed]
22. Belli, H.M.; Yang, A.E.T.; Bresee, C.S.; Hartmann, M.J.Z. Variations in vibrissal geometry across the rat mystacial pad: Base diameter, medulla, and taper. *J. Neurophysiol.* **2016**, *117*, 1807–1820. [CrossRef] [PubMed]
23. Voges, D.; Carl, K.; Klauer, G.J.; Uhlig, R.; Schilling, C.; Behn, C.; Witte, H. Structural Characterization of the Whisker System of the Rat. *J. Sens.* **2012**, *12*, 332–339. [CrossRef]

Article

Tacsac: A Wearable Haptic Device with Capacitive Touch-Sensing Capability for Tactile Display

Oliver Ozioko [1], William Navaraj [2], Marion Hersh [3] and Ravinder Dahiya [1,*]

[1] Bendable Electronics and Sensing Technologies (BEST) Group, University of Glasgow, Glasgow G12 8QQ, UK; Oliver.Ozioko@glasgow.ac.uk

[2] Department of Engineering, Nottingham Trent University, Clifton Campus, Nottingham NG11 8NS, UK; william.navaraj@ntu.ac.uk

[3] Biomedical Engineering, University of Glasgow, Glasgow G12 8LP, UK; Marion.Hersh@glasgow.ac.uk

* Correspondence: Ravinder.Dahiya@glasgow.ac.uk

Received: 21 July 2020; Accepted: 21 August 2020; Published: 24 August 2020

Abstract: This paper presents a dual-function wearable device (Tacsac) with capacitive tactile sensing and integrated tactile feedback capability to enable communication among deafblind people. Tacsac has a skin contactor which enhances localized vibrotactile stimulation of the skin as a means of feedback to the user. It comprises two main modules—the touch-sensing module and the vibrotactile module; both stacked and integrated as a single device. The vibrotactile module is an electromagnetic actuator that employs a flexible coil and a permanent magnet assembled in soft poly (dimethylsiloxane) (PDMS), while the touch-sensing module is a planar capacitive metal-insulator-metal (MIM) structure. The flexible coil was fabricated on a 50 µm polyimide (PI) sheet using Lithographie Galvanoformung Abformung (LIGA) micromoulding technique. The Tacsac device has been tested for independent sensing and actuation as well as dual sensing-actuation mode. The measured vibration profiles of the actuator showed a synchronous response to external stimulus for a wide range of frequencies (10 Hz to 200 Hz) within the perceivable tactile frequency thresholds of the human hand. The resonance vibration frequency of the actuator is in the range of 60–70 Hz with an observed maximum off-plane displacement of 0.377 mm at coil current of 180 mA. The capacitive touch-sensitive layer was able to respond to touch with minimal noise both when actuator vibration is ON and OFF. A mobile application was also developed to demonstrate the application of Tacsac for communication between deafblind person wearing the device and a mobile phone user who is not deafblind. This advances existing tactile displays by providing efficient two-way communication through the use of a single device for both localized haptic feedback and touch-sensing.

Keywords: actuator; tactile sensor; deafblind communication; tactile display

1. Introduction

Tactile displays enable people's interaction with the environment by creating tactile sensation on the skin as haptic feedback which is perceived and interpreted by the brain. Haptics, which involves both tactile (cutaneous) and kinesthetic (force) feedback, plays a great role in the way we communicate, interact with and perceive the environment [1–4]. In the area of robotics and virtual reality, it's viewed as real and simulated touch interactions between robots and humans [5]. This means that these interactions may be accomplished by human, machine or both and the environments can be real or virtual. This is recently changing the way humans interact with information and communicate ideas [6,7]. In case of haptic visual aids for deafblind people, the interaction is not accompanied by vision or hearing.

The human sense of touch is well developed and there is often a minimal perceived intensity for various stimuli like pressure, temperature, heat, and vibration. Humans are able to recognize common objects by touch within 1–2 s and the skin is very sensitive to light pressure [8]. Although there are different mechanoreceptors, tactile reception results from a combination of all the receptors in a particular skin area. The threshold of tactile perception depends on several factors like location, contact area, type of stimulus, duration, age and even hormone levels [9–14]. In [12], it was observed that the most significant age difference in vibrotactile detection threshold was found in the glabrous finger with the difference being less pronounced at lower frequencies (5–10 Hz) and more pronounced as the frequency increases to 300 Hz. Studies have equally shown that the perceived intensity of stimulation is determined by both the depth of penetration and the rate of skin indentation [8,15]. The large numbers of touch receptors present in our skin allow us to obtain rich information through haptic interaction by touching the objects around [16–21]. So, the importance of touch sensing [22] is evident in situations where other sensing modalities such as vision are insufficient or unavailable [23]. An example is the interaction by the hearing and visually impaired people in the real-world environment [24–27]. Through various forms of tactile stimulations such as skin stretching [28,29], vibration, force or painless electric shock, they substitute the hearing and vision impairments. Thus, they substantially rely on the tactile sensing [30] as well as tactile feedback (e.g., using actuators in tactile displays) to explore and manipulate objects around them [6,7,31]. Similarly, combination of tactile sensing [32] and tactile feedback is needed in areas such as minimal invasive surgery [2,5] and virtual reality [33,34] to improve the user's haptic interaction experience. Therefore, a tactile display with an inherent ability to provide touch feeling as well as the vibrotactile feedback within these limits would be advantageous. Further, considering the close contact of such displays with the curvy human body, they are desired to be soft and flexible [35,36].

Several assistive tactile displays with a single actuator or array of actuators have been developed for the purpose of providing tactile feedback. Early research in this area mainly focused on sensory substitution for the blind using Braille as well as camera-based types such as opticon [37] which reproduces texts in tactile form. However, the application domain has recently expanded to fields such as robotics, [38] haptics, teleoperation, virtual reality, video games, and prosthetics etc. Although all tactile feedback presents certain information to the user, the difference between tactile information for sensory substitution and information for robotic surgery is obvious. Here, we have grouped these existing tactile displays as: (a) the exploration types (ET), in which the user explores the surface of objects as in Braille, and (b) the localized stimulation types (LST), in which the actuator is positioned on the palm or elsewhere on the body to exert localized stimulation such as vibration, or skin stretch. The latter is popularly used for assistive purposes [27,39,40].

Table 1 summarizes different types of tactile displays and the actuation technologies explored so far. These include electromagnetic, electrostatic, dielectric elastomer (DE), piezoelectric [41], pneumatic, rheological fluids, and shape-memory alloy (SMA) [41]. Each of these technologies has advantages and disadvantages with regards to the development of tactile displays. Electrostatic actuators which are based on the attraction of two forces are widely used due to their low power requirements and fast actuation speed but suffer limited spatial resolution and performs better on dry fingers. Dielectric elastomers are excellent smart materials for actuation and are widely explored for a number of applications including soft robotics and tactile displays for assistive purposes. However, the common drawback of DE actuators is the requirement of very high voltage (usually in the order of kV). SMA provides high force and large displacement but has characteristic slow response time and suffers hysteric effect. Another technology that provides high force and large displacement is the pneumatic-based tactile displays, but they are often quite bulky and the lack of portability poses a problem for wearability. In comparison with these technologies, the electromagnetic principle used here to develop the integrated sensor and actuator offers high precision and displacement at comparatively low voltages (Table 1), which are important for wearable applications. Additionally, none of the tactile displays (e.g., Braille displays and deafblind smart gloves) reported so far have the dual integrated capability of tactile sensing and vibrotactile feedback.

Table 1. Summary of selected tactile displays and their actuation technology.

Technology	Tactile Sensing	Disp. (μm)	Type	Current Req. (mA)	Voltage Req.	Freq (Hz)	Ref.
Piezoelectric	No	0.257	ET	<250	80 Vpk	250	[41]
Electromagnetic		1000	ET	0.250	5 V	N/A	[42,43]
Electrostatic	No	N/A	ET	<200	2000 Vp–p	100–400	[44]
Electroactive Polymer	No	680	LST	N/A	5 kV/mm	0.1–300	[45]
Thermal		61	ET	N/A	28.7 V	N/A	[46]
Electrotactile	No	N/A	ET	N/A	9.3–63.4 V	20	[47]
ERM Motor	No	NA	LST	72	5 V	208	[39,48]
SMA	No	up to 2000	ET	N/A	120 mA	2	[49,50]
Pneumatic	No	560	ET	N/A	N/A	0.2–20	[51]
Electrorheological Fluid	No	700	LST	N/A	4 kV	10	[52]
Electromagnetic	Capacitive	377	LST	100 mA	3–5 V	10–200	This Work

N/A—Not Available. ET-Exploration Type; LST—Localized Stimulation Type; ERM—Eccentric Rotation Mass; SMA—Shape Memory Alloy.

Here we present a device (Tacsac) with both capacitive tactile sensing capability and vibrotactile feedback for application in tactile displays (LST); e.g., assistive communication devices by the deafblind people who otherwise use tactile communication methods such as deafblind manual alphabets [53]. It consists primarily of two modules; (1) a capacitive touch sensing module fabricated using flexible printed circuit board; and (2) an actuation module based on electromagnetic principle which is capable of providing vibrotactile feedback at a range of frequencies (10 Hz–200 Hz) within the range perceivable by human. A mobile app was also developed and utilized for the demonstration of its potential application as a means of communication. In particular, the presented device could be used by deafblind to communicate with people without vision/hearing impairment via a smartphone as we demonstrate here (Figure 1). The capacitive touch sensing layer serves as a touch interface for sending message to the mobile phone user (via the developed mobile app), while the vibrotactile module is used as a means of interpreting the message to the user.

Further, the Tacsac device is advantageous as it utilizes a skin contactor which directly stimulates the skin of the user. This enables the provision of localized vibration rather than full-body vibration like popular conventional commercial vibration motors used so far in deafblind communication devices.

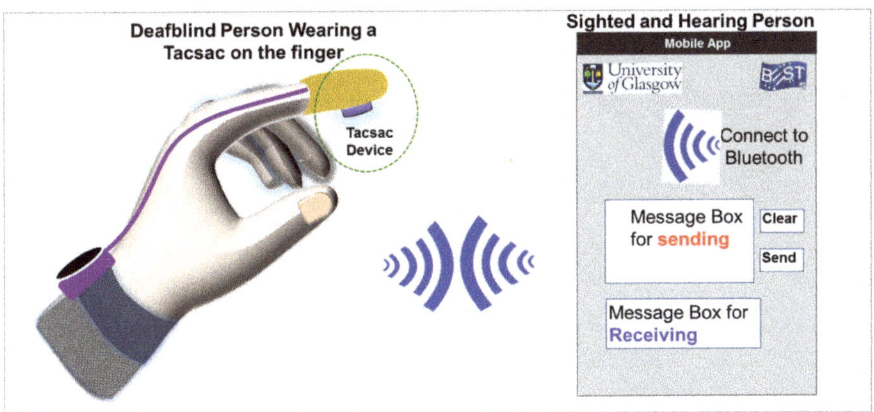

Figure 1. Application of the Tacsac device.

2. Materials and Methods

2.1. Device Structure and Operating Principle

Figure 2 shows the structure and working principle of the Tacsac device. It comprises two main modules: (a) the touch-sensing module, and (b) the vibrotactile module. The touch sensing module was designed with a simple metal-insulator-metal capacitive configuration with capacitance expressed as:

$$C = \varepsilon_r \varepsilon_0 A/d \quad (1)$$

where C is the capacitance, which changes when the sensor is touched, ε_r is the relative permittivity of the dielectric material, ε_0 is the permittivity of free space, A is the area of metal plates and d is the distance between the two conducting metal plates. The actuation module uses electromagnetic principle and is driven by the interaction of a spiral coil and a permanent magnet. As a principal component of the actuator, the coil determines the magnetic field produced. When a uniform square current pulse flows through the coil at a particular frequency, a pulsating magnetic field is produced along the axis, which exerts a force on the tiny magnet of the actuator and leading to periodic actuation of the top layer. The choice of the parameters of the coil is very key in the realization of a spiral coil which is capable of giving the desired results. Here we present how different parameters of the coil were calculated.

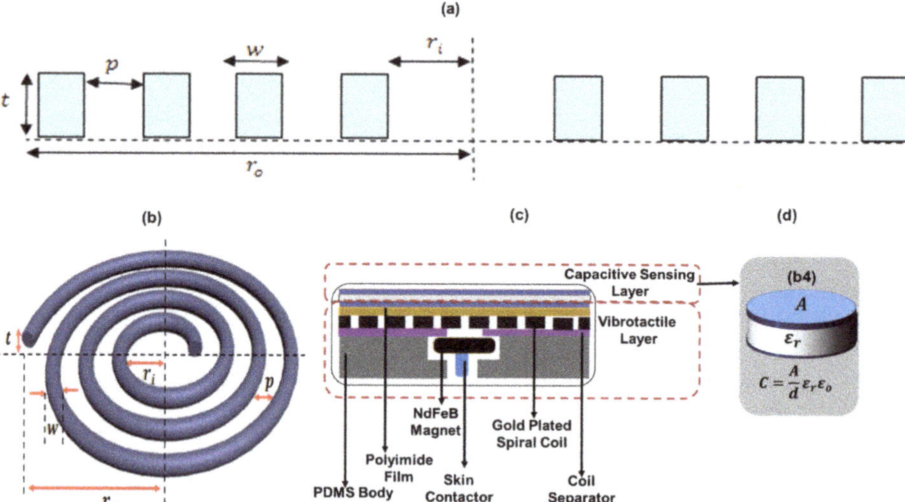

Figure 2. The structure and working principle of Tacsac (**a**) 2D axisymmetric structure of the spiral coil (**b**) Structure of the spiral coil (**c**) Structure of the Tacsac device. (**d**) Capacitive Sensing Layer.

Any wire carrying current creates a magnetic field in the region surrounding it and the relationship between the current and the accompanying magnetic field is given by Biot Savart's Law. If a spiral coil carrying current I is considered to be made up of n number of concentric circular loops with inner radius r_i, outer radius r_o and thickness t (Figure 2a,b), then if $n = N$ where N = number of turns of the spiral.

Utilizing the 2D-axisymmetric model of the spiral coil as shown in Figure 2a (adapted from [54]), the relationship between the internal and external radii of the spiral coil is given as:

$$r_o - r_i = N(p + w) \quad (2)$$

where p is the pitch of the coil, w is the width of the coil conductor. The coil aspect ratio is also an important consideration in as different coil aspect ratio could affect the magnetic field produced at the center of the coil which provides the required force of actuation when it interacts with a magnet. Defining $\alpha = w/t$ as the coil aspect ratio we have:

$$r_o - r_i = N(p + \alpha t) \tag{3}$$

Given the length of the spiral coil as $l = N2\pi r_o$, we derived the total magnetic field generated at the center by the N-turns of the spiral as:

$$B_{Center\ of\ Spiral} = \sum_{1}^{N} \frac{\mu_0 I}{2(N(\alpha t + p))} \ln\left(\frac{r_o}{r_o - N(\alpha t + p)}\right) \tag{4}$$

where, μ_0 is the relative permeability of free space ($4\pi \times 10^{-7}$) *[H/m]*. Equation (4) depicts the relationship between the coil parameters and shows how varying these parameters affect the magnetic field generated at the center of the actuating coil.

2.2. Device Fabrication

This section presents the fabrication of the Tacsac device, which involves three main tasks: (1) fabrication of the spiral coil, which is the primary element of the integrated actuator that provides the vibrotactile feedback; (2) fabrication of the capacitive sensing layer; and (3) Integration of the vibrotactile actuator and the capacitive sensing layer to realize Tacsac.

2.2.1. Fabrication of the Spiral Coil

The spiral coil is the primary element of the actuating layer and was fabricated using the Lithographie Galvanoformung Abformung (LIGA, Figure 3) process [55], which is needed to produce structures with high aspect ratio. A Plassys MEB 550S Electron Beam Evaporator system (Plassys, Glasgow, UK) was used to deposit a 20 nm/80 nm NiCr/Au on a flexible 50 µm polyimide sheet. Following this was the spinning of an AZ4562 photoresist (Clariant GmbH, Wiesbaden, Germany) at 2000 rpm for ~3 s, and the sample left at room temperature for ~30 min prior to baking on a hot plate at ~100 °C for 10 min. The baked sample was left again at room temperature for 30 min to allow evaporation of the solvent. This was followed by an exposure of the sample to ultraviolet (UV) light for ~60 min following a standard lithography technique. An AZ826 developer was utilized to develop the exposed sample for ~10 min, after which it was rinsed in reverse osmosis water.

In order to increase the thickness of the resulting spiral coil, we electroplated it using a litre of non-cyanide gold electroplating solution. The solution was prepared by first heating 250 mL of RO water at 50 °C using a hotplate. The water was then removed from the hotplate and 60 mL of solution containing Na_2SO_3 and tripotassium citrate ($K_3C_6H_5O_7$) was added to the water. Following this, 5 mL of brightener (containing arsenic salt) was added. Next is the addition of 50 mL of gold complex solution which contains gold sulphite. Finally, the resultant solution was adjusted to reach 1liter by adding more RO water.

The coil sample was then electroplated for ~45 min using the prepared non-cyanide gold complex solution [56] to realize a coil with ~17 µm thickness. The unwanted gold layer was then etched using conventional gold etchant for ~15 s, thus exposing only the NiCr seed layer. The sample was annealed at 350 °C in a furnace for ~20 min under a nitrogen atmosphere to increase the strength of electroplated metal and avoid undesirable lift-off during etching of the seed layer. The NiCr seed layer was then was removed using Nichrome etchant to realize the required spiral coil as shown in Figure 3h.

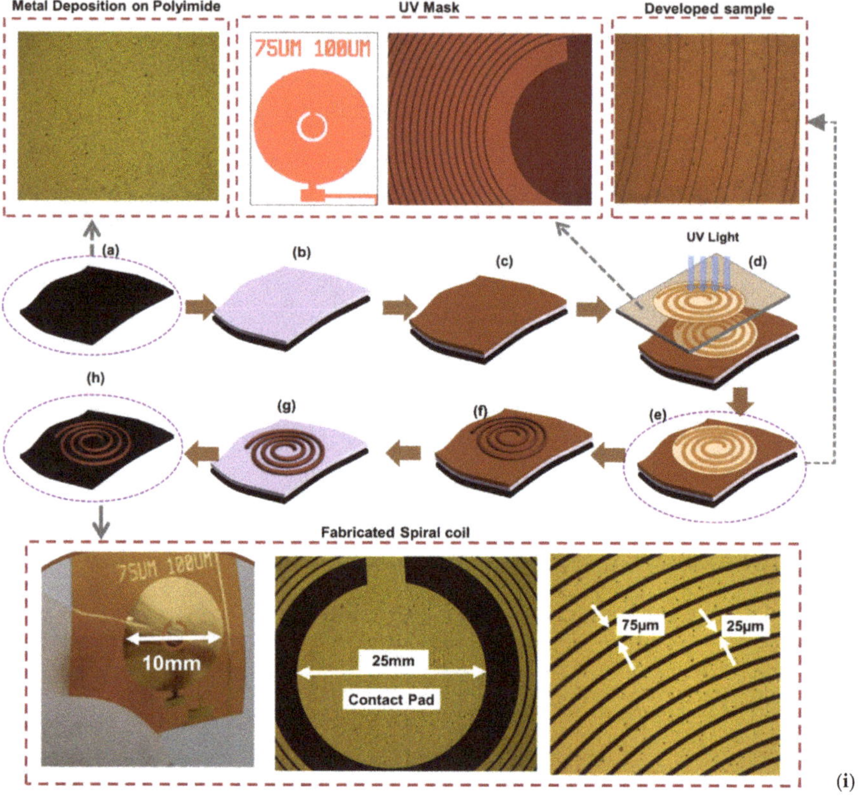

Figure 3. Fabrication steps for realization of the coil: (**a**) Initial flexible substrate; (**b**) Gold deposition; (**c**) Spin-coating of photoresist; (**d**) Exposure of photoresist; (**e**) Developing the photoresist; (**f**) Electroplating the coil; (**g**) lift-off the photoresist; (**h**) Etching of the seed layer; and (**i**) Fabricated Coil.

2.2.2. Fabrication of the Capacitive Sensing Layer

The touch-sensing module was realized following the steps depicted in Figure 4a. It was fabricated using a facile planar capacitive structure realized by bonding two layers of single-sided flexible printed circuit board (FPCB) with ~35 µm of copper on a polyimide substrate. This was achieved by sandwiching a PVC film between the non-conducting surface of one FPCB with the conducting copper surface of the other, and so the polyimide substrate of the top FPCB and the PVC serves as the dielectric layer for the capacitive touch sensor (Figure 4a). The FPCB and PVC were cut using the Silhouette Cameo 2 blade cutter (Silhouette, Lindon, UT, USA, Figure 4(a3,a5)). The software of the Silhouette Cameo has a built-in library of different materials with options of editing their properties. The pattern to be cut was first designed in a graphic software and then transferred to the Silhouette Cameo software for cutting Figure 4(a1). The Silhouette Cameo was set to be able to cut only the required portions of the sheet. To do this, the speed, force, and blade position of the blade cutter was set to 5, 20, and 10 cm·s^{-1} respectively. This was followed by attachment of a plain FPCB on a sticky 12 in × 12 in cutting mat, from which the designed pattern was cut out Figure 4(a3). Similar steps were followed to cut a PVC film (Figure 4(a4,a5)). After cutting, the unwanted parts were removed revealing only the pattern, as shown in Figure 4(a5). The cut piece of PVC was sandwiched between two cut pieces of FPCB as shown in Figure 4(a6). This was then followed by the attachment of a fine copper wire to serve as electrodes Figure 4(a7).

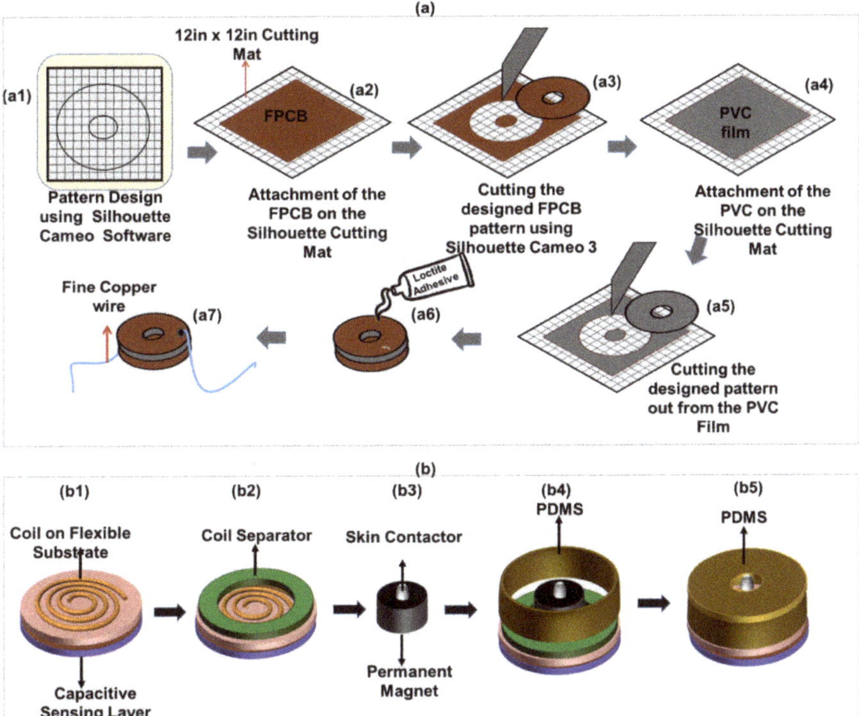

Figure 4. (**a**) Fabrication steps for the sensing layer; (**a1**) Design of the pattern using CAD software; (**a2**) attachment of the FPCB on the 12 in × 12 in cutting mat; (**a3**) Cutting of two layers of the pattern using the Silhouette Cameo 3 blade cutter; (**a4**) Attachment of the PVC film on the 12 in × 12 in cutting mat (**a5**) Cutting of the PVC film using the Silhouette Cameo 3 blade cutter; (**a6**) Bonding of the FPCB and PVC layer using Loctite adhesive; (**a7**) Soldering of a fine copper wire on both layers of the FPCB to serve as electrode; and (**b**) Fabrication steps for integration of the capacitive sensing layer and vibrotactile actuator(**b1**) Attachment of the touch-sensing layer to the coil; (**b2**) Integration of the coil separator; (**b3**) Attachment of skin contactor to the permanent magnet; (**b4**) Integration of the PDMS packaging; (**b5**) Final packaging of the actuator using PDMS cover.

2.2.3. Device Integration

The steps for the realization of the touch-sensitive actuator (Tacsac) are shown in Figure 4b. The main components of the actuator include the touch sensitive layer, the coil, a permanent magnet, skin contactor, and PDMS packaging. The integration and packaging was carefully designed to reduce damping of the actuation/vibration. The actuator has an overall diameter of ~1.5 cm and was designed to fit appropriately into the user's finger. It was realized layer by layer as shown in Figure 4b. A cylindrical mould with 1.5 cm and 1.9 cm inner and outer diameter respectively was designed and realized with 3D printer for the PDMS packaging. The mould was designed for the body of the actuator and meant to realize a PDMS packaging of diameter 1.5 cm and height 0.4 cm, with a hole of 1mm inner diameter for the movement of the skin contactor.

For the realization of the actuator, 10:1 PDMS (Sylgard 184, Sigma-Aldrich, Gillingham, UK) comprising of mixture of pre-polymer base and crosslinking agent was prepared and poured into the mould, and then cured at 80 °C in the oven for 15 min. The same PDMS was also used to attach the touch-sensitive layer to the coil and cured for 10 min Figure 4(b1). The coil separator was then attached to the coil substrate Figure 4(b2) using Loctite transparent adhesive (Amazon UK, Slough, UK).

The permanent magnet used here was a 2 mm thick N42 grade neodymium magnet from E-Magnets (E-Magnets, Berkhamsted, UK), comprising of ~29–32% neodymium, 64.2–68.5% iron and 1–1.2% E-Boron (NdFeB). A 1 mm^2 skin contactor (made of polylactic acid (PLA) plastic) was attached to the permanent magnet using a Loctite transparent adhesive, and both placed on the coil separator (Figure 4(b3)). The actuator was then packaged with the cured PDMS described earlier in this section (Figure 4(a4,a5)).

2.3. Device Characterization

This section describes the procedure for characterizing the actuation (vibrotactile) and sensing (capacitive) capabilities of the Tacsac device. The actuation characterization (displacement and amplitude characteristics at different frequencies) of the actuator in response to varying current pulse inputs was carried out by employing custom-made optical lever technique (Figure 5).

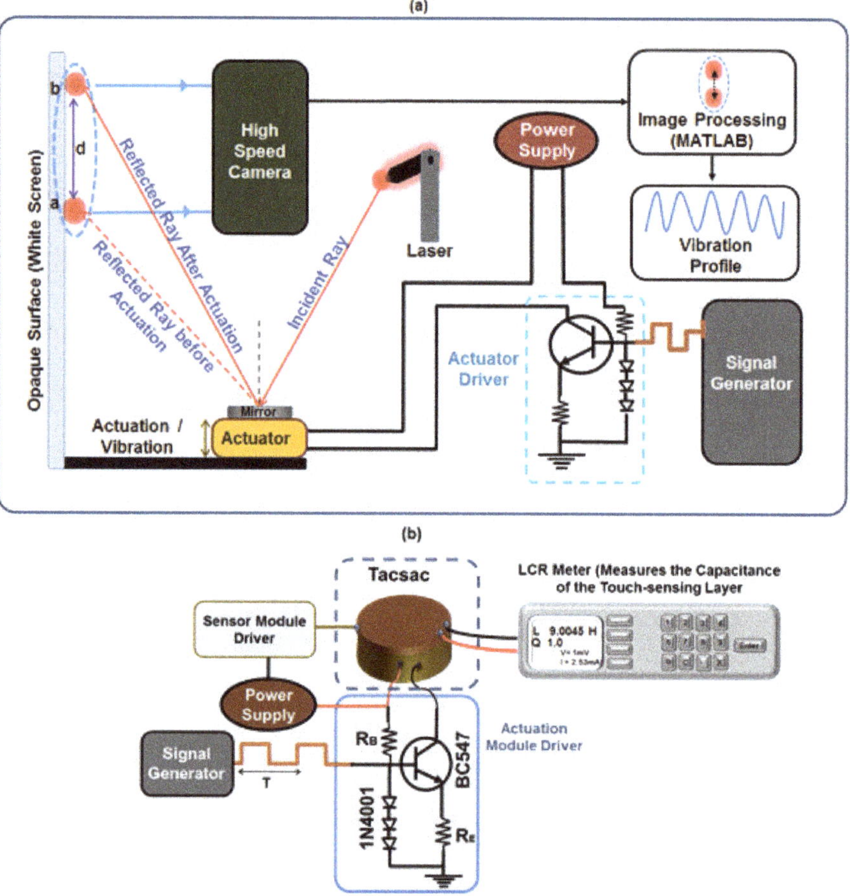

Figure 5. (**a**) Setup for Characterization of the Actuation Module (**b**) Setup for the characterization of sensing module of the fabricated Tacsac with and without actuation.

In this technique, a pointed laser ray is directed onto a reflective material (mirror) placed on the actuator and the reflected ray properly directed on to an opaque screen where the actuation is measured as a magnified value of the original displacement (Figure 5). Before the experiment, the laser and the actuator were all firmly attached to a stable optical table and the ray adjusted to obtain a sharp image on the screen. The actuator was then connected to a power supply, and a signal generator via a custom-made constant current source (Figure 5a). During the experiment, the actuator was driven with different currents (60, 90, 120, 150, and 180 mA) and at different frequencies ranging from 10 Hz to 200 Hz in order to understand the behavior of the actuator over this range which is detectable by human. To record displacement at any time, the camera is used to capture the motion of the ray on the screen when the actuator is turn on at the given current and frequency. In our experiment, the motion of the reflected laser spot on the opaque screen during the vibration of the actuator was recorded as a video using a high-speed camera with 960 frames per second (fps). This enabled the amplitude and displacement characteristics of Tacsac to be effectively captured in real time for different vibration frequencies ranging from 10 Hz to 200 Hz. Further, the recorded videos were processed (with appropriate calibration) using digital signal processing by a MATLAB program to obtain the dynamic response of the Tacsac device.

The capacitive touch-sensing layer was equally characterized to understand how the device responds when touched and when not touched. Force sensing was not studied here since this is not key for the intended application. This characterization was carried out using an E4980AL precision LCR meter (Keysight, Santa Rosa, CA, USA) and a LabVIEW 2018 Robotics v18.0f2 (National Instruments, Austin, TX, USA) program installed in a computer for automatic reading of the capacitance values (Figure 5). Capacitive values were read with the LabVIEW program for the case when actuator was touched and the vibration OFF and then ON. Similar setup was used to drive the actuator in Figure 5 at 150 Hz using a square wave of 5 Vp-p, 50% duty cycle and zero offset (Figure 5). While the actuator was vibrating, the capacitive sensing layer was touched at different intervals using a finger and the change in capacitance was recorded using the LCR meter via the LabVIEW program.

3. Results and Discussion

Figure 6a shows the captured laser images for recorded amplitude of the actuator oscillation at 70 Hz and 100 Hz. Similar images captured for other frequencies were processed in MATLAB to determine the actual dynamic response of the actuator. Figure 6b shows the normalized transient response profiles of the actuator as a function of elapsed time for a peak pulse current of 180 mA. The actuator displacement was measured for various frequencies ranging from 10 Hz to 200 Hz with equal intervals of 10 Hz which is well within the range detected by human. At low frequencies it can be observed that, as the magnet switches from one position to another, it undergoes damped oscillations before coming to stabilize in that particular position. For frequencies above 50 Hz, the magnet switches smoothly from its position before coming to a steady state. This results in the actuator not getting sufficient time to restore to its equilibrium state before the subsequent input current pulse. The switching pulse may result in constructive or destructive interference with the damped oscillations, resulting in an oscillatory behavior.

Figure 6c shows the absolute amplitude (displacement) of the actuator for different frequencies (within the range detectable by human) at constant current of 180 mA, with the maximum displacement observed at 70 Hz. As the frequency increases further, the displacement drops significantly owing to the inherent mechanical impedance during oscillations. The resonance frequency of the oscillator is in the range of 60–70 Hz. The slight increase of actuator displacement at 120 Hz and 190 Hz is likely related to the first and second overtones of the resonance frequency between 120–130 Hz and 180–190 Hz respectively. Our control experiments at different currents and fixed frequency showed that the actuator response is similar, except proportional increase in amplitude (Figure 6d). During this experiment, the current in the spiral coil was increased to observe its effect on the actuation and to choose the optimum coil current for perceivable vibration. This is because, the magnetic field generated along the axis of the coil increases with pulsed current passing through the coil, leading to stronger repulsive impulse force on the magnet to displace it from its mean position. The maximum input current to the coil was restricted to 180 mA owing to the dimensional limit of the casing of the actuator on the amplitude of actuation. Increasing the current to very large values (>200 mA) could increase the chances of the coil getting hot as result of Joule heating. At a constant current of 180 mA, the actuator was able to give an off-plane displacement of 0.377 mm with negligible Joule heating effect. The transient response of the actuator was equally recorded for various coil currents (60, 90, 120, 150 and 180 mA) at a fixed frequency of 10 Hz (Figure 6d). This showed similar oscillating pattern in each case with current and displacement having direct proportionality. This means that 60 mA gave the least peak to peak displacement while 180 mA gave the highest in this case (Figure 6d). The actuator has a response time ~93 ms (at 180 mA, for as frequency as low as 10 Hz) which is similar to that of conventional vibration motors (~50–140 ms).

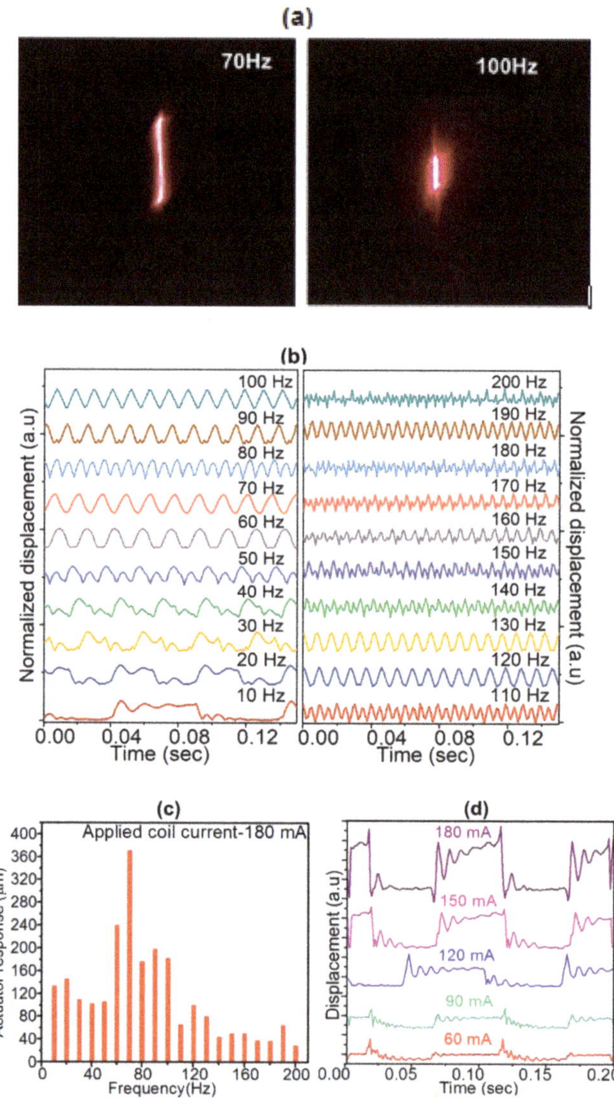

Figure 6. (**a**) Recorded amplitude of the actuator oscillation at 70Hz and 100Hz (**b**) Normalized transient displacement response of the actuator and (**c**) Peak actuator displacement as a function of applied frequency at a current of 180 mA (**d**) Transient response of the actuator at different currents and input frequency of 10 Hz.

The response of the sensor module characterization, recorded using an LCR meter and a LabVIEW program, is shown in Figure 7a,b. Capacitance values were read when the actuator vibration was OFF and ON. The result shows average $\Delta C/Co \sim 0.35$ for both cases. This means that the sensing layer is able to respond quite similar both when the actuator is ON or OFF. The results were similar other frequencies tested and so only the response of the actuator at frequency of 150 Hz is presented as shown in Figure 7.

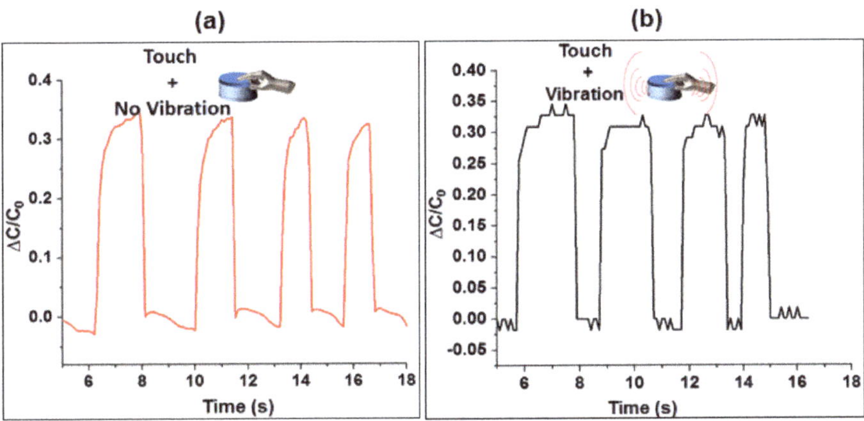

Figure 7. Response of the capacitive sensing layer with and without actuation; (**a**) Sensor output when touched and vibration is OFF; (**b**) Sensor's output when touched and vibration is ON.

4. Application

Figure 8 shows the block diagram and result of the actuator with the mobile app developed in this work. This is to demonstrate one of the applications of Tacsac, which is wireless communication between (1) a deafblind person and a hearing-and-sighted person who uses a mobile phone, or (2) deafblind-to-deafblind people, both of whom wear the device. The overall communication system comprises four main modules (1) the fabricated Tacsac device, (2) the control module including the drive and readout circuits, (3) the wireless module, and (4) the developed mobile app.

The deafblind user employs the actuator to communicate with a mobile phone held by the sighted and hearing person. Messages are sent by the deafblind person using the capacitive sensing layer of the actuator, and then received via the vibrotactile actuator in the form of vibration. This could also be adapted for use as a Morse code communication device for deafblind people. Morse code is one of the methods used by deafblind people for communication and this device could be of benefit for users of this communication method. In this case, messages sent from the mobile app would be converted to Morse codes in the form of vibration of varying frequencies and duration. To send messages in the form of Morse code, deafblind people could tap the capacitive sensing layer, which would be decoded and converted to text messages by the mobile app. In comparison to Braille for instance, Morse code has advantage in terms of wearability and simplicity and with only a single device (integrated sensor and actuator), a two-way communication can be established. Braille introduces some level of complexity in terms of the number of devices required to represent the six dots of Braille.

However, in this work we have only demonstrated the use of Tacsac to successfully send and receive message from the developed mobile app with a single fabricated actuator. This demonstration was achieved using a mobile app which communicates with the actuator via an HC-05 Bluetooth module (Amazon UK, Slough, UK). When the app is launched, the user presses the connect button (Figure 8) to connect with the actuator and a connection status is displayed (Figure 8). To communicate from the app to the Tacsac device, the user types a number in the message box of the app and sends it via Bluetooth. When the actuator receives this information, it vibrates accordingly. For instance, when the number "1" is sent from the app, the actuator vibrates once, while it vibrates twice when the number "2" is sent. To demonstrate the communication from Tacsac to the mobile app, the user touches the capacitive sensing layer, this touch is then sensed, and the information sent to the mobile app via Bluetooth (Supplementary Materials Video S1). When the mobile app receives this information, it displays the word "Touched" or "None" otherwise (see message receipt status in Figure 8).

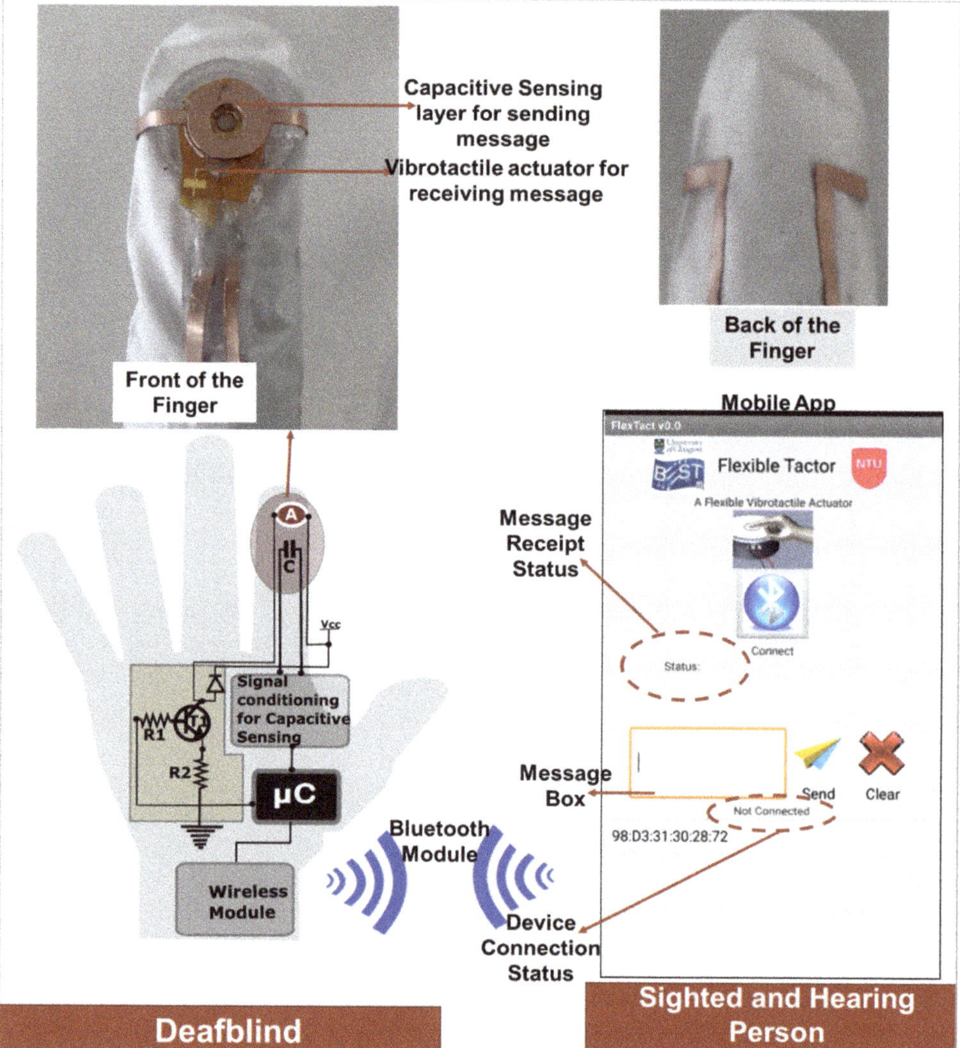

Figure 8. Application of Tacsac for wireless communication of deafblind people with mobile phone users.

5. Conclusions

In this work, an assistive haptic device for application in tactile displays (such as smart assistive gloves) is presented. The actuation is based on electromagnetic principle and the device also has capacitive touch-sensing capability. Considering the state of the art, this is a step forward given that majority of the existing tactile displays do not have the ability for both touch sensing and vibrotactile feedback. The addition of touch-sensitivity to the tactile displays is advantageous for simultaneous sending and receiving of information—particularly for deafblind people as it will provide a close-loop communication system. The sensor and actuator combination show good performance and the potential to be used in a tactile display for deafblind communication. The integrated actuator provides a 0.377 mm displacement and its measured vibration profiles showed its capability to provide perceivable vibration within a wide range of perceivable frequencies (10 Hz to 200 Hz). A developed app was used to send and receive messages to/from the actuator, demonstrating its application for communication between deafblind people and mobile phone users who have hearing and vision. This work finds application in Morse code communication for deafblind people where the user is able to compose messages based on Morse code and then send it to mobile phone user. Future work will involve the fabrication of array of such integrated sensor and actuator for application in other methods of deafblind communication (e.g., British deafblind manual alphabet).

Supplementary Materials: The following are available online at http://www.mdpi.com/1424-8220/20/17/4780/s1, Video S1: Tacsac communicating with a custom-made mobile app.

Author Contributions: Conceptualization, O.O., and R.D, Methodology, W.N., O.O and R.D; Software W.N. and O.O.; Analysis, O.O., W.N., R.D and M.H.; resources, R.D and M.H.; Writing—original draft preparation, W.N. and O.O.; Writing—review and editing, R.D., M.H; O.O and W.N.; Supervision, R.D; Funding acquisition, R.D and M.H. All authors have read and agreed to the published version of the manuscript.

Funding: This work was supported in part by Engineering and Physical Sciences Research Council (EPSRC) Engineering Fellowship for Growth neuPRINTSKIN (EP/R029644/1), European Commission through PH-CODING (H2020-FETOPEN-2018- 829186) and Tertiary Education and Trust Fund (TETFund) Nigeria.

Conflicts of Interest: The authors declare no conflict of interest.

References

1. Culbertson, H.; Schorr, S.B.; Okamura, A.M. Haptics: The Present and Future of Artificial Touch Sensation. *Annu. Rev. Control Robot. Auton. Syst.* **2018**, *1*, 385–409. [CrossRef]
2. Abiri, A.; Pensa, J.; Tao, A.; Ma, J.; Juo, Y.-Y.; Askari, S.J.; Bisley, J.; Rosen, J.; Dutson, E.P.; Grundfest, W.S. Multi-Modal Haptic Feedback for Grip Force Reduction in Robotic Surgery. *Sci. Rep.* **2019**, *9*, 5016. [CrossRef]
3. Dahiya, R.; Yogeswaran, N.; Liu, F.; Manjakkal, L.; Burdet, E.; Hayward, V.; Jörntell, H. Large-Area Soft e-Skin: The Challenges Beyond Sensor Designs. *Proc. IEEE* **2019**, *107*, 2016–2033. [CrossRef]
4. Ozioko, O.; Karipoth, P.; Hersh, M.; Dahiya, R. Wearable Assistive Tactile Communication Interface Based on Integrated Touch Sensors and Actuators. *IEEE Trans. Neural Syst. Rehabil. Eng.* **2020**, *28*, 1344–1352. [CrossRef]
5. Okamura, A.M. Haptic Feedback in Robot-Assisted Minimally Invasive Surgery. *Curr. Opin. Urol.* **2009**, *19*, 102–107. [CrossRef]
6. Sreelakshmi, M.; Subash, T.D. Haptic Technology: A comprehensive review on its applications and future prospects. *Mater. Today Proc.* **2017**, *4*, 4182–4187. [CrossRef]
7. Srinivasan, M.A.; Basdogan, C. Haptics in virtual environments: Taxonomy, research status, and challenges. *Comput. Graph.* **1997**, *21*, 393–404. [CrossRef]
8. Goethals, P. *Tactile Feedback for Robot Assisted Minimally Invasive Surgery: An Overview*; Department of Mechanical Engineering KU Leuven: Leuven, Belgium, 2008.
9. Hale, K.S.; Stanney, K.M. Deriving haptic design guidelines from human physiological, psychophysical, and neurological foundations. *IEEE Comput. Graph. Appl.* **2004**, *24*, 33–39. [CrossRef]
10. Gescheider, G.A.; Verrillo, R.T.; McCann, J.T.; Aldrich, E.M. Effects of the menstrual cycle on vibrotactile sensitivity. *Percept. Psychophys.* **1984**, *36*, 586–592. [CrossRef]
11. Verrillo, R.T. Effects of aging on the suprathreshold responses to vibration. *Percept. Psychophys.* **1982**, *32*, 61–68. [CrossRef]
12. Venkatesan, L.; Barlow, S.M.; Kieweg, D. Age- and sex-related changes in vibrotactile sensitivity of hand and face in neurotypical adults. *Somatosens. Mot. Res.* **2015**, *32*, 44–50. [CrossRef]
13. Wang, Y.; Millet, B.; Smith, J.L. Designing wearable vibrotactile notifications for information communication. *Int. J. Hum. Comput. Stud.* **2016**, *89*, 24–34. [CrossRef]
14. Jones, L.A.; Lederman, S.J. *Human Hand Function*; Oxford University Press: New York, NY, USA, 2006.

15. Lederman, S.J.; Browse, R.A. The Physiology and Psychophysics of Touch. In *Sensors and Sensory Systems for Advanced Robots*; Dario, P., Ed.; Springer: Berlin/Heidelberg, Germany, 1988.
16. Tiwana, M.I.; Redmond, S.J.; Lovell, N.H. A review of tactile sensing technologies with applications in biomedical engineering. *Sens. Actuators A Phys.* **2012**, *179*, 17–31. [CrossRef]
17. Wan, Y.; Wang, Y.; Guo, C.F. Recent progresses on flexible tactile sensors. *Mater. Today Phys.* **2017**, *1*, 61–73. [CrossRef]
18. Dahiya, R. E-Skin: From Humanoids to Humans. In *Proceedings of the IEEE*; IEEE: Piscataway Township, NJ, USA, 2019; Volume 107, pp. 247–252.
19. Taube Navaraj, W.; García Núñez, C.; Shakthivel, D.; Vinciguerra, V.; Labeau, F.; Gregory, D.H.; Dahiya, R. Nanowire FET Based Neural Element for Robotic Tactile Sensing Skin. *Front. Neurosci.* **2017**, *11*. [CrossRef]
20. Soni, M.; Dahiya, R. Soft eSkin: Distributed touch sensing with harmonized energy and computing. *Philos. Trans. R. Soc. A* **2020**, *378*, 20190156. [CrossRef]
21. Dahiya, R.S. Epidermal electronics—flexible electronics for biomedical applications. In *Handbook of Bioelectronics: Directly Interfacing Electronics and Biological Systems*; Iniewski, K., Carrara, S., Eds.; Cambridge University Press: Cambridge, UK, 2015; pp. 245–255.
22. Dahiya, R.S.; Valle, M. *Robotic Tactile Sensing: Technologies and System*; Springer: New York, NY, USA, 2013.
23. Hersh, M. Mobility technologies for blind, partially sighted and deafblind people: Design issues. In *Mobility of Visually Impaired People*; Springer: Berlin/Heidelberg, Germany, 2018; pp. 377–409.
24. Hersh, M.A. Mobility Technologies for Blind, Partially Sighted and Deafblind People: Design Issues. In *Mobility of Visually Impaired People: Fundamentals and ICT Assistive Technologies*; Pissaloux, E., Velazquez, R., Eds.; Springer International Publishing: Cham, Switzerland, 2018; pp. 377–409.
25. Ogrinc, M.; Farkhatdinov, I.; Walker, R.; Burdet, E. Horseback riding therapy for a deafblind individual enabled by a haptic interface. *Assist. Technol.* **2017**, *30*, 1–8.
26. Dammeyer, J. Deafblindness: A Review of the Literature. *Scand. J. Public Health* **2014**, 554–562. [CrossRef]
27. Sorgini, F.; Caliò, R.; Carrozza, M.C.; Oddo, C.M. Haptic-assistive technologies for audition and vision sensory disabilities. *Disabil. Rehabil. Assist. Technol.* **2018**, *13*, 394–421. [CrossRef]
28. Aggravi, M.; Pausé, F.; Giordano, P.R.; Pacchierotti, C. Design and evaluation of a wearable haptic device for skin stretch, pressure, and vibrotactile stimuli. *IEEE Robot. Autom. Lett.* **2018**, *3*, 2166–2173. [CrossRef]
29. Schorr, S.B.; Quek, Z.F.; Romano, R.Y.; Nisky, I.; Provancher, W.R.; Okamura, A.M. Sensory substitution via cutaneous skin stretch feedback. In Proceedings of the 2013 IEEE International Conference on Robotics and Automation, Karlsruhe, Germany, 6–10 May 2013; pp. 2341–2346.
30. Caporusso, N.; Biasi, L.; Cinquepalmi, G.; Trotta, G.F.; Brunetti, A.; Bevilacqua, V. A wearable device supporting multiple touch-and gesture-based languages for the deaf-blind. In Proceedings of the International Conference on Applied Human Factors and Ergonomics, Los Angeles, CA, USA, 17–21 July 2017; pp. 32–41.
31. Srinivasan, M.; Cutkosky, M.; Howe, R.; Salisbury, J. Human and Machine Haptics. *RLE Prog. Rep.* **1999**, *147*, 1–39.
32. Navaraj, W.; Dahiya, R. Fingerprint-Enhanced Capacitive-Piezoelectric Flexible Sensing Skin to Discriminate Static and Dynamic Tactile Stimuli. In *Advanced Intelligent Systems*; WILEY-VCH Verlag GmbH & Co. KGaA: Weinheim, Germany, 2019.
33. Chen, X.; Hu, J. A review of haptic simulator for oral and maxillofacial surgery based on virtual reality. *Expert Rev. Med. Devices* **2018**, *15*, 435–444. [CrossRef] [PubMed]
34. Ge, J.; Wang, X.; Drack, M.; Volkov, O.; Liang, M.; Cañón Bermúdez, G.S.; Illing, R.; Wang, C.; Zhou, S.; Fassbender, J.; et al. A bimodal soft electronic skin for tactile and touchless interaction in real time. *Nat. Commun.* **2019**, *10*, 4405. [CrossRef] [PubMed]
35. Dahiya, R.; Navaraj, W.T.; Khan, S.; Polat, E.O. Developing Electronic Skin with the Sense of Touch. *Inf. Disp.* **2015**, *31*, 6–10. [CrossRef]
36. Dahiya, R.; Akinwande, D.; Chang, J.S. Flexible Electronic Skin: From Humanoids to Humans [Scanning the Issue]. In *Proceedings of the IEEE*; IEEE: Piscataway Township, NJ, USA, 2019; Volume 107, pp. 2011–2015.
37. Seo, D.G.; Cho, Y.-H. Resonating tactile stimulators based on piezoelectric polymer films. *J. Mech. Sci. Technol.* **2018**, *32*, 631–636. [CrossRef]
38. Noguchi, T.; Nagai, S.; Kawamura, A. Electromagnetic Linear Actuator providing High Force Density per Unit Area without Position Sensor as a Tactile Cell. *IEEJ J. Ind. Appl.* **2018**, *7*, 259–265. [CrossRef]
39. Zárate, J.J.; Shea, H. Using Pot-Magnets to Enable Stable and Scalable Electromagnetic Tactile Displays. In *IEEE Transactions on Haptics*; IEEE: Piscataway Township, NJ, USA, 2017; Volume 10, pp. 106–112.

40. Hyun, U.K.; Hyun Chan, K.; Jaehwan, K.; Sang-Youn, K. Miniaturized 3 × 3 array film vibrotactile actuator made with cellulose acetate for virtual reality simulators. *Smart Mater. Struct.* **2015**, *24*, 055018. [CrossRef]
41. Phung, H.; Nguyen, C.T.; Jung, H.; Nguyen, T.D.; Choi, H.R. Bidirectional tactile display driven by electrostatic dielectric elastomer actuator. *Smart Mater. Struct.* **2020**, *29*, 035007. [CrossRef]
42. Mitsuhiro, S.; Tsubasa, I.; Shinji, U.; Takaaki, M.; Kazuo, S. Fabrication of a bubble-driven arrayed actuator for a tactile display. *J. Micromechanics Microengineering* **2008**, *18*, 065012. [CrossRef]
43. Kitamura, N.; Chim, J.; Miki, N. Electrotactile display using microfabricated micro-needle array. *J. Micromechanics Microengineering* **2015**, *25*, 025016. [CrossRef]
44. Ozioko, O.; Taube, W.; Hersh, M.; Dahiya, R. SmartFingerBraille: A tactile sensing and actuation based communication glove for deafblind people. In Proceedings of the 2017 IEEE 26th International Symposium on Industrial Electronics (ISIE), Edinburgh, UK, 19–21 June 2017; pp. 2014–2018.
45. Gollner, U.; Bieling, T.; Joost, G. Mobile Lorm Glove—Introducing a Communication Device for Deaf-Blind People. In *TEI 2012*; Association for Computing Machinery: New York, NY, USA, 2012; pp. 127–130.
46. Sapra, P.; Parsurampuria, A.K.; Muralikrishnan, S.; Gupta, V.; Karthikeyan, H.; Bhagavatheesh, K.; Venkatesan, A.; Valiyaveetil, S.; Balakrishnan, M.; Rao, P.V.M. Refreshable Braille Display Using Shape Memory Alloy With Latch Mechanism. In Proceedings of the 13th ASME/IEEE International Conference on Mechatronic and Embedded Systems and Applications, Cleveland, OH, USA, 6–9 August 2017; p. V009T07A040. [CrossRef]
47. Matsunaga, T.; Totsu, K.; Esashi, M.; Haga, Y. Tactile display using shape memory alloy micro-coil actuator and magnetic latch mechanism. *Displays* **2013**, *34*, 89–94. [CrossRef]
48. Wu, X.; Kim, S.H.; Zhu, H.; Ji, C.H.; Allen, M.G. A Refreshable Braille Cell Based on Pneumatic Microbubble Actuators. *IEEE J. Microelectromechanical Syst.* **2012**, *21*, 908–916. [CrossRef]
49. Mazursky, A.; Koo, J.-H.; Yang, T.-H. Design, modeling, and evaluation of a slim haptic actuator based on electrorheological fluid. *J. Intell. Mater. Syst. Struct.* **2019**, *30*, 2521–2533. [CrossRef]
50. Hathazi, A. Developing and implementing a curriculum for children with deafblindness/multisensory impairment. *Educatia* **2010**, *21*, 145–150.
51. Do, T.N.; Phan, H.; Nguyen, T.Q.; Visell, Y. Soft Electromagnetic Actuators: Miniature Soft Electromagnetic Actuators for Robotic Applications. *Adv. Funct. Mater.* **2018**, *28*, 1800244. [CrossRef]
52. Caporusso, N.; Trizio, M.; Perrone, G. Pervasive assistive technology for the deaf-blind need, emergency and assistance through the sense of touch. In *Pervasive Health*; Springer: Berlin/Heidelberg, Germany, 2014; pp. 289–316.
53. Ozioko, O.; Hersh, M.A. Development of a Portable Two-way Communication and Information Device for Deafblind People. In Proceedings of the 13th AAATE International Conference, Budapest, Hungary, 9–12 September 2015; pp. 518–525.
54. Busch-Vishniac, I.J. The case for magnetically driven microactuators. *Sens. Actuators A Phys.* **1992**, *33*, 207–220. [CrossRef]
55. Genolet, G.; Lorenz, H. UV-LIGA: From development to commercialization. *Micromachines* **2014**, *5*, 486–495. [CrossRef]
56. Schlesinger, M.; Paunovic, M. *Modern Electroplating*; John Wiley & Sons: Hoboken, NJ, USA, 2011; Volume 55.

MDPI
St. Alban-Anlage 66
4052 Basel
Switzerland
Tel. +41 61 683 77 34
Fax +41 61 302 89 18
www.mdpi.com

Sensors Editorial Office
E-mail: sensors@mdpi.com
www.mdpi.com/journal/sensors

www.ingramcontent.com/pod-product-compliance
Lightning Source LLC
LaVergne TN
LVHW070112170726
843515LV00007B/1674

* 9 7 8 3 0 3 6 5 0 4 2 4 7 *